A BUILDING CONTRACT CASEBOOK

A BUILDING CONTRACT CASEBOOK

VINCENT POWELL-SMITH
LLM, DLitt, FCIArb

AND

MICHAEL FURMSTON
TD, BCL, MA, LLM

BSP PROFESSIONAL BOOKS

OXFORD LONDON EDINBURGH

BOSTON PALO ALTO MELBOURNE

First published 1984 by
Granada Publishing
(ISBN 0 246 12108 4)
First edition (revised) published 1987 by
BSP Professional Books

British Library
Cataloguing in Publication Data
Powell-Smith, Vincent
A building contract casebook.
1. Building — Contracts and specifications
— Great Britain — Cases
I. Title   II. Furmston, M.P.
344.103′7869   KD1641.A7

ISBN 0–632–02109–8

BSP Professional Books
Editorial offices:
Osney Mead, Oxford OX2 0EL
  (*Orders:* Tel. 0865 240201)
8 John Street, London WC1N 2ES
23 Ainslie Place, Edinburgh EH3 6AJ
52 Beacon Street, Boston
  Massachusetts 02108, USA
667 Lytton Avenue, Palo Alto
  California 94301, USA
107 Barry Street, Carlton
  Victoria 3053, Australia

Set by V & M Graphics Ltd
Aylesbury, Bucks
Printed and bound in Great Britain

Contents

Preface

Most modern building contracts are entered into on standard form terms, such as JCT 80, and this has engendered the belief that the printed document contains all the contractual obligations. This belief is erroneous because the law of building contracts is firmly fixed in the mainstream of the general law.

The cases to be found in this work explain some of the principles which are especially applicable to building contracts. The approach is practical rather than academic, this being reflected in both treatment and choice of cases. We have concentrated on basic principles, and our aim has been to illustrate the application of the law to particular and sometimes special fact situations, which arise in the construction industry. It must, however, be remembered that parties are often free to depart from general principles, many provisions in the standard forms being designed toward this end. Individual cases must for this reason be read both for their facts and for the relevant contractual provisions.

We hope that the book will be of use to building employers, architects, quantity surveyors, contractors and sub-contractors alike. We hope, too, that students will find it of help and interest. A study of the case law shows the development and application of the law in practice. The law is not a series of unadorned rules; rather, it is a living and vital subject. The book is meant to be used in conjunction with the standard texts on building contracts and is an adjunct to them.

The selection of cases is our own. The problem has been what to leave out rather than what to put in, and this has caused us much anxious thought. Moreover, although each of us assumed primary responsibility for particular sections, the result is a truly joint effort, involving much discussion and consultation.

Some Commonwealth decisions are included either because the point at issue does not appear to have been decided in any reported English case or because, in our view, the Commonwealth court has approached the matter in a way more readily understandable to this book's intended readers. Many of the cases abstracted here are not easily found in the usual run of law reports available to members of the construction team: others are old favourites. If a further edition of this book is called for, no doubt new cases will need to be added, for an ever-increasing number of construction industry disputes are reaching the courts to the benefit of not only the legal profession but, hopefully, to that of the construction industry as a whole. Some may disagree and retort:

'Physicians of the utmost fame
Were called at once; but when they came,
They answered, as they took their fees,
"There is no cure for this disease".'
(Hilaire Belloc 1870-1953)

The need for a reprint so soon after publication has proved the need for this book and has enabled us to include additional cases as an Appendix, with cross-references to the main text. For the most part these are taken from *Construction Law Reports*, of which we are joint editors. The text is thus updated to include material available to us up to 1 August 1986. We are grateful to Miss Linda Mulcahy, LLB for her help in connection with the Appendix.

VINCENT POWELL-SMITH

MICHAEL FURMSTON

Acknowledgments

We are grateful to Mr John F. Mulvehill, LLB, Chief Executive of Dudley Metropolitan Borough, for making available to us the transcript of the decision of the Court of Appeal in *Dudley Corporation* v. *Parsons & Morrin Ltd*, and to Mr G.V. Semain, Chief Executive of Thurrock Borough Council for a similar courtesy in respect of the judgments in *Thomas Bates & Son Ltd* v. *Thurrock Borough Council*. Dr John Parris, PhD, LLB, was also good enough to supply us with transcripts of a few unreported cases. We also wish to thank Mr J.R.J. Palmer, who prepared the index, and undertook the preparation of the table of cases, and Mrs Elizabeth Powell-Smith for help with the reading of the proofs.

V.P.-S.
M.P.F.

PART I

FORMATION OF THE CONTRACT

Chapter 1

Letters of Intent

Usually, a letter of intent merely expresses an intention to enter into a contract in the future, and creates no contractual liability. It is a question upon the facts of each case whether a letter of intent gives rise to any and, if so, what liability.

British Steel Corporation v. Cleveland Bridge & Engineering Co. Ltd

QUEEN'S BENCH DIVISION   (1981), 24 BLR 94

CBE were involved in the construction of a bank in Saudi Arabia, and needed to be supplied with cast steel nodes, for which supply they negotiated with BSC. On 21 February 1979, CBE sent the following letter of intent to BSC:

'Samma Bank: Damman

We are pleased to advise you that it is [our] intention to enter into a sub-contract with your company, for the supply and delivery of the steel castings which form the roof nodes on this project . . . We understand that you are already in possession of a complete set of our node detail drawings and we request that you proceed immediately with the works pending the preparation and issuing to you of the official form of sub-contract'.

Consequent upon this letter BSC started work. On 27 February 1979 CBE sent a telex to BSC listing the sequence in which they required the nodes to be delivered, this being the first intimation that CBE required the nodes to be manufactured in a particular order. Despite the problems this caused, BSC continued to manufacture the nodes.

CBE did not send 'the official form of sub-contract' to BSC, nor did the parties agree on price or delivery dates. Ultimately CBE did not pay BSC for the delivered nodes.

BSC claimed £229,832.70 from CBE as the price of the nodes, the claim being based in contract and alternatively upon a *quantum meruit*. CBE admitted that the goods were sold and delivered to them, and admitted liability in part. This admission was subject to a plea of set-off totalling £867,735.68, which CBE counter-claimed on the ground that BSC, in breach of contract, delivered the nodes late and out of sequence.

BSC's primary contention was that they were entitled to be paid on a *quantum meruit* because, if there was no binding contract between the parties, there was no legal basis for CBE's counter-claim. CBE argued that the agreement between the parties was comprised in the letter of 21 February, the telex concerning the delivery sequence, and BSC's conduct in proceeding with the manufacture of the nodes.

Held: No contract had come into existence between the parties on the basis of the letter of intent and BSC's performance, but BSC were entitled to be paid upon a *quantum meruit*.

> ROBERT GOFF J: [The] question whether in a case such as the present any contract has come into existence must depend on a true construction of the relevant communications which have passed between the parties, and the effect (if any) of their actions pursuant to those communications. There can be no hard and fast answer to the question whether a letter of intent will give rise to a binding agreement; everything must depend on the circumstances of the particular case. In most cases where work is done pursuant to a request contained in a letter of intent, it will not matter whether a contract did or did not come into existence; because if the party who has acted on the request is simply claiming payment, his claim will usually be based upon a *quantum meruit*, and it will make no difference whether that claim is contractual or quasi-contractual. Of course, a *quantum meruit* claim (like the old actions for money had and received, and for money paid) straddles the boundaries of what we now call contract and restitution; so the mere

framing of a claim as a *quantum meruit* claim, or a claim for a reasonable sum, does not assist in classifying the claim as contractual or quasi-contractual. But where, as here, one party is seeking to claim damages for breach of contract, the question whether any contract came into existence is of crucial importance.

As a matter of analysis the contract (if any) which may come into existence following a letter of intent may take one of two forms – either there may be an ordinary executory contract, under which each party assumes reciprocal obligations to the other; or there may be what is sometimes called an 'if' contract, i.e. a contract under which A requests B to carry out a certain performance and promises B that, if he does so, he will receive a certain performance in return, usually remuneration for his performance. The latter transaction is really no more than a standing offer which, if acted upon before it lapses or is lawfully withdrawn, will result in a binding contract.

The former type of contract was held to exist by Judge Fay in *Turriff Construction Ltd* v. *Regalia Knitting Mills Ltd* (1971), and it is the type of contract for which [CBE] contended in the present case. Of course, as I have already said, everything must depend on the facts of the particular case; but certainly, on the facts of the present case – and, as I imagine, on the facts of most cases – this must be a very difficult submission to maintain. It is only necessary to look at the terms of CBE's letter of intent in the present case to appreciate the difficulties. In that letter, the request to BSC to proceed immediately with the work was stated to be 'pending the preparation and issuing to you of the official form of sub-contract', being a sub-contract which was plainly in a state of negotiation, not least on the issues of price, delivery dates, and the applicable terms and conditions. In these circumstances, it is very difficult to see how BSC, by starting work, bound themselves to any contractual performance. No doubt it was envisaged by CBE at the time they sent the letter that negotiations had reached an advanced stage, and that a formal contract would soon be signed; but since the parties were still in a state of negotiation, it is impossible to say with any degree of certainty what the material terms of the contract would be. I find myself quite unable to conclude that, by starting work in these circumstances, BSC bound themselves to complete the work. In the course of argument, I put to [counsel] the question whether BSC were free at any time, after starting work, to cease work; his submission was that they were not free to do so, even if negotiations on the terms of the formal contract broke down completely. I find this submission to be so repugnant to common sense and the commercial realities that I am unable to accept it. It is perhaps revealing that, on 4 April 1979, BSC did indeed state that they were not prepared to proceed with the contract until they had an agreed specification – a reaction which, in my judgment, reflected not only the commercial, but also the legal, realities of the situation.

I therefore reject CBE's submission that a binding executory contract came into existence in this case. There remains the question whether, by reason of BSC carrying out work pursuant to the request contained in CBE's letter of intent, there came into existence a contract by virtue of which BSC were entitled to claim reasonable remuneration; i.e., whether there was an 'if' contract of the kind I have described. In the course of argument, I was attracted by this alternative – really on the basis that, not only was it analytically possible, but also that it could provide a vehicle for certain contractual obligations of BSC concerning their performance – e.g., implied terms as to the quality of goods supplied by them. But the more I have considered the case, the less attractive I have found this alternative. The real difficulty is to be found in the factual matrix of the transaction, and in particular the fact that the work was being done *pending* a formal sub-contract the terms of which were still in a state of negotiation. It is, of course, a notorious fact that, when a contract is made for the supply of goods on a scale and in circumstances such as the present, it will in all probability be subject to standard terms, usually the standard terms of the supplier. Such standard terms will frequently legislate, not only for the liability of the seller for defects, but also for the damages (if any) for which the seller will be liable in the event not only of defects in the goods but also of late delivery. It is a commonplace that a seller of goods may exclude liability for consequential loss, and may agree liquidated damages for delay. In the present case, an unresolved dispute broke out between the parties on the question whether CBE's or BSC's standard terms were to apply – the former providing no limit to the seller's liability for delay, and the latter excluding such liability altogether. Accordingly when in a case such as the present the parties are still in a state of negotiation, it is impossible to predicate what liability (if any) will be assumed by the seller, e.g., for defective goods or late delivery, if a formal contract should be entered into. In these circumstances, if the buyer asks the seller to commence work 'pending' the parties entering into a formal contract, it is difficult to infer from the seller acting upon that request that he is assuming any responsibility for his performance, except such responsibility as will rest upon him under the terms of the contract which both parties confidently anticipate they will shortly enter into. It would be an extraordinary result if, by acting on such a request in such circumstances, the buyer were to assume an unlimited liability for his contractual performance, when he would never assume such liability under any contract which he entered into.

For these reasons, I reject the solution of the 'if' contract. In my judgment, the true analysis of the situation is simply this. Both parties confidently expected a formal contract to eventuate. In these circumstances, to expedite performance under that anticipated contract, one requested the other to commence the contract work, and the other complied with that request. If thereafter – as anticipated – a contract was entered into, the work done as requested will be treated as having been performed under that contract; if, contrary to their expectation, no contract was entered into, then the performance of the work is not referable to any contract of which the terms can be ascertained, and the law simply imposes an obligation on the party who made the request to pay a reasonable sum for such work as has been done pursuant to that request, such an obligation sounding in quasi-contract or, as we now say, in restitution. Consistently with that solution, the party making the request may find himself liable to pay for work which he would not have had to pay for as such if the anticipated contract had come into existence, e.g. preparatory work which will, if the contract is made, be allowed for in the price of the finished work: cf. *William Lacey (Hounslow) Ltd* v. *Davis* (1957). This solution moreover accords with authority, e.g. the decision in *Lacey* v. *Davis* (above); the decision of the Court of Appeal in the unreported case of *Sanders & Forster Ltd* v. *A. Monk & Co. Ltd* (1980), though that decision rested in part on a concession; and the crisp dictum of Parker J in *O.T.M. Ltd* v. *Hydranautics* (1981), when he said of a letter of intent that 'its only effect would be to enable the defendants to recover on a *quantum meruit* for work done pursuant to the direction' contained in the letter. I only wish to add to this part of my judgment the footnote that, even if I had concluded that in the circumstances of the present case there was a contract between the parties and that that contract was of the kind I have described as an 'if' contract, then I would still have concluded that there was no obligation under that contract on the part of BSC to continue with or complete the contract work, and therefore no obligation on their part to complete the work within a reasonable time. However, my conclusion in the present case is that the parties never entered into any contract at all.

In the course of his argument, [Counsel] submitted that, in a contract of this kind, the price is always an essential term in the sense that, if it is not agreed, no contract can come into existence. [He] relied upon a dictum of Lord Denning MR in *Courtney & Fairbairn Ltd* v. *Tolaini Brothers (Hotels) Ltd* (1975), to the effect that the price in a building contract is of fundamental importance. I do not however read the Master of the Rolls' dictum as stating that in every building contract the price is invariably an essential term, particularly as he expressly referred to the substantial size of the contract then before the Court. No doubt in the vast majority of business transactions, particularly those of substantial size, the price will indeed be an essential term, but in the final analysis it must be a question of construction of the particular transaction whether it is so. This is plain from the familiar trilogy of cases – *May & Butcher Ltd* v. *The King* (1929); *W.N. Hillas and Co. Ltd* v. *Arcos Ltd* (1932); and *Foley* v. *Classique Coaches Ltd* (1934), – which show that no hard and fast rule can be laid down but that the question in each case is whether, on a true construction of the relevant transaction, it was consistent with the intention of the parties that even though no price had been agreed a reasonable price should be paid. In the present case, however, I have no doubt whatsoever that, consistently with the view expressed by the Master of the Rolls in *Courtney* v. *Tolaini Brothers*, the price was indeed an essential term, upon which (among other essential terms) no final agreement was ever reached.

It follows that BSC are entitled to succeed on their claim; and that CBE's set-off and counterclaim must fail.

Turriff Construction Ltd and Turriff Ltd v. Regalia Knitting Mills Ltd

QUEEN'S BENCH DIVISION   (1971), 9 BLR 20

Regalia invited Turriff to tender for the design and construction of a new factory at Corby. The tender was finalised in May 1969. At a meeting on 2 June 1969, Regalia informed Turriff that they wished them to carry out the project based on completion in 1972. Turriff asked for 'an early letter of intent . . . to cover [them] for the work [they] will now be undertaking'. On 17 June 1969 Regalia sent the following letter to Turriff:

'As agreed at our meeting of 2 June 1969 it is [our] intention to award a contract to [you] to build a factory including production, stores, offices, and canteen facilities to be built in four continuous phases.

Phase I to be on a fixed price basis as agreed and phases II, III and IV to be calculated on the same basis as phase I and completed by 1972. The commencing date to be 1 August and the terms of payment to be negotiated on a monthly form against bills of quantity supplied by [our] surveyor.

All this to be subject to obtaining agreement on the land, and leases with the Corby Development Corporation, full building and bye-law consent, and the site investigation being undertaken by [other contractors].

The whole to be subject to agreement on an acceptable contract.'

Once planning permission was received, Turriff began their detailed design work and started to negotiate with Regalia's architect about the form of contract. For various reasons, work had still not begun on site in December 1969 and Turriff asked for an interim payment of £3,500 in respect of design costs. They then learned that the project was to be abandoned, and on 12 February 1970 Regalia wrote repudiating liability on the basis that 'no contract exists between our two companies since we were, in fact, careful when writing to you on 17 June 1969 to explain that the whole matter was "subject to agreement on an acceptable contract" . . .'. Turriff issued a writ claiming the cost of work done from the date of the meeting held on 2 June 1969.

Held: Regalia was liable for the work carried out. At the meeting of 2 June 1969, Turriff had offered to carry out the preparatory work provided Regalia assumed liability, and had indicated that they would regard a letter of intent as an acceptance of their offer. The letter of intent sent on 17 June 1969 did not negative liability. There was an ancillary contract covering the interim costs.

JUDGE EDGAR FAY QC: [Counsel for the plaintiffs] put his claim in two ways. First, he said the services were rendered generally at a time when both parties believed that Turriff would be awarded the contract. In support of this proposition he cited the case of *William Lacey (Hounslow) Ltd* v. *Davis* (1957). This is a claim in quasi-contract and raises difficult questions of fact and law . . . but I accept [the defendant's] submission . . . that the parties had at the material time addressed their minds to the question of liability in issue and this feature takes the matter out of the realm of quasi-contract and into that of contract. Since the parties discussed the matter the question is whether their discussion resulted in some and if so what contractual obligation. . . . [Turriff say] the defendants assumed an obligation to pay for the services the subject of the action by their agreement to provide and subsequent provision of a Letter of Intent for the reasons expressed at the meeting of 2 June 1969. . . . [This] argument appears at first sight to share [Turriff's] view that some legal virtue attaches to a document called a Letter of Intent. As I understand it such a letter is no more than the expression in writing of a party's present intention to enter into a contract at a future date. Save in exceptional circumstances it can have no binding effect. However that does not dispose of the matter because much more occurred in this case than a mere request for a letter of intent and the giving of such a letter. . . .

[The judge then reviewed the evidence about the events between 2 and 17 June 1969 and found, amongst other things, that Turriff had made known to Regalia that they wanted an assurance that they would be paid for the preparatory work even if the project were abandoned, and that they would regard a letter of intent as being that assurance. The letter of intent was accepted by Turriff as that assurance, and they carried out the preparatory work in consequence of it. Accordingly he held that Regalia was liable.]

Chapter 2

Costs of Tendering

The cost of tendering is generally borne by the contractor, but if the contractor performs additional services at the employer's request, he may be entitled to reasonable payment.

William Lacey (Hounslow) Ltd v. Davis

QUEEN'S BENCH DIVISION   [1957] 2 All ER 712

The plaintiff tendered for the re-construction of war-damaged premises belonging to the defendant, who led the plaintiff to believe that it would receive the contract. At the defendant's request, the plaintiff calculated the timber and steel requirements for the building, and prepared various further schedules and estimates which the defendant made use of in negotiations with the War Damage Commission. Eventually, the plaintiff was informed that the defendant intended to employ another builder to do the work. In fact, the defendant sold the premises.

The plaintiff claimed damages for breach of contract, and, alternatively, remuneration on a *quantum meruit* in respect of the work done by it in connection with the reconstruction scheme.

Held: Although no binding contract had been concluded between the parties, a promise should be implied that the defendant would pay a reasonable sum to the plaintiff in respect of the services rendered. Judgment was entered for the plaintiff in the sum of £250 13s 5d.

BARRY J: I am unable to see any valid distinction between work done which has to be paid for under the terms of a contract erroneously believed to be in existence, and work which was to be paid for out of the proceeds of a contract which both parties erroneously believed was about to be made. . . . In both cases, when the beliefs of the parties were falsified, the law implies an obligation . . . to pay a reasonable price for the services which had been obtained.

Chapter 3

Tenders and Estimates

The contractor's offer to carry out the works is called a Tender. It must be definite and unambiguous in its terms if its acceptance is to conclude an agreement enforceable by the law as a contract.

Peter Lind & Co. Ltd v.
Mersey Docks & Harbour Board

QUEEN'S BENCH DIVISION   [1972] 2 Lloyd's Rep 234

The plaintiffs submitted to the defendants alternative tenders for the construction of a freight terminal, one on a fixed-price basis, and the other on a 'cost-plus' basis. The defendants purported to accept 'your tender', but did not specify which one. The plaintiffs carried out the work and claimed payment on a *quantum meruit*.

Held: There was no concluded contract because the defendants' purported acceptance did not specify which tender was being accepted. The plaintiffs were entitled to payment on a *quantum meruit*.

Normally an estimate will be held to be a firm offer in law, and its acceptance will result in a binding contract.

Crowshaw v. Pritchard and Renwick

QUEEN'S BENCH DIVISION   (1899), 16 TLR 45

The plaintiff wrote to the defendant enclosing a drawing and specification, and inviting a tender for some alteration work. The defendant replied:

> 'ESTIMATE – Our estimate to carry out the sundry alterations to the above premises, according to the drawings and specifications, amounts to £1,230.'

The plaintiff wrote next day accepting the defendant's 'offer to execute' the works, but later the defendant refused to go ahead. He contended that by using the word 'estimate' he did not intend it as an offer to do the work, and also that there was a trade custom that a letter in this form was not to be treated as an offer.

Held: The 'estimate' was an offer which had been accepted by the plaintiff and the defendant was liable in contract.

BIGHAM J: It has been suggested that there is some custom or well-known understanding that a letter in this form is not to be treated as an offer. There is no such custom, and, if there is, it is contrary to law.

Chapter 4

Incorporation of Documents

A reference in a contractual document to the contract being subject to conditions 'available on request' and brought to the notice of the other party is sufficient to incorporate the current edition of those conditions into the contract.

Smith v. South Wales Switchgear Ltd

HOUSE OF LORDS   [1978] 1 All ER 18

S had overhauled SWS's electrical equipment for some years. On 18 March 1970 SWS wrote to S asking him to carry out an overhaul of the equipment in July 1970 and S replied that he would do so. On 27 May SWS sent S a 'purchase note' requesting the overhaul 'subject to . . . our General Conditions of Contract 24001, obtainable on request'. S replied stating that he had given instructions for the work to be carried out. He did not request a copy of the conditions, although on 1 July 1970 a copy of the 1969 conditions was sent to him. There were two other versions of the conditions, including one dated March 1970.

Held: The reference in the purchase order to 'General Conditions . . . available on request' was sufficient to incorporate the general conditions as revised in March 1970. The reference clearly indicated how the terms might be ascertained by S. The conditions referred to were those current at the date of the contract because it was common knowledge that such conditions were revised from time to time. Had S asked for a copy of the general conditions, he could reasonably expect to have received the latest edition.

LORD KEITH: I consider that the contract was concluded by the appellants receiving the respondents' original purchase order dated 27 May 1970, coupled with their letter to the respondents dated 16 July 1970 saying that they had instructed the work to be carried out . . . The purchase order referred to the respondents' 'General Conditions of Contract 24001 obtainable on request.' It thus clearly indicated the manner in which the terms of these conditions could be ascertained, and that was sufficient in law, unconditional acceptance having followed, for their incorporation into the contract. A question however arises as to what in the circumstances is the meaning to be attributed to 'General Conditions of Contract 24001, obtainable on request', it having emerged that the original terms of these conditions were revised by the respondents in January 1969 and again in March 1970. In my opinion the meaning reasonably to be attributed is that the conditions referred to are the current conditions, that is to say, the conditions as revised in March 1970. It is common experience that the general conditions of contract of various undertakers are revised from time to time, and anyone requesting a copy of such conditions would reasonably expect to receive the current up-to-date edition. Further, that is the edition which a responsible employee of the undertaker clearly ought to send to an inquirer. Viewing the matter objectively therefore, I am of opinion that the parties contracted on the basis of the respondents' General Conditions of Contract 24001 as revised in March 1970. I do not consider that the reference in the purchase order to 'General Conditions of Contract 24001', without any mention of revision, precludes the relevant conditions from being the latest revised version. All three versions in fact bore the number 24001. The situation does not differ from that which would exist had the reference simply been to 'General Conditions of Contract' without any number. It would be unreasonable to suggest that wherever such conditions are revised the fact of revision must necessarily be indicated on any contractual reference to them. Any reference to conditions can only be understood, in the mind of an ordinary reasonable man, as a reference to the conditions currently in force. The circumstance that on 1 July 1970 the respondents sent to the appellants a purchase amendment order having printed on the reverse a copy of the general conditions as revised in January 1969 does not, in my view, affect the matter. This occurred at a time when both parties were erroneously attributing the number NQ492122 to contractual work which had been completed at the end of 1969, and engaged in fixing the price to be paid for that work. In that situation the January 1969 version of the general conditions was quite rightly attributed to the particular work in question.

The terms of a standard form contract may be incorporated into the contract by reference.

Killby & Gayford Ltd v. Selincourt Ltd

COURT OF APPEAL (1973), 3 BLR 104

A contract was made by an exchange of letters. In a letter dated 11 February 1972 the architects asked the contractors to price alteration work. The letter concluded:

'Assuming that we can agree a satisfactory contract price between us, the general conditions and terms will be subject to the normal standard form of RIBA contract'.

The contractors submitted a written estimate on 6 March 1972, to which the architect replied that he was accepting 'on behalf of the client . . . your estimate dated 6 March 1972 and I shall be obliged if you will accept this letter as your formal instruction to proceed with the work.' No JCT contract was ever signed, but the contractors proceeded with the work.

Held: The exchange of letters incorporated the current JCT form.

Even where the parties use loose or inappropriate language when referring to standard form contracts the court will endeavour to make sense of vague references to them. However, the court will not create a contract for the parties, but will seek to apply their presumed intention.

Brightside Kilpatrick Engineering Services v. Mitchell Construction (1973) Ltd

COURT OF APPEAL   (1975), 1 BLR 62

The plaintiffs were nominated as sub-contractors under JCT 63. The sub-contract was placed by the defendants on an order form dated 24 March 1972, which stipulated that the sub-contract documents should consist of a 'Standard Form of Tender dated 9.11.71.', a specification, 'Conditions of Contract' and a 'Schedule of Facilities'. The order continued that 'The Form of Contract with the Employer is RIBA 1963 Edition, July 1971 Revised', and concluded 'The Conditions applicable to the Sub-Contract shall be those embodied in RIBA as above agreement'. Printed references to the 'green' form of sub-contract had been deleted.

Held: The words should be read as referring to those clauses in the main contract which referred to matters concerning nominated sub-contractors. Clause 27 of JCT 63 regulated the nominated sub-contract relationship, and the only way to give a sensible meaning to the relationship was to read into it the terms of the 'green' form.

BUCKLEY LJ: It is not an easy point, for, as so often happens when parties try to import into their contractual relationships terms in other documents which do not easily fit those relationships, it is difficult to know what is the proper construction to put upon the documents in the case.

The contention for the defendants has been that the statement in the document of 24 March 1972 that 'The Conditions applicable to the Sub-Contract with you shall be those embodied in RIBA as above' is a clear reference to the main contract, which was the RIBA 1963 form of contract, and that the effect of that is to import into the contractual relationship between the contractors and the sub-contractors all the terms of the head contract *mutatis mutandis* substituting for references to the building owner references to the contractor, and substituting for references to the contractor references to the sub-contractor, and that that contract as so modified rules the relationship between the contractor and the sub-contractor accordingly, at any rate so far as the terms of that contract as so modified are capable of applying as a practical matter to the relationship between a contractor and sub-contractor.

I say 'capable of applying as a practical matter' because . . . when one comes to read the head contract adapted *mutatis mutandis* in the way I have indicated, many of its clauses are already quite inappropriate and incapable of being given a sensible effect. That that is the position, to my mind, is a strong indication that one should approach interpreting the order form of 24 March 1972 in this respect in this way with the greatest circumspection, particularly as so construing the document brings it into conflict with the terms of the tender document, which, as I have said, contemplated that the parties would enter into a contract between contractor and sub-contractor in the green form and not in a form produced by modifying the RIBA contract.

As I have said, it is not easy to know precisely what these words were intended to mean, but I am much inclined to the view that upon their true construction what they say, and all that they say, is this: that the sub-contractual relationship between the contractor and the sub-contractor shall be such as to be consistent with all those terms in the head contract that specifically deal with matters relating to sub-contractors, so that there shall be no conflict between the head contract and the sub-contract.

It is to be observed that the words, 'The Conditions applicable to the Sub-Contract with you shall be those embodied in RIBA as above' do not say, in terms at any rate, that the sub-contract between the contractor and the sub-contractor shall be in the RIBA form, or the RIBA form suitably adapted, or anything of that kind, but that some conditions applicable to the sub-contract shall be found in the RIBA contract. In my judgment, the best way to give a sensible interpreation to those words is to read them as referring to those clauses in the head contract which relate to matters concerning sub-contractors.

Modern Building (Wales) Ltd v. Limmer & Trinidad Co. Ltd

COURT OF APPEAL (1975), 14 BLR 101

The plaintiffs were nominated sub-contractors to the defendants, the sub-contract having been placed by an order which read:

'To supply adequate labour, plant and machinery to carry out [the sub-contract works] . . . in full accordance with the appropriate form for nominated sub-contractors (RIBA 1965 edition). . . . All as your quotation'.

There was no RIBA sub-contract form, nor was there any RIBA contract in a 1965 edition. The main contract was in JCT Standard Form, 1963 edition.

Held: The words used in the order should be read as describing the then current 'green' form of sub-contract, which was thereby incorporated. The words in brackets in the order were inappropriate and could be disregarded.

BUCKLEY LJ: I think we have to consider whether, on the language of the order of 18 December, the arbitration clause in the 'green' form was or was not imported into the contract between the parties. . . . [Counsel] has said that there are various ways in which that order can be construed which would result in its being held that the arbitration clause was not so imported. First of all, he says that the words within the brackets, '(RIBA 1965 edition)' are really nonsensical, and the whole of the importation clause in the order form, if I can so describe it, is therefore of no effect at all, upon the basis, I think, that it must be read as a whole, and if it purports to import a non-existent form of contract, then it has no effect. In my judgment, the answer to that argument is this: that this is a clear case of *falsa demonstratio*. The words 'the appropriate form for nominated sub-contractors', if read in the light of the evidence . . . are clearly capable of identifying the document that was intended to be referred to. . . . There is language perfectly appropriate to describe the 'green' form. It is not language which ought to be read as intended to cut down and restrict the identifying operation of the preceding words. It is just an added description which on investigation turns out to be factually inaccurate. . . . I think the right way to construe this order form is to ignore the words in brackets altogether . . . and to accept that the reference to the appropriate form for sub-contractors is a reference to the 'green' form, that being the only form to which it is suggested that those words could apply and the form to which it is said that anybody in the trade would understand them as applying. . . .

[Counsel] has suggested that the written contract contains insufficient indication as to how various matters that are left blank in that 'green' form contract, and in particular the appendix to that contract, which relates to such matters as the completion period, the retention money and so forth, should be filled in. Where the parties by an agreement import the terms of some other document as part of their agreement those terms must be imported in their entirety, in my judgment, but subject to this that if any of the imported terms in any way conflict with the expressly agreed terms, the latter must prevail over what would otherwise be imported. Here it is not disputed that the written contract between the parties, consisting of the quotation and the order, contains all the essential terms of the contract, and, in my judgment, the 'green' form of contract must be treated as forming part of the written contract, subject to any modifications that may be necessary to make the clauses in the 'green' form accord in all respects with the express terms agreed between the parties.

Terms can be incorporated into a contract by a course of dealing between the parties as well as by express reference. Terms can also be incorporated on the basis of the common understanding of the parties.

British Crane Hire Corporation Ltd v. Ipswich Plant Hire Ltd

COURT OF APPEAL   [1974] 1 All ER 1059

The parties were engaged in the plant hire business. A drag line crane was hired by telephone. Hiring charges were agreed, but nothing was said about the other terms. After delivery, the plaintiff owners sent to the defendants the then current 'General Conditions for the Hiring of Plant', but the defendants failed to sign and return the acceptance note.

Held: As both parties were in the plant hire industry, and both used the same standard conditions when hiring plant, they must be treated as having assumed that the contract was made on those terms.

LORD DENNING MR: In support of the course of dealing, the plaintiffs relied on two previous transactions in which the defendants had hired cranes from the plaintiffs. Each was on a printed form which set out the hiring of a crane, the price, the site, and so forth; and also setting out the conditions the same as those here. There were thus only two transactions many months before and they were not known to the defendants' manager who ordered this crane. In the circumstances I doubt whether those two transactions would be sufficient to show a course of dealing.

In *Hollier* v. *Rambler Motors (AMC) Ltd* (1972), Salmon LJ said he knew of no case 'in which it has been decided or even argued that a term could be implied into an oral contract on the strength of a course of dealing (if it can be so called) which consisted at the most of three or four transactions over a period of five years.' That was a case of a private individual who had had his car repaired by the defendants and had signed forms with conditions on three or four occasions. The plaintiff there was not of equal bargaining power with the garage company which repaired the car. The conditions were not incorporated.

But here the parties were both in the trade and were of equal bargaining power. Each was a firm of plant hirers who hired out plant. The defendants themselves knew that firms in the plant-hiring trade always imposed conditions in regard to the hiring of plant; and that their conditions were on much the same lines. The defendants' manager (who knew the crane), was asked about it. He agreed that he had seen these conditions or similar ones in regard to the hiring of plant. He said that most of them were, to one extent or another, variations of a form which he called 'the Contractors' Plant Association form'. The defendants themselves (when they let out cranes) used the conditions of that form. The conditions on the plaintiffs' form were in rather different words, but nevertheless to much the same effect. . . .

From [the] evidence it is clear that both parties knew quite well that conditions were habitually imposed by the supplier of these machines: and both parties knew the substance of those conditions. In particular that, if the crane sank in soft ground, it was the hirer's job to recover it; and that there was an indemnity clause. In these circumstances, I think the conditions on the form should be regarded as incorporated into the contract. I would not put it so much on the course of dealing, but rather on the common understanding which is to be derived from the conduct of the parties, namely, that the hiring was to be on the terms of the plaintiffs' usual conditions.

As Lord Reid said in *McCutcheon* v. *David MacBrayne Ltd* (1964) . . .: 'The judicial task is not to discover the actual intentions of each party: it is to decide what each was reasonably entitled to conclude from the attitude of the other.' It seems to me that, in view of the relationship of the parties, when the defendants requested this crane urgently and it was supplied at once – before the usual form was received – the plaintiffs were entitled to conclude that the defendants were accepting it on the terms of the plaintiffs' own printed conditions. . . .

The parties may attempt to contract by exchanging letters, estimates, orders, and acknowledgments, each of which purports to be on the usual terms of the party producing it. This gives rise to the 'battle of the forms'. Different views have been expressed as to who wins this 'battle'. Probably the correct approach is to see whether one party has by words or conduct accepted the other's terms. In the absence of such acceptance there is probably no contract.

Butler Machine Tool Co. Ltd v. Ex-Cell-O-Corporation (England) Ltd

COURT OF APPEAL   [1979] 1 All ER 965

On 23 May 1969 the sellers offered to sell machine tools to the buyers for £75,535, delivery ten months forward. The offer took the form of a written quotation subject to terms which 'shall prevail over any terms and conditions in the buyer's order'. These terms included a price variation clause. The buyers replied on 27 May 1969 placing an order for the machinery, subject to their own terms, which did not include a price variation clause. There was a tear-off acknowledgment of receipt of order attached. This said: 'We accept your order on the terms and conditions stated thereon'. The sellers completed and signed the acknowledgment and returned it to the buyers with a letter saying that the order was 'being entered in accordance with our revised quotation of 23 May for delivery in 10/11 months'. After delivery, the sellers claimed increased costs, which the buyers refused to pay.

Held: The buyers' order of 27 May was a counter-offer which destroyed the offer made in the seller's quotation of 23 May. By completing and returning the acknowledgment slip, which was said to be on the buyers' terms and conditions, the sellers had accepted the counter-offer on those terms, which did not include a price variation clause. The sellers' accompanying letter referring to the quotation of 23 May was irrelevant: it merely referred to the price and identity of the machine-tools.

LORD DENNING MR: In most cases where there is a 'battle of forms' there is a contract as soon as the last of the forms is sent and received without objection being taken to it. . . . The difficulty is to decide which form, or which part of which form, is a term or condition of the contract. In some cases the battle is won by the man who fires the last shot. He is the man who puts forward the latest terms and conditions: and, if they are not objected to by the other party, he may be taken to have agreed to them. . . . In some cases, however, the battle is won by the man who gets the blow in first. If he offers to sell at a named price on the terms and conditions stated on the back and the buyer orders the goods purporting to accept the offer on an order form with his own different terms and conditions on the back, then, if the difference is so material that it would affect the price, the buyer ought not to be allowed to take advantage of the difference unless he draws it specifically to the attention of the seller. There are yet other cases where the battle depends on the shots fired on both sides. There is a concluded contract but the forms vary. The terms and conditions of both parties are to be construed together. If they can be reconciled so as to give a harmonious result, all well and good. If differences are irreconcilable, so that they are mutually contradictory, then the conflicting terms may have to be scrapped and replaced by reasonable implication.

LAWTON LJ: The modern commercial practice of making quotations and placing orders with conditions attached, usually in small print, is indeed likely, as in this case, to produce a battle of forms. The problem is how should that battle be conducted? The view taken by the judge was that the battle should extend over a wide area and the court should do its best to look into the minds of the parties and make certain assumptions. In my judgment, the battle has to be conducted in accordance with set rules. It is a battle more on classical 18th century lines when convention decided who had the right to open fire first rather than in accordance with the modern concept of attrition.

The rules relating to a battle of this kind have been known for the past 130-odd years. They were set out by the then Master of the Rolls, Lord Langdale, in *Hyde* v. *Wrench* (1840), and Lord Denning

MR has already referred to them; and, if anyone should have thought they were obsolescent, Megaw J in *Trollope & Colls Ltd* v. *Atomic Power Construction Ltd* (1962) called attention to the fact that those rules are still in force.

When those rules are applied to this case, in my judgment, the answer is obvious. The sellers started by making an offer. That was in their quotation. The small print was headed by the following words: 'General. All orders are accepted only upon and subject to the terms set out in our quotation and the following conditions. These terms and conditions shall prevail over any terms and conditions in the buyer's order.' That offer was not accepted. The buyers were only prepared to have one of these very expensive machines on their own terms. Their terms had very material differences in them from the terms put forward by the sellers. They could not be reconciled in any way. In the language of article 7 of the Uniform Law on the Formation of Contracts for the International Sale of Goods they did materially alter the terms set out in the offer made by the sellers.

As I understand *Hyde* v. *Wrench* (1840) and the cases which have followed, the consequence of placing the order in that way, if I may adopt Megaw J's words, was 'to kill the quotation'. It follows that the court has to look at what happened after the buyers made their counter-offer. By letter dated 4 June 1969 the sellers acknowledged receipt of the counter-offer, and they went on in this way: 'Details of this order have been passed to our Halifax works for attention and a formal acknowledgment of order will follow in due course.' That is clearly a reference to the printed tear-off slip which was at the bottom of the buyers' counter-offer. By letter dated 5 June 1969 the sales office manager at the sellers' Halifax factory completed that tear-off slip and sent it back to the buyers.

It is true, as counsel for the sellers has reminded us, that the return of that printed slip was accompanied by a letter which had this sentence in it: 'This is being entered in accordance with our revised quotation of 23 May for delivery in 10/11 months.' I agree with Lord Denning MR that, in a business sense, that refers to the quotation as to the price and the identity of the machine, and it does not bring into the contract the small print conditions on the back of the quotation. Those small print conditions had disappeared from the story. That was when the contract was made. At that date it was a fixed price contract without a price escalation clause.

As I pointed out in the course of argument to counsel for the sellers, if the letter of 5 June which accompanied the form acknowledging the terms which the buyers had specified had amounted to a counter-offer, then in my judgment the parties never were ad idem. It cannot be said that the buyers accepted the counter-offer by reason of the fact that ultimately they took physical delivery of the machine. By the time they took physical delivery of the machine, they had made it clear by correspondence that they were not accepting that there was any price escalation clause in any contract which they had made with the plaintiffs.

Where the express terms of the contract are clear and unambiguous the courts will not imply a term simply to extricate a party from difficulties.

Trollope & Colls Ltd v. North-West Metropolitan Regional Hospital Board

HOUSE OF LORDS   (1973), 9 BLR 60

Building works were to be carried out in three phases, for each of which there was a separate contract sum and set of conditions, all based on JCT terms. The date of commencement of Phase III was to be fixed by reference to the completion date of Phase I. Phase III was to be completed by a specified date. Phase I was delayed by 59 weeks, 57 of which were allowed under the extension of time clause, and in effect the period for completing Phase III was reduced from 30 to 16 months. The contractors contended that a term for extending the date for completion of Phase III should be implied.

Held: No such term could be implied. The contract was clear and free from ambiguity in giving the date for completion of Phase III.

> LORD PEARSON: The basic principle [is] that the court does not make a contract for the parties. The court will not even improve the contract which the parties have made for themselves, however desirable the improvements might be. The court's function is to interpret and apply the contract which the parties have made for themselves. If the express terms are perfectly clear and free from ambiguity, there is no choice to be made between different possible

meanings: the clear terms must be applied even if the court thinks some other terms would have been more suitable. An unexpressed term can be implied if and only if the court finds that the parties must have intended that term to form part of their contract: it is not enough for the court to find that such a term would have been adopted by the parties as reasonable men if it had been suggested to them: it must have been a term that went without saying, a term *necessary* to give business efficacy to the contract, a term which, though tacit, formed part of the contract which the parties made for themselves.

The relevant express term is entirely clear and free from ambiguity: the date for completion of phase III is the date stated in the appendix to [the] conditions. . . . That term in itself can have only one meaning.

The standard forms contain no technical information about the work to be done. The contract will consist, therefore, of a combination of the contract conditions and other documents such as drawings, specifications and bills of quantities which describe the works. Sometimes the various contractual documents contain provisions which appear inconsistent with each other. Normally the courts in such a situation would attach more weight to documents such as bills which have been produced for the particular contract. The effect of JCT 80, clause 2.2.1 [JCT 63, clause 12(1)], which provides that 'nothing contained in the contract bills shall override or modify the application or interpretation of that which is contained in the Articles of Agreement, the Conditions or the Appendix' is that the printed provisions prevail over any special clauses in the bills. This reverses the ordinary rule that written words prevail over printed words in standard form contracts. Accordingly, the bills cannot modify the interpretation of the contract conditions.

North-West Metropolitan Regional Hospital Board v. T.A. Bickerton & Son Ltd

HOUSE OF LORDS   (1970), 7 BLR 107

For the facts and holding see p. 334.

LORD HODSON: I have not referred to [a] clause in the bills of quantities (which supports this construction) in view of the difficulty presented by condition 12(1) which makes it clear that, subject to immaterial exceptions, the conditions cannot be over-ridden, modified or affected by what is contained in the bills.

English Industrial Estates Corporation v. George Wimpey & Co. Ltd

COURT OF APPEAL   (1972), 7 BLR 122

For the facts and decision see p. 77.

STEPHENSON LJ: To apply the general principle that type should prevail over print seems to me to contradict the express provision of clause 12 that the reverse is to be true of this particular contract: the special conditions in type are to give way to the general conditions in print. The words in clause 12 'or affect in any way whatsoever the application or interpretation of that which is contained in these conditions' seem to mean that in this contract the bills of quantities are 'Contract Bills' in so far as they deal with 'the quality and quantity of the work included in the contract sum', but that in so far as they state conditions of the contract they have no effect on the printed conditions. In so far as they repeat or copy printed conditions or amendments which have been added to the printed conditions, as they do at pages 3 to 7, they add nothing to those conditions. In so far as they introduce further contractual obligations, as they do at page 9, they may add obligations which are consistent with the obligations imposed by the conditions, but they do not affect them by overriding or modifying them or in any other way whatsoever.

It follows from a literal interpretation of clause 12 that the court must disregard – or even reverse – the ordinary and sensible rules of construction and that the first of the documents comprising the works Wimpeys offered in their tender of 17 December 1968 to carry out expressly prevents the court from looking at the second of those documents to see what the first of them means. But that is because the second document is . . . a hybrid document and part of it deals with matters which should have been incorporated in the first; and that part can only live with clause 16 if it has . . . a different subject matter.

Those building owners who put forward the RIBA contract ought, in my judgment, to see either that clause 12 is struck out or amended or that all contractual terms which should be read together with the printed conditions are not left in the bills of quantities only but are incorporated in the conditions so that they cannot be shut out by clause 12 from being so read. Otherwise building owners have only themselves to blame if they find that they are saddled with such a liability as they have successfully contested in this case.

See also *Bramall & Ogden Ltd* v. *Sheffield City Council* (1983), Appendix, p.378.

Chapter 5

Acceptance of the Tender

An unambiguous acceptance of a tender by or on behalf of the employer and communicated to the contractor results in a binding contract, but there is no binding contract unless the parties are agreed on all essential terms.

Courtney & Fairbairn Ltd v. Tolaini Brothers (Hotels) Ltd

COURT OF APPEAL   (1975), 2 BLR 97

The defendants wished to develop a site by building a motel complex. They approached the plaintiffs who asked that if they (the plaintiffs) arranged the necessary finance for the project, the defendants would 'be prepared to instruct [their] Quantity Surveyor to negotiate fair and reasonable contract sums in respect of the . . . projects as they arise . . . based upon agreed estimates of the net cost of the work and general overheads with a margin for profit of 5%'. The defendants replied stating that they 'agreed to the terms specified'. The plaintiffs arranged the finance, but were not awarded the contract. The question was whether there was a binding and enforceable contract between the parties.

Held: There was no enforceable contract because there was no agreement about the price or any method of ascertaining it. The price was a matter of fundamental importance to the contract.

LORD DENNING MR: I can find no agreement on the price or any method by which the price was to be calculated. The agreement was only an agreement to 'negotiate' fair and reasonable contract sums. The words of the letter are 'your Quantity Surveyor to *negotiate* fair and reasonable contract sums in respect of each of the three projects as they arise'. Then there are words which show that estimates had not yet been agreed, but were yet to be agreed. The words are: 'These [the contract sums] would, incidentally be based upon agreed estimates of the net cost of work and general overheads with a margin for profit of 5 per cent.' Those words show that there were no estimates agreed and no contract sums agreed. All was left to be agreed in the future. It was to be agreed between the parties themselves. If they had left the price to be agreed by a third person such as an arbitrator, it would have been different. But here it was to be agreed between the parties themselves.

Now the price in a building contract is of fundamental importance. It is so essential a term that there is no contract unless the price is agreed or there is an agreed method of ascertaining it, not dependent on the negotiations of the two parties themselves. In a building contract both parties must know at the outset, before the work is started, what the price is to be, or, at all events, what agreed estimates are. No builder and no employer would ever dream of entering into a building contract for over £200,000 without there being an estimate of the cost and an agreed means of ascertaining the price.

In the ordinary course of things the architects and the quantity surveyors get out the specification and the bills of quantities. They are submitted to the contractors. They work out the figures and tender for the work at a named price: and there is a specified means of altering it up or down for extras or omissions and so forth, usually by means of an architect's certificate. In the absence of some such machinery, the only contract which you might find is a contract to do the work for a reasonable sum or for a sum to be fixed by a third party. But here there is no such contract at all. There is no machinery for ascertaining the price except by negotiation. In other words, the price is still to be agreed. Seeing that there is no agreement on so fundamental a matter as the price, there is no contract.

But then this point was raised. Even if there was not a contract actually to build, was not there a contract to negotiate? In this case Mr Tolaini did instruct his quantity surveyor to negotiate, but the negotiations broke down. It may be suggested that the quantity surveyor was to blame for the failure of the negotiations. But does that give rise to a cause of action? There is very little guidance in the book about a contract to negotiate. It was touched on by Lord Wright in *Hillas & Co. Ltd* v. *Arcos Ltd* (1932) where he said:

'There is then no bargain except to negotiate, and negotiations may be fruitless and end without any contract ensuing'. Then he went on: '. . . yet even then, in strict theory, there is a contract (if there is good consideration) to negotiate, though in the event of repudiation by one party the damages may be nominal, unless a jury think that the opportunity to negotiate was of some appreciable value to the injured party'. That tentative opinion by Lord Wright does not seem to me to be well founded. If the law does not recognise a contract to enter into a contract (when there is a fundamental term yet to be agreed) it seems to me it cannot recognise a contract to negotiate. The reason is because it is too uncertain to have any binding force. No court could estimate the damages because no one can tell whether the negotiations would be successful or would fall through; or if successful, what the result would be. It seems to me that a contract to negotiate, like a contract to enter into a contract, is not a contract known to the law. . . . I think we must apply the general principle that when there is a fundamental matter left undecided and to be the subject of negotiation, there is no contract. So I would hold that there was not any enforceable agreement in the letters between the plaintiff and the defendants. I would allow the appeal accordingly.

Where the parties use unclear or ambiguous language, the court will do the best it can to give a sensible meaning to their expressed intention, but it will not make a contract for them where none exists.

Constable Hart & Co. Ltd v. Peter Lind & Co. Ltd

COURT OF APPEAL   (1978), 9 BLR 1

CH were sub-contractors to PL. The sub-contract was in the FCEC standard form and incorporated CH's quotation which provided that it would 'remain Fixed Priced until 3 June 1975; any work carried out after this date to be negotiated' and this was confirmed by PL's purchase order which read 'Price fixed until 3 June 1975 – £434,732.29p. less 5 per cent discount'. The works were not completed by 3 June 1975 and no agreement was reached thereafter upon the terms on which CH were to be paid for work executed after that date.

Held: The sub-contract was subject to an implied term that the rates for work carried out after 3 June 1975 would be reasonable rates.

SIR DOUGLAS FRANK QC (whose judgment was affirmed by the Court of Appeal): [Counsel] for the claimants submitted that where parties have agreed that building work be carried out and have concluded a contract but where the price has not been agreed for part of the work, the law implies that a reasonable sum be paid. He referred me to *Foley* v. *Classique Coaches Ltd* (1934), *F. & G. Sykes (Wessex) Ltd* v. *Fine Fare Ltd* (1967), *Liverpool City Council* v. *Irwin* (1977). However, those cases only go to establish the proposition that where a contract is silent as to the price, the court, in order to save the contract, will say that there is an implied term that the price is to be a reasonable price. They do not assist in determining a formula for a reasonable price or in the case where there is more than one possible formula, for ascertaining a price. He also referred to *Shell UK Ltd* v. *Lostock Garage Ltd* (1976), and in particular to Lord Denning's two broad categories of implied terms. It was accepted by him and by . . . the respondents that this case falls within the second category, that is, it is not of common occurrence and that the implication is based on an imputed intention by the parties from their actual circumstances. These are what he described as the 'officious bystander' type of case, and in such cases a term is not to be implied on the ground that it would be reasonable but only when it is necessary and can be formulated with sufficient degree of precision. Therefore the question I have to decide is what advice would have been given by an officious bystander, albeit one with experience of civil engineering contracts, to and accepted by the parties had they applied their minds to the term which should have been incorporated in the contract. . . .

[The respondent's counsel] said that the words in the letter of 8 June 1973 'any works carried out after this date to be negotiated' are too uncertain to be given any meaning unless there is a yardstick by which they are to be agreed, and therefore one is led back to the implied term. That term, he said, must be determined in the light of the expressed terms. He emphasised that this is a lump sum contract, that is, at a fixed price for defined works, and that is the price referred to in clause 5. He pointed out that the contract provides for the valuation of variations in the work at specified rates and there is no similar provision for additional cost for work carried out after 3 June 1975 and so it can only be paid for as an additional sum under sub-clause (5) of clause 15. Thus there is no means of paying for the additional cost other than as an additional sum without modifying the price and that, he argued, cannot be done.

One difficulty I find with [that] argument is that it seems to me only to go to the time at which the additional payment is to be made and not to the method of its quantification. I think I must go back to June 1973 to see what was in the minds of the parties and I may here say that although [counsel] contended to the contrary, I find that the words 'any work carried out after this date to be negotiated' were embodied in the contract. It is clear that the claimants in arriving at their quotation had done so on the footing that the work would be completed by about 3 June 1975. Indeed, the period of completion specified in the contract was approximately 75 per cent of the main carriageway to be constructed in 1974, and the remainder in 1975. It is also clear that they had quoted on the basis of their priced bills and that their quotation had been accepted on that basis, and their quotation was expressed to form part of the contract. Thus, that the quotation was based on the priced bills and on the footing that the price would hold good until 3 June 1975 was recognised by both parties. It seems to me to follow naturally that had the parties then been asked the basis for payment after that date they would have replied that appropriate adjustments either up or down would have to be made to the prices in the bills and I think that it was when the claimants referred to the work to be negotiated they meant that there would have to be negotiations for the adjustment of the prices in the bills. This is a matter mainly of impression from the documents and all the surrounding circumstances. I do not think that the fact that the price is defined in the contract vitiates this conclusion. It seems to me that the terms of the letter of 8 June 1973 are broad enough to include adjustment of the price as defined and that is the conclusion I have reached.

Esmil Ltd v. Fairclough Civil Engineering Ltd

COURT OF APPEAL   (1981), 19 BLR 129

E were sub-contractors to F under 'Model Form "A" Home – with erection' terms, on the basis of a fixed price for a period of 18 months and thereafter with price variation using the BEAMA formula. This was not designed to operate only after the expiry of a fixed price period. The price clause was expressed to be 'subject to amendment at the time of entering into a formal contract and will embody the provisions of sub-clause 1.2.5(a) or 1.2.5(b) functional on the financial ramifications of partial fixed/variable and wholly variable cost contracts'. The formal contract merely incorporated the tender documents leaving the price clause unamended and unqualified, except by this statement. The fixed price period expired on 7 April 1975; E's work was programmed for completion by 1 July 1975, but in fact work continued until 1977. The question was: what were E's rights to additional payment for work done after 7 April 1975?

Held: The note to the price clause was almost incomprehensible but had to be given a sensible meaning. The intention of the parties was clear and E were entitled to an adjustment of the price to reflect the increased costs after the expiry of the fixed price period in respect of all their work which properly fell to be performed thereafter.

DONALDSON LJ: It is a wise and elementary precaution to agree expressly upon the terms of a contract before undertaking its performance. The parties to this appeal failed to do so. . . . The resulting controversy was wholly predictable. . . . The parties were agreed that they had been sufficiently *ad idem* for there to be an enforceable contract and that the contract documents evidenced the terms of that contract. The task can thus be labelled one of 'construction', bearing in mind that this word covers 'the action of framing, devising or forming' as well as that of 'construing' (*Shorter Oxford English Dictionary*). Plainly it was more concerned with the former than the latter. . . .

A tender and acceptance may amount to a contract, even though the acceptance refers to a formal contract to be drawn up afterwards.

Lewis v. Brass

COURT OF APPEAL (1877) 3 QBD 667

An architect invited tenders for building work. The defendant's tender read: 'I hereby agree to execute complete, within the space of 26 weeks from the day of receiving instructions to commence, the whole of the work required to be done . . . for the sum of £4,193'. The architect replied: 'I am instructed by my client . . . to accept your tender of £4,193 for works as above referred to. The contract will be prepared by [Lewis's solicitors], and I have no doubt it will be ready for signature in the course of a few days'. The defendant had made a mistake in his tender and tried to withdraw it.

Held: The tender and acceptance formed a contract.

Where one party makes an unambiguous offer to the other party he is bound by his offer when accepted, even if it contains a mistake of which the other party is unaware.

Centrovincial Estates PLC v. Merchant Investors Assurance Co. Ltd

COURT OF APPEAL 4 March 1983

C let several floors of an office building to an Industrial Training Board, which in turn sub-let one floor to M, at a yearly rent of £68,320. The sub-lease contained a rent review clause providing for the rent to be increased to the current market rental value. On 22

June 1982, C's solicitors wrote to M inviting them to agree the figure of £65,000 a year as the appropriate rental figure at the review date. M replied on 23 June accepting the offer. Some days later, the solicitors telephoned M saying that the letter of 22 June had contained an error and that C had intended to propose a figure of £126,000 a year. Subsequently, the solicitors wrote to M inviting them to treat the letter of 22 June as if it had quoted £126,000 a year. M refused to agree to the suggested correction and maintained that a binding agreement had been concluded by the two letters of 22 and 23 June.

Held: A binding agreement had been so concluded. It was contrary to well-established principles of law to suggest that the offeror under a bilateral contract could withdraw an unambiguous offer, after it had been properly accepted, merely because he had made a mistake which the other party neither knew nor could reasonably have known at the time when he accepted it.

An offer may be accepted by conduct, as where a contractor starts work on receipt of an order.

A. Davies & Co. (Shopfitters) Ltd v. William Old Ltd

QUEEN'S BENCH DIVISION   (1969), 67 LGR 395

A main contract in JCT 63 form was made between the employer and the defendants. The architect wrote to the plaintiffs, inviting them to tender as nominated sub-contractors, and saying that payment would be made on architect's certificates through the main contractor. The architect accepted the plaintiffs' tender, informing them that the main contractor would place an order with them. The main contractor's order introduced a new term, namely a 'pay when paid' clause. The sub-contractor started work without protest. The employer became insolvent, leaving unpaid sums due to the main contractor for work carried out by the plaintiffs, who now claimed payment from the main contractor.

Held: The sub-contractor had accepted the main contractor's terms by starting work. The order was a counter-offer, which had been accepted by conduct. The plaintiffs' action failed.

If the contractor makes a unilateral mistake in his tender price, or there are individual errors in pricing, multiplication, or addition, he is bound by such errors unless, before acceptance, the employer or the architect discovers the error and realizes that it is not intentional.

W. Higgins Ltd v. Northampton Corporation

CHANCERY DIVISION   [1927] 1 Ch. 128

H contracted with the local authority to erect 58 houses. H completed the tender incorrectly. The result was that H thought he was tendering for the erection of the houses at £1670 a pair, when in fact it was £1613 because of the way in which H had made up the Bills.

Held: H was bound by his mistake and could not claim to have the contract set aside or rectified.

ROMER J: The result is very unfortunate for the plaintiff and an extreme hardship on him. But can I rectify the contract so as to make it give effect to what was undoubtedly his intention when he entered into the contract? It appears to me that I cannot. [In] my opinion the parties were not labouring under a mutual mistake. I think the contract expresses what the defendants intended it to express. At any rate, it has not been shown to me that the defendants were labouring under any mistake, and of course the onus is upon the plaintiff to prove that fact, if fact it were. . . .

In these circumstances, with the utmost regret, I come to the conclusion that I cannot give the plaintiff the relief that he desires, and unless, even at the eleventh hour, the defendants can make some concession to the plaintiff, I am afraid that the ratepayers will have benefited by the bona fide and serious mistake made on the part of the plaintiff, which was really brought about by the carelessness of some official of the defendant corporation in drawing up the original bill of quantities. If he had made it clear in that bill of quantities, as he ought to have done, that the 50 yards of chimneys were already included in the 321 yards of wall, Mr Higgins would never have made the mistake that has resulted in this action.

If the employer discovers an error in the contractor's tender he cannot accept the mistaken tender so as to create a contract where the contractor's mistake is as to the terms of the offer itself and not merely as to the motive or underlying assumptions on which the offer is based.

McMaster University v. Wilchar Construction Ltd

ONTARIO HIGH COURT (1971), 22 DLR (3d) 9

The employer purported to accept a tender knowing that the contractor had omitted the entire first page of his bid. That page included an intended price fluctuations clause.

Held: The contractor was not liable.

THOMPSON J: There is not the slightest doubt in my mind that the real reason the plaintiff purported to accept Wilchar's tender was in the hope that it might be able to recover the penalty of the bid bond, knowing full well as early as 3 October that Wilchar had made a mistake in its tender and that it would refuse to enter into a contract unless the mistake were remedied. To me this is patently a case where the offeree, for its own advantage, snapped at the offeror's offer well knowing that the offer was made by mistake.

[Counsel for McMaster] concedes that Wilchar had, by mistake, not included the escalator clause as a term of its tender, but he argues that this was not a mistake of a fundamental character such as to vitiate the tender and that it was a mistake merely in the motive or the reason for making the offer.

I am not prepared to accede to such an argument. In a construction contract the price is always a fundamental term of the contract. In fact it is the very *quid pro quo* of such a contract. In the instant case, for the contractor the mistake meant a loss of thousands of dollars as contrasted with a profit in its absence; for the contractee, it meant an advantage of some $16,000.

Were it not that a provision or stipulation as to price were a fundamental term of a contract, then such decisions of the courts, ranging over a number of years, as in *Webster* v. *Cecil* (1861); *Garrard* v. *Frankel* (1862); *Hartog* v. *Colin and Shields* (1939) and *Bennett* v. *Adams River Lumber Co.* (1910) have all been in vain. . . .

In my view, this is truly a case of unilateral mistake. While perhaps inadvertent, it was an honest mistake, and one which Wilchar hastened to make known to the plaintiff as soon as it was discovered; one which was made known to the plaintiff long before the tender or offer was accepted. Moreover, I conclude that from the circumstances, the plaintiff must, in any event, be taken to have known of the mistake before acceptance of the offer. In this context, it should be stressed that one is taken to have known that which would have been obvious to a reasonable person in the light of the surrounding circumstances: see *Hartog* v. *Colin and Shields* (1939).

With full knowledge of the mistake, and the evidence as to this is irrefragable, apart from the concession of counsel, the plaintiff purported to accept Wilchar's tender, despite the protest of Wilchar and its endeavour to amend it.

There can be little doubt that there was no real agreement between the parties and that this is but a bold attempt by the plaintiff to force the defendant Wilchar to fulfil a promise in a sense which the plaintiff knew that Wilchar did not intend it and to which its mind did not assent: and so I hold.

To put it simply, this is a case where one party intended to make a contract on one set of terms and the other intended to make it upon another set of terms, with the result that there is lack of consensus. The parties were not *ad idem*. The existing circumstances prevented the formation of a contract.

I further hold that in any event, the circumstances were such that it would be unconscionable, unfair and unjust to permit the plaintiff to maintain the contract in the light of the conduct of its project manager, conduct which I find was not only fraudulent in that more broad equitable sense of the word, but which went even further.

Wilchar would be entitled to the equitable right of rescission, in the event that a contract had been consummated, on the basis that under the circumstances, the contract was voidable for fundamental mistake in its formation. Under such conditions, such a mistake is a good defence to any action brought to enforce such an alleged contract or to obtain damages for its breach: see *Raffles* v. *Wichelhaus* (1864).

Whether a bill rate applies to extra work will depend on the construction of the bill.

Dudley Corporation v. Parsons & Morrin Ltd

COURT OF APPEAL   8 April 1959

A contract for the building of a school was in the RIBA 1939 Form, with quantities. The contract terms in issue were essentially the same as those of JCT 63 and JCT 80. The contractors priced an item for excavating 750 cu. yds. in rock at £75, i.e. two shillings a cube. This was a gross underestimate, although it was not known whether rock would be met. In carrying out the excavations described in the drawings and bills, the contractors excavated a total of 2230 cu. yds. of rock. The architect valued the work at two shillings a cube for 750 cu. yds., and the balance at £2 a cube. £2 was not unreasonable if no other price applied. The employer disputed this amount. The arbitrator found 'no sufficient evidence . . . that the price of two shillings per yard cube . . . was a mistake'.

The question before the court was: 'Whether on the true construction of the Bills . . . the item "750 yards cube. *Extra only* for excavating in rock, including use of compressors etc. (Provisional) 2/-,£75" . . . refers and is applicable only to the three items immediately preceding the same (relating to the excavation of heating basement) or to all the items in respect of excavation preceding the same . . . or to some other and, if so, what items'.

Held: The contractors were entitled only to an extra two shillings per cu. yd. for all excavation in rock over the quantity provisionally allowed in the priced bills.

PEARCE LJ: Under the heading 'Excavator' there are first some general observations and specifications with regard to excavation. There follow seventeen particular items of excavation to be priced by the contractor. Then comes the extra item in dispute. It is in perfectly general terms. In my view, according to the natural reading of the document, that extra relates generally to all the previous items of excavation should it prove necessary in any of them to excavate rock. Further, that extra has no words limiting it to the foregoing three items. The letters 'do.' for 'ditto' have been very freely and very carefully used throughout wherever they are needed to link one item to the foregoing item. But here no such link is inserted. This very strongly confirms the natural construction namely, that the extra is of general application to all the preceding items.

I appreciate that there is considerable force in the argument so clearly put forward by [counsel for the defendant] which prevailed in the learned judge's view, but, with all respect, I do not feel that they suffice to overcome the matters I have mentioned. Moreover, the case shows that no one knew before hand whether rock would be encountered, there had been no trial drills, there was no reason to suppose that rock would be found in the area of the heating chamber in particular and no reason why the extra should be limited to the excavation in that area. If the extra relates only to that area, it would provisionally allow for 750 cu. yds. of rock in an area of 804 cu. yds. of soil, a rock percentage of well over 90 per cent. If, on the other hand, it related to the other items of excavation, it would provisionally allow for 750 cu. yds. of rock in more widely spread areas totalling 2,915 cu. yds., which is a rock percentage of just over twenty-five. This seems to me more likely.

In my view, the actual financial result should not affect one's view of the construction of the words. Naturally, one sympathises with the contractor in the circumstances, but one must assume that he chose to take the risk of greatly under-pricing an item which might not arise, whereby he lowered the tender by £1,425. He may well have thought it worth while to take that risk in order to increase his chances of securing the contract.

Where work has commenced before a contract is concluded, but the parties later agree on terms, the terms so agreed will have retrospective effect.

Trollope & Colls Ltd v. Atomic Power Constructions Ltd

QUEEN'S BENCH DIVISION   [1962] 3 All ER 1035

In February 1959 the plaintiffs submitted to the defendants a tender for carrying out, as sub-contractors, certain civil engineering work. Negotiations then took place and were continuing when, in June 1959, the plaintiffs began work at the defendants' request, and on the basis of a letter of intent. On 11 April 1960, the form of the general conditions was finally agreed between the parties, the judge holding that on that date the parties were then agreed on all the essential terms. Disputes arose as to whether the terms of the contract so agreed, and in particular the provisions as to variations, governed the rights of the parties for work done prior to 11 April 1960.

Held: The contract finally concluded on 11 April 1960 governed the rights of the parties as to prior work.

MEGAW J: So far as I am aware, there is no principle of English law which provides that a contract cannot in any circumstances have retrospective effect, or that, if it purports to have, in fact, retrospective effect, it is in law a nullity. If, indeed, there were such a principle, there would be many important mercantile contracts which would, no doubt to the consternation of the parties, be nullities. Frequently, in large transactions a written contract is expressed to have retrospective effect, sometimes lengthy retrospective effect: and this in cases where the negotiations on some of the terms have continued up to almost, if not quite, the date of the signature of the contract. The parties have meanwhile been conducting their transactions with one another, it may be for many months, on the assumption that a contract would ultimately be agreed on lines known to both the parties, though with the final form of various constituent terms of the proposed contract still under discussion. The parties have assumed that when the contract is made – when all the terms have been agreed in their final form – the contract will apply retrospectively to the preceding transactions. Often, as I say, the ultimate contract expressly so provides. I can see no reason why, if the parties so intend and agree, such a stipulation should be denied legal effect. Take, as an example, a simple case. Suppose that a contract for the sale of goods is under negotiation. The offeror has said: 'I will sell you one thousand tons of coal on such and such terms'. While the detailed terms of sale are still under negotiation, the offeree asks for the delivery of five hundred tons and the offeror makes delivery, both parties intending and anticipating that this delivery will count against the contract quantity if and when the contract is made, or perhaps believing wrongly that the contract has already been made. The final terms are then agreed, the offer is accepted, and the contract is made. Even if, in the actual acceptance by the offeree, no express reference is made to the antecedent delivery of the five hundred tons, I should have thought that there would be little room for doubt that the contract was intended to govern, and in law did govern, that antecedent delivery; and that neither party could successfully assert that there was no contract, or that the five hundred tons was delivered on a *quantum meruit* basis; or that the whole one thousand tons still fell to be delivered after the contract. Of course, the position would be different if no contract were ultimately made, as, for example, by the offeror's withdrawal of the offer before acceptance. So here. If a contract was made on 11 April 1960, and if the contract expressly provided, or should in law be assumed to have provided, that its terms as then agreed were to apply retrospectively to previous acts of the parties done since the date of the tender in anticipation of the making of such a contract at a later date, I see no reason why, in law, effect should not be given to such a provision. It is true that such retrospectivity may give rise to difficulties for one or other or both of the parties, where things have been done at a time when certain of the terms which ultimately and retrospectively become contractual were in a state of flux. Thus, it might turn out that something done by a party before the making of the contract had been done otherwise than in accordance with a term as subsequently agreed. That is a matter which may have to be taken into account as tending to militate against the implication of such an agreement as to retrospective effect. It would not, however, be sufficient to negative the existence of a contract with retrospective effect, if the assent of the parties to such a contract were clearly established.

In the present case, so far as I have seen, a retrospective effect is nowhere expressly stated or stipulated. Is it to be implied into the contract of 11 April 1960, if contract there was? That, I conceive, is the crucial question on this issue.

It has been said on many occasions that terms are not to be implied merely because they are desirable, or merely because the parties, if they had considered the question, would probably or as reasonable men have agreed such terms. Terms can only be implied where, to use the common phrase, they are necessary in order to give 'business efficacy' to the contract. In the present case, in one sense at least, this term of retrospectivity is necessary to give business efficacy to the contract, if contract there was. The submission of counsel for the plaintiffs is in my view right to this extent, that, in the absence of retrospective effect to the variations clauses, an agreement made on 11 April 1960, would not only be devoid of business efficacy but also would not be capable of being a contract at all, because it would pre-suppose a state of affairs (the original specifications and price) which had long since ceased to be realistic or practicable or within the true intentions of either party in relation to the work to be done under an agreement. On the other hand, I do not think that a term such as this can be implied simply for the purpose of upholding the existence of a contract, unless it can clearly be seen that it conforms with what the parties truly intended and with what they both would have accepted as a matter of course had the question been raised in the course of the negotiations or at the moment of the making of the supposed contract. But if those factors are present, in my judgment it is right and necessary that such a term should be treated as being implied.

I am satisfied from all the circumstances that both parties, in all that they did in the course of the negotiations, in the defendants' requests or instructions to the plaintiffs to carry out the work as varied, and in the plaintiffs' acceptance of those instructions, were doing so on the understanding and in the anticipation that, if a contract were made, and whenever it was made, that contract would apply to and govern what was meanwhile being done by the parties. I am satisfied that if, on 11 April 1960 (still assuming that a contract was, otherwise, made on that date) the question had been raised, both parties would have said, as a matter of course: 'This contract is to be treated as applying, not only to our future relations, but also to what has been done by us in the past since the date of the tender in the anticipation of the making of this contract'.

PART II
PERFORMANCE OF THE CONTRACT

Chapter 6

The Contractor's Obligations

(i) DESIGN

In the classic form of building contract, the employer and his professional advisers assume responsibility for the design of the building, and the contractor agrees to build the building as designed. In such a contract, the contractor does not in principle undertake liability for the design.

Bower v. Chapel-en-le-Frith Rural District Council

KING'S BENCH DIVISION   (1911), 9 LGR 339

The plaintiffs were the successful tenderers for the erection of a waterworks for the Council. The contract was for a lump sum and contained specifications and bills of quantities. The plaintiffs were required by the contract to buy a windmill tower and pump from a named supplier at a named price and to fix them. The windmill proved hopelessly defective. The Council claimed that the plaintiffs must replace it with an efficient one. The plaintiffs argued that they were not responsible as they had played no part in the choice of the windmill. It was not argued that the default was in any way due to defective installation.

Held: The plaintiffs were not liable.

LAWRENCE J: The question is: Were the plaintiffs liable for the inefficient working of the mill, or did their liability cease when they properly erected it under the contract? Another way of putting the question is: Was the plaintiffs' contract to make the mill answer its purpose; or was it a contract to do the work in accordance with the specification and plans? I need hardly say that in my judgment the answer to each of these questions should be in favour of the plaintiffs on the ground that I have already given, namely, that the mill was chosen by the defendants, and it was the mill the plaintiffs were bound by their contract to erect, and which they did erect, and that the plaintiffs had no means or power of objecting to the mill, or altering it, or doing anything to it. It was no part of their contract to guarantee in any way that the mill would be efficient.

[The Court of Appeal ordered a new trial on the ground that the judge had misunderstood the way which the parties intended to present the case, but expressed no opinion on the question of the law involved.]

Cable (1956) Ltd v. Hutcherson Brothers Pty. Ltd

HIGH COURT OF AUSTRALIA　　　　　　　　　　　　(1969) 43 ALJR 321

The respondents successfully tendered for the design, supply, and installation of a bulk storage and handling plant. This was to be built on reclaimed harbour land for the appellant. A formal contract was entered into, under which the drawings of the proposed installations, prepared by the respondents, had been approved by the appellant's consulting engineer. These drawings showed ring beam foundations for storage bins. Subsidence occurred when the bins were erected and filled.

Held: The promise of the respondent was to carry out the work agreed in the drawings.

BARWICK CJ: The principal matter to be resolved is the effect upon the promise of the respondent of the incorporation in the contract of the terms of the specification. The competing views are that of the appellant that the promise of the respondent was to produce a result, namely, storage bins of given holding capacity which with the ancillary plant would operate and carry out the appellant's scheme or project, that is to say, receive, hold and discharge the volume of the stated mineral to the specified extent at the specified rate: and that of the respondent, that the promise was to do only the work described in the agreed drawings, the appellant being itself satisfied upon its own engineer's advice that work so done would operate and carry out the scheme or project it proposed. In my opinion, the choice between the contentions can be quite securely made without reference to decided cases or indeed any attempt to find an analogy in any of them or in the facts of any of them. No doubt at times the question whether the promise of a builder is to produce a result or merely to do specific work is difficult to answer. In reaching a conclusion, the fact that it can be seen that reliance is placed upon the skill and judgment of the builder may on occasions be an important if not a decisive consideration. Here, there is no doubt that the respondent knew the result which the appellant sought to achieve and it might well be that at the time of tender an area of the design of the installation was left to the tenderer. Indeed, the respondent tendered a price to include 'design' (see the letter of tender) though oddly enough the word 'design' is not found in the basis of tender clause in the specification. The respondent did in fact propose the design of the foundation for the storage bins and prepared the first drawings for them. But, by the time the agreement was made the 'design' of the foundations had been agreed, after participation of the appellant's consulting engineer in the determination of its form. That agreed design, extrapolated in a drawing, was expressed to be part of the contract works (see reference to drawing No. 155/8 in the articles of agreement). But the appellant says that notwithstanding that fact so much may be drawn from the inclusion of the specification in the articles of agreement, that the responsibility for the suitability of the design to the appellant's mutually known purpose remained with the respondent. It is therefore necessary and important to observe what is brought out of the specifications prepared for tendering into the operation of the articles of agreement.

The description of the 'works' in the portion of the articles of agreement which I set out earlier in these reasons includes a reference to the specifications. The respondent's promise in clause 1 of the articles is to do the work shown in the drawings and described by or referred to in the specification and conditions. The conditions refer to the works shown in the drawings and described in the specification: but the contract conditions are paramount (clause 1(c)). However, in any case the specification is only incorporated insofar as it describes the 'works', presumably, the works which are to be found in or which form part of the works found in the drawings. It seems to me that the part of the specification most relied upon by the appellant, namely the basis of tender, does not perform or relate to the function of describing the works. In terms it relates only to the footing on which the tenderer should make his bid. Whether or not the building owner will exact a promise to make good that footing is quite another matter, particularly where as here, it is throughout intended that the mutual rights and obligations of the parties shall be exclusively expressed in a formal manner. So far from the basis of tender clause in the specification being a description of work, it seems to me to have become at the time of the execution of the contract no more than an historical step in a negotiation which preceded the making of the agreement. It is not reflected in the terms of that agreement either expressly or by implication. But out of deference to the submissions which were pressed upon us, let me say that, even if one moved the whole of the clause 'Basis of tender' into the operative part of the articles of agreement, the result would, in my opinion, be the same. Nothing can be made, in my opinion, of the use of the word 'turnkey'. It is not a term of art and, even if it could be taken to mean that the works must be handed over as a going concern, I would not have thought that in the context of these articles the word or expression meant that the builder warranted the efficacy of the works he had agreed to erect. The same reasoning would, in my opinion, deny that the acceptance of all responsibility for supply and erection and efficient operation of the project – itself a very inapt word in any case in the circumstances – involved any undertaking as to the suitability of the agreed design. Lastly, the paramountcy of the contract conditions, which do describe the works agreed to be done, would prevent the clause of the specification having the submitted effect, if in truth it could be construed as imposing the obligations the appellant seeks to draw from it.

In my opinion, for all these reasons I conclude that the respondent promised no more than to carry out the specified work in a workmanlike manner. This, according to the arbitrator it did.

In practice the contractor may be under a measure of design liability for various reasons. The contractor's obligations as to materials and workmanship often include a measure of liability for what may be called 'second order design', i.e. detailed decisions as to suitability of materials and methods of construction.

Reg. Glass Pty Ltd v. Rivers Locking Systems Pty Ltd

HIGH COURT OF AUSTRALIA   (1968) 120 CLR 576

The appellants employed the respondents, who were specialists in security installations, to fit a number of doors at their shop premises. The quotation for one door read:

> 'Rear single door: supply, fit and hang core door to suit opening, steel sheet outside area, fit Rivers 4 point non-break-out locking system operated by key from inside, hinge, stops, etc.'

The respondents fitted the door upon an existing wooden frame made of Pacific maple, a softwood. Thieves later broke into the shop by forcing the door away from the frame with a jemmy.

Held: Although the door had been installed in compliance with the express terms of the quotation, the respondents were in breach of an implied term that they would supply fit and hang a door which would provide reasonable protection against persons seeking to break in.

BARWICK CJ: By common law, however, there was a term to be implied in the contract. The business of the defendant was to provide burglar-proof protection and the plaintiff unquestionably relied upon its skill and ability to supply, fit and hang a door which would provide reasonable protection against persons seeking to break in when the locking devices were in operation. . . . It would, for instance, have been a plain breach of the contract if the locking devices had done no more than penetrate the wooden jambs and lintel so that pressure against the door could have split the wood and pushed the door in. Similarly, if the defendant had fixed the steel sheeting to the outside area of the door in such a way as it could have readily been removed, e.g., by exposed screws, although there would have been compliance with the express terms of the contract, the implied term would have been broken. The problem, as we see it therefore, is whether the Supreme Court was correct in setting aside the finding of the learned trial judge that the manner of the hanging of the door was such that the door did not provide reasonable protection against persons seeking to enter the shop.

It is increasingly common for the contractor expressly to assume some liability for design, especially under 'package deal' or 'design and build' contracts.

Greaves & Co. (Contractors) Ltd v. Baynham Meikle & Partners

COURT OF APPEAL   (1975), 4 BLR 56

The plaintiff contractors were employed under a 'package deal' contract to build a warehouse which was to be used to store oil drums. The drums were to be kept on the first floor of the warehouse and to be moved about by fork-lift trucks. In fact the warehouse as constructed was defective in that the movement of the fork-lift trucks caused vibrations which caused the floor to crack. The plaintiffs admitted liability to the employers, and the present action concerned the plaintiffs' claim against the defendant structural engineers, whom they had engaged to design the warehouse.

LORD DENNING MR: [As] between the owners and the contractors, it is plain that the owners made known to the contractors the purpose for which the building was required, so as to show that they relied on the contractors' skill and judgment. It was therefore the duty of the contractors to see that the finished work was reasonably fit for the purpose for which they knew it was required. It was not merely an obligation to use reasonable care. The contractors were obliged to ensure that the finished work was reasonably fit for the purpose.

That appears from the recent cases in which a man employs a contractor to build a house: *Miller* v. *Cannon Hall Estates Ltd* (1931); *Hancock* v. *B.W. Brazier (Anerley) Ltd* (1966). It is a term implied by law that the builder will do his work in a good and workmanlike manner; that he will supply good and proper materials; and it will be reasonably fit for human habitation. Similarly in this case the contractors undertook an obligation towards the owners that the warehouse should be reasonably fit for the purpose for which, they knew, it was required, that is as a store in which to keep and move barrels of oil. In order to get the warehouse built, the contractors found they needed expert skilled assistance, particularly in regard to the structural steel work. . . .

What was the cause of this cracking of the floors? The structural engineers said that it was due to the shrinkage of the concrete for which they were not responsible. There was nothing wrong, they said, with the design which they produced. But the judge did not accept that view. He found that the majority of the cracks were caused by vibration and not by shrinkage. He held that the floors were not designed with sufficient strength to withstand the vibration which was produced by the stacker trucks.

On those findings the first question is: what was the duty of the structural engineers towards the contractors? The judge found that there was an implied term that the design should be fit for the use of loaded stacker trucks, and that it was broken. Alternatively, that the structural engineers owed a duty of care in their design, which was a higher duty than the law in general imposes on a professional man; and that there was a breach of that duty. . . .

What then is the position when an architect or an engineer is employed to design a house or a bridge? Is he under an implied warranty that, if the work is carried out to his design, it will be reasonably fit for the purpose? Or is he only under a duty to use reasonable care and skill? This question may require to be answered some day as matter of law. But in the present case I do not think we need answer it. For the evidence shows that both parties were of one mind on the matter. Their common intention was that the enginer should design a warehouse which would be fit for the purpose for which it was required. That common intention gives rise to a term implied *in fact*.

In exceptional circumstances, it may be possible to infer that the employer looked to the contractor to advise him that the design was faulty.

Brunswick Construction Ltee v. Nowlan

SUPREME COURT OF CANADA   (1974), 21 BLR 27

N engaged an architect to design a house and then contracted with the appellants to build the house according to the architect's design. The design was faulty and made no sufficient provision for ventilation of the roof space and timbers. The result was a serious attack of rot. No architect was engaged to supervise the construction.

Held: The appellants were liable.

RITCHIE J: In my opinion a contractor of this experience should have recognized the defects in the plans which were so obvious to the architect subsequently employed by the respondents, and, knowing of the reliance which was being placed upon it, I think the appellant was under duty to warn the respondents of the danger inherent in executing the architect's plans, having particular regard to the absence therein of any adequate provision for ventilation.

(ii) MATERIALS AND WORKMANSHIP

In a building contract, unless the parties have agreed to the contrary, the builder impliedly agrees:

(a) *that he will do his work in a good and workmanlike manner;*
(b) *that he will supply good and proper materials; and*
(c) *(where the contract is to build a house) that the house will be reasonably fit for human habitation. In other cases, this obligation is to ensure that the completed structure will be reasonably fit for its intended purpose.*

Hancock v. B.W. Brazier (Anerley) Ltd

[1966] 2 All ER 901

LORD DENNING MR: In the autumn of 1958 the builders, B.W. Brazier (Anerley) Ltd were proposing to develop an estate of some forty houses known as the Haven Close Estate in Swanley. They laid it out in plots. Each prospective purchaser could choose his own plot. The three plaintiffs did so. Each of them selected a plot and paid a sum down to get it allotted to him; but no contract was made at that time for the erection of the houses. In the next three months, between October 1958 and January 1959 the defendants did a good deal of work on the foundations. They put in some hardcore and then put a layer of four inch concrete above it. They started to build the houses. All this before any contracts were signed. After the work had been done written contracts were entered into by the purchasers with the builders. By clause 1 the purchaser contracted to buy the house for £2,750. Clause 9 said: 'The vendor [Braziers] will prior to completion at its own cost and charge and in a proper and workmanlike manner erect build and complete on the above freehold land a messuage or dwelling-house in accordance with the plan and specification supplied to the purchaser subject only to such variations as may be ordered by the purchaser.' The plan and specification were in the ordinary form providing for the brick-work, the rooms, and so forth. There was nothing in the specification about the foundations, except that the site concrete was to be four inches thick and laid to required levels. On the plan there was this: 'four-inch site concrete on hardcore'.

The builders went on with their work. By May 1959 the houses were substantially finished. Completion took place on 20 May. The purchasers went into occupation. All appeared well; but two years later there was trouble in one house. Three or four years later in another house. Then in other houses. The trouble was this: the floors were cracking and breaking up. It was investigated. The cause was found in the hardcore which had been put into the foundations. It was not just the ordinary kind of hardcore which comes out of demolished buildings, such as bricks, flints, chalk and earth. There was quite a lot of sodium sulphate in this hardcore. We have been shown some of this sodium sulphate. The blocks of it look very much like flints or chalk. Anyone might easily mistake it for ordinary hardcore. Somebody must have got it from some old dump from a factory, must have thought it was stones and must have used it for hardcore. Yet it turns out to be a dangerous substance to use as hardcore. I say it is a 'dangerous substance' because, when it is dry, it soaks up water whenever it gets the chance, and then it expands with almost irresistible force. If it is already damp, it will soak up more water and expand. If it gets into contact with concrete, it brings about a chemical reaction which may cause the concrete to disintegrate. Hence the damage caused to these houses. It was the sodium sulphate in the hardcore that caused the floors to break up. The purchasers have been put to much expense to get it put right.

Let me say at once that all this trouble was in no way the fault of the builders. They had no reason to suspect the hardcore. They bought it in good faith. The defects were not apparent. No one could tell by looking at this hardcore that it had sodium sulphate mixed in with it. The builders used all reasonable care, skill and judgment. Counsel for the builders says that in these circumstances they were not guilty of any breach of contract. The entire obligation of the builders, was contained in clause 9 of the contract. That was a clause whereby they had to use only all reasonable care and skill and judgment in their work. It did not mean that they warranted the suitability of the materials which they used. He relies on the principle that where a matter is covered by express provision in a contract, it is not for the court to make an implication on the same matter.

It is quite clear from *Lawrence* v. *Cassell* (1930) and *Miller* v. *Cannon Hill Estates Ltd* (1931), that when a purchaser buys a house from a builder who contracts to build it, there is a threefold implication: that the builder will do his work in a good and workmanlike manner; that he will supply good and proper materials; and that it will be reasonably fit for human habitation. Sometimes this implication, or some part of it, may be excluded by an express provision, as for instance in *Lynch* v. *Thorne* (1956). The specification there expressly provided that the walls were to be nine-inch brick walls. The work was done with good materials and workmanship and exactly in accordance with the specification; but the walls did not keep out the driving rain. The builder was held not liable. The question in this case is whether the threefold implication is excluded by clause 9. I think that it is not, for this simple reason: clause 9 deals only with workmanship. It does not deal with materials. The quality of the materials is left to be implied; and the necessary implication is that they should be good and suitable for the work. I am quite clear that it is implied in the contract that the hardcore must be good and proper hardcore, in the same way as the bricks must be good and proper bricks. I know that the builders were not at fault themselves. Nevertheless this is a contract: it was their responsibility to see that good and proper hardcore was put in. As it was not put in, they are in breach of their contract. If it is any consolation to them, they can try and get hold of their suppliers and sue them if they can prove it against them; but they have to take responsibility so far as the purchasers are concerned.

In principle implied terms may be varied by agreement so as to impose a higher or lower duty on the contractor. (Any reduction of the contractor's liability might now be subject to the requirement of 'reasonableness' under the Unfair Contract Terms Act 1977 and Supply of Goods and Services Act 1982.) Such an agreement may be arrived at either from express words or as a necessary deduction from the circumstances.

Steel Company of Canada Ltd v. Willand Management Ltd

SUPREME COURT OF CANADA   (1966), 58 DLR(2d) 595

The respondent roofing contractor had bid for, and been awarded, a number of roofing contracts on buildings being constructed for the appellants. The roofs of the buildings were to be covered by a fluted steel deck, and the roofing membrane (a layer of insulation and roofing felt) was to be attached to the steel deck by a compound of fire-resistant material known as 'Curadex'. As part of the contract the respondent guaranteed 'for a period of five years that all work above specified will remain weather tight and that all material and workmanship employed are first-class and without defect'.

The roof suffered damage in high winds and the evidence was that this was due to the 'Curadex' proving an unsatisfactory way of fixing the membrane to the steel deck.

Held: The guarantee extended to damage to the roof arising out of faulty design, and the appellant was therefore entitled to recover the cost of repairing the roof.

RITCHIE J: It accordingly appears to me that the question which lies at the heart of this appeal is whether the responsibility for the results of using 'Curadex' rests upon the appellant who prescribed it or upon the respondent who applied it, and in this regard it seems to me to be of first importance to consider the circumstances under which this adhesive came to be included in the specifications. . . .

It therefore appears to me that when [a director] signed the tenders on behalf of the respondent, although he had had no actual experience in the use of 'Curadex' on sloping roofs, [he] was, as the result of lengthy discussions with the appellant's officers and of his having previously used the product on flat roofs, fully aware of the factors necessary to enable him to decide whether or not this adhesive was a first class material for its intended use, and whether it was one which his company was prepared to guarantee to remain 'weather tight' for a period of five years. . . .

In construing the guarantee as he did [the trial judge], was clearly influenced by the fact that he did not think that it would have been reasonable for the defendant to have expected, and the plaintiff to have given, an absolute guarantee against the elements when neither had had any experience with the capacity of 'Curadex' to perform properly on the sloping steel deck.

In this regard it is, however, to be remembered that the respondent is an experienced contractor specializing in the roofing business and that it was bidding in competition with several other roofing contractors. Under these circumstances the language employed by Cockburn CJ, in *Stadhard* v. *Lee* (1863), which was quoted with approval in the Exchequer Court of Canada in *Jones* v. *The Queen* (1877), appears to me to be particularly pertinent: 'It frequently happens, in the competition which notoriously exists in the various departments of business, that persons anxious to obtain contracts submit to terms which, when they come to be enforced, appear harsh and oppressive. From the stringency of such terms escape is often sought by endeavouring to read the agreement otherwise than according to its plain meaning. But the duty of a court in such cases is to ascertain and to give effect to the intention of the parties as evidenced by the agreement, and though, where the language of the contract admit of it, it should be presumed that the parties meant only what was reasonable, yet, if the terms are clear and unambiguous the court is bound to give effect to them without stopping to consider how far they may be reasonable or not.'

In construing the guarantee by supplying the words 'in so far as the plans and specifications with which it had to comply would allow' it appears to me that the Courts below have tacitly accepted the proposition that no matter how experienced a contractor may be in a particular field, he nevertheless bears no responsibility for the employment of defective material in the work which he has undertaken, provided that it is a material which has been selected by the owner and included in the specifications. This proposition finds support in the judgment of Vann J, in the *MacKnight* case, (1950), in a passage which was expressly adopted by Hughes J, which reads as follows: 'The defendant [i.e., the owner] specifically selected both material and design and ran the risk of a bad result. If there was an implied warranty of sufficiency, it was made by the party who prepared the plan and specifications, because they were its work, and in calling for proposals to produce a specified result by following them, it may fairly be said to have warranted them adequate to produce that result.'

I cannot accept this proposition which appears to me to run contrary to a long line of decisions in England starting with *Thorn* v. *Mayor and Commonalty of London* (1876), which have been followed in this country . . . and the effect of which is summarized in part in Hudson's *Building & Engineering Contracts* (1959), 8th edition, page 147 where it is said: 'Sometimes, again, a contractor will expressly undertake to carry out work which will perform a certain duty or function in conformity with plans and specifications, and it turns out that the work constructed in accordance with the plans and specifications will not perform that duty or function. It would appear that generally the express obligation to construct a work capable of carrying out the duty in question overrides the obligation to comply with the plans and specifications, and the contractor will be liable for the failure of the work notwithstanding that it is carried out in accordance with the plans and specifications. Nor will he be entitled to extra payment for amending the work so that it will perform the stipulated duty.'

The agreement to furnish a written guarantee 'that all work above specified will remain weather tight' for five years in my view constitutes at the very least an express undertaking 'to carry out work which will perform a certain . . . function in conformity with plans and specifications' and in accordance with the principles stated in the paragraph last above cited, I think that it follows that when a work so constructed does not perform the function which the contractor agreed that it would perform, the contractor is liable for the failure of the work and is not entitled to extra payment for repairing it 'so that it will perform the stipulated duty'.

In the course of his reasons for judgment, Hughes J expresses the following view: 'It would seem . . . that from this evidence that the defendant corporation was taking a calculated risk in specifying the adhesive designed and required to fasten the roofing membrane to a roof of new design, and it would seem that they knew this to be the case.'

In my opinion the evidence discloses that both parties were fully alerted to any limitations which may have attached to the use of 'Curadex' as an adhesive on these roof decks and in view of the fact that neither of them had had any experience in using it on sloping roofs, I think that some risk was involved. This may have been the reason why the appellant required the contractors who were tendering on the work to provide the guarantee in question, but whatever the reason may have been, it appears to me that any risk involved in the undertaking was accepted by those who were prepared to tender in accordance with specifications which included the requirement of providing a written guarantee that all material employed in the work was first class and without defect, and that 'all work . . . specified' would remain weather tight for a period of five years.

Young and Marten Ltd v. McManus Childs Ltd

HOUSE OF LORDS   (1969), 9 BLR 77

The respondents, builders engaged in developing a housing estate at Gerrards Cross, sub-contracted the roofing of the houses to the appellants. The roofing contract called for the use of 'Somerset 13' tiles, which were manufactured only by JB. The appellants obtained supplies of tiles from their own suppliers in London who in turn obtained them from JB. Some of the tiles supplied had a defect, apparently due to faulty manufacture, which was not discoverable by any reasonable inspection by the appellants. After less than twelve months the tiles began to disintegrate and the respondents claimed the cost of the re-roofing of the houses. The appellants argued that, as they had not chosen the type of tile and could not have discovered the defect, they were not liable.

Held: This argument was rejected.

LORD REID: This is a contract for the supply of work and materials and this case raises a general question as to the nature and extent of the warranties which the law implies in such a contract. As regards the contractor's liability for the work done there is no dispute in this case: admittedly it must be done with all proper skill and care. The question at issue relates to his liability in respect of material supplied by him under the contract. The appellants maintain that the warranty in respect of materials is similar to that in respect of work, so that, if the selection of material and of the person to supply it is left to the contractor, he must exercise due skill and care in choosing the material and the person to supply it. But where, as in this case, the material and the supplier were chosen by the respondents, the appellants maintain that there was no warranty as to the fitness or quality of the tiles. The respondents admit that, if it is held that the choice of this type of tile was theirs and theirs alone, there can be no implied warranty that this type of tile was fit for the contract purpose. But they say that there still was a warranty that the tiles would be of good quality and that that warranty must be implied notwithstanding the fact that they left no choice to the appellants in selecting the person who was to supply the tiles. If that is right then the respondents must succeed. The loss was not caused by Somerset 13 tiles being unsuitable for the contract purpose: it was caused by the tiles which were supplied being of defective quality.

There is not very much authority on this matter so it may be well first to consider it as a question of principle. In my view no warranty ought to be implied in a contract unless it is in all the circumstances reasonable. If authority be required for that proposition I find it in the judgment of the Exchequer Chamber in *Readhead* v. *Midland Railway Co* (1869): 'Warranties implied by law are for the most part founded on the presumed intention of the parties, and ought certainly to be founded on reason, and with a just regard to the interests of the party who is supposed to give the warranty, as well as of the party to whom it is supposed to be given.' I take first the general question of the contractor's liability where the material which he is required to use can be obtained from any one of several suppliers and the choice of suppliers is left to him. There is no doubt that in every case he is bound to make a proper inspection of the material before using it, and he will be liable if the loss is caused by the use of material which reasonable inspection would have shown to be defective. The question is whether he warrants the material against latent defects.

There are in my view good reasons for implying such a warranty if it is not excluded by the terms of the contract. If the contractor's employer suffers loss by reason of the emergence of the latent defect, he will generally have no redress if he cannot recover damages from the contractor. But if he can recover damages the contractor will generally not have to bear the loss: he will have bought the defective material from a seller who will be liable under section 14(2) of the Sale of Goods Act because the material was not of merchantable quality. And if that seller had in turn bought from someone else there will again be liability so that there will be a chain of liability from the employer who suffers the damage back to the author of the defect. Of course the chain may be broken because the contractor (or an earlier buyer) may have agreed to enter into a contract under which his supplier excluded or limited his ordinary liability under the Sale of Goods Act. But in general that has nothing to do with the employer and should not deprive him of his remedy. If the contractor chooses to buy on such terms he takes the risk of having to bear the loss himself if the goods prove to be defective.

Moreover many contracts for work and materials closely resemble contracts of sale: where the employer contracts for the supply and installation of a machine or other article, the supply of the machine may be the main element and the work of installation be a comparatively small matter. If the employer had bought the article and installed it himself he would have had a warranty under section 14(2), and it would be strange that the fact that the seller also agreed to install it should make all the difference.

The speciality in the present case is that these tiles were only made by one manufacturer. So the contractor had to buy them from him or from someone who bought from him. Why should that make any difference? It would make a difference if that manufacturer was only willing to sell on terms which excluded or limited his ordinary liability under the Sale of Goods Act, and that fact was known to the employer and the contractor when they made their contract. For it would be unreasonable to put on the contractor a liability for latent defects when the employer had chosen the supplier with knowledge that the contractor could not have recourse against him. If the manufacturer's disclaimer of liability caused him to supply the goods at a cheaper price, as in theory at least it should, the employer ought not to get the benefit of a cheap price as well as a warranty from the contractor.

A more difficult case would be where the employer and contractor had no reason to suppose, when they made their contract, that the manufacturer would refuse to sell subject to a seller's ordinary liabilities in respect of the goods which he sells. But I need not consider that case now because there is no suggestion that [JB] had refused to sell except on terms which limited their ordinary liability in respect of latent defects in their tiles. No doubt there will be some cases where, although the contractor had a right of recourse against the manufacturer, he cannot in fact operate that right. The supplier may have become insolvent, or, as in the present case, the action against the contractor may be so delayed that he has no time left in which to sue his supplier. But these cases must be relatively few and it would seem better that the contractor should occasionally have to suffer than that the employer should very seldom have any remedy at all. It therefore seems to me that general principles point strongly to there being an implied warranty of quality in this case.

Gloucestershire County Council v. Richardson

HOUSE OF LORDS   [1969] 1 AC 480

The council engaged the respondents under a RIBA contract (1939 edition, 1957 revision) as main contractors for the erection of extensions to a technical college. The architect nominated C as supplier of concrete units, and sent to the respondents the quotation from C, instructing the respondents to accept it, which they did. The quotation contained clauses limiting C's liability if the goods supplied proved defective. Under the terms of the main contract the respondents were entitled to make reasonable objection to nominated sub-contractors but not to nominated suppliers.

The concrete units proved defective and this led to great delays. In due course the respondents purported to exercise their right to determine the contract on the ground that the work had been delayed for more than one month 'by reason of architect's instructions', and the appellants in turn claimed damages for wrongful repudiation. It was conceded that the respondents were not entitled to terminate if they were in breach of an implied warranty as to the quality of the concrete units.

Held: The respondents were not in breach of any implied warranty.

LORD PEARCE: My Lords, the contractor in any particular field of business, when he engages to do certain work and supply material, impliedly warrants that the materials will be of good quality, unless the particular circumstances of the case show that the parties intended otherwise. To find the intention one must consider the express terms of the contract and any admissible surrounding circumstances.

Here the parties entered into a Royal Institute of British Architects contract, a complicated and sophisticated document. There is no express acceptance by either party of liability for the quality of the nominated materials. The contractor must comply with the instructions of the architect. He must accept the architect's nomination in respect of certain sub-contractors and nominated suppliers. No nominated sub-contractor, however, can be employed (clause 21) if the contractor makes reasonable objection to him or if, inter alia, the sub-contractor will not enter into a sub-contract indemnifying the contractor against claims for negligence of the sub-contractor and 'against the same obligations in respect of the sub-contract as those for which the contractor is liable in respect of this contract.' These words seem to make it clear that the contractor is accepting liability in respect of work done by the nominated sub-contractor.

The situation with regard to nominated suppliers, however, is noticeably different. The clause (22) which deals with nominated suppliers follows directly on that which deals with nominated sub-contractors. It provides no veto on the ground of the contractor's reasonable objections, nor on the ground of the nominated supplier refusing to indemnify the contractor. This omission cannot, I think, be unintentional. It seems, in contrast to clause 21, to point to an intention that the contractor is not undertaking liability for materials provided by a nominated supplier. Otherwise he must surely have been given, as in the case of a nominated sub-contractor, an opportunity of making reasonable objections, and a right to insist on an indemnity from the supplier.

It would not be unreasonable for the parties to intend that an employer should take the responsibility for materials provided by nominated suppliers. They have been selected, without giving the contractor any right to express views, by the employer's own expert architect who has decided that the nominated goods are suitable for the purpose and who has made the preliminary arrangements with the suppliers either before or during the main contract. The contractor is simply instructed to obtain his supplies from the nominated supplier. It is the employer who, through his architect, alone arranges the price, which is liable to be reflected in the quality and who alone can insist on tests and checks of quality. All the circumstances of the nomination appear actually to exclude any reliance on the contractor's skill and judgment. And, though the contractor receives a profit on the nominated supply, it is a controlled profit and he has certain duties to perform such as co-ordinating the delivery with the work and doing his best to see that there are no delays (see clause 18(vii)).

On the other hand, if the contractor is not liable for material provided by nominated suppliers, the employer is left without a remedy for faulty material. For the contract, by clause 22, indicates clearly that the nominated supplier is in contractual relation with the contractor only and, although the employer is paying for the nominated materials, he pays the contractor for them and the contractor pays the nominated supplier. Thus to hold that the contractor is not liable for nominated supplies is to go against one of the important reasons for the general rule that there is a warranty of good quality in materials supplied under a contract for labour and materials, namely that the employer should have a remedy against the contractor who can in turn enforce it against the supplier with whom the fault lies.

Yet, in spite of this important fact, I think that the contrast between the wording of clause 21 and that of clause 22 persuades one to the view that the contract shows an intention to exclude a warranty by the contractor in respect of nominated supplies. And the particular circumstances of the case fortify this view. The employers (or their architect) employed a skilled engineer to advise in respect of the columns in question. The engineer, without any consultation with the contractor, prepared detailed designs of the columns, chose out [C] as suitable suppliers and gave them such instructions as he considered were necessary to specify the composition of the columns. The architect obtained a quotation from [C] which was lower than the amount quoted by another firm and also substantially lower than the P.C. item, and which contained a substantial limitation of the purchaser's right of recourse in the event of the columns being defective. The architect presumably was satisfied with this and without discussing the matter with the contractor, instructed him to accept the quotation.

The employers, through their architect, having directed the contractor to buy from a manufacturer who had substantially limited his own liability, it would not be reasonable to suppose that the parties were intending the contractor to accept an unlimited liability, which might have been very great, for those columns over whose manufacture he had no control whatsoever. It was suggested in argument by the appellants that one should assume some implied intention that there will be liability on the contractor limited to the curtailed right of recourse provided under [C's] quotation. If so, one would have to imply in respect of each nominated supplier that there is a liability limited to the right of recourse provided in each particular sub-contract of supply. But I see no ground for such a complicated implication which, if it existed, must surely have been embodied in express terms in clause 22. In my opinion, the limitation of recourse against the nominated supplier and the other particular circumstances in respect of the supply of the columns provide confirmation of the intention indicated by the contract itself, namely, that any warranty by the contractor in respect of the quality of the nominated supplies was to be excluded.

LORD WILBERFORCE: The situation thus created was one of a special and complex character, differing greatly from that which arose in *Young and Marten Ltd* v. *McManus Childs Ltd* (1969). There the employer nominated a brand article to be supplied by the manufacturer with no limitation on the contractor's freedom to contract with the manufacturer as he thought fit. The contractor could, and it would be the expectation that he would, or at least it would be his responsibility if he did not, deal with the manufacturer on terms attracting the normal conditions or warranties as to quality or fitness.

But here, the design, materials, specification, quality and price were fixed between the employer and the sub-supplier without any reference to the contractor: and so far from being expected to secure conditions or warranties from the sub-supplier, he had imposed upon him special conditions which severely restricted the extent of his remedy. Moreover, as reference to the main contract shows, he had no right to object to the nominated supplier, though, by contrast, the contract does provide a right to object to a nominated sub-contractor if the latter does not agree to indemnify him against his liability under the contract.

In these circumstances, so far from there being a good reason to imply in the contract (which throughout its elaborate 27 clauses makes no express provision as regards this matter) a condition or warranty binding the contractor in respect of latently defective goods, the indications, drawn from the conduct of the contracting parties, are strongly against any such thing. It would, indeed, be most unjust if when the employer has (possibly to his own advantage as reflected in the price) limited the contractor's right or recourse as severely as he has (since consequential damage or loss in such cases as this may be very great and much in excess of the cost of replacement) he should be given by implication an unlimited right to recover damages from the contractor.

Norta Wallpapers (Ireland) Ltd v. John Sisk & Sons (Dublin) Ltd

IRISH SUPREME COURT   (1977), 14 BLR 49

The claimants engaged the respondents as main contractors to build a factory for making wallpaper and nominated H as sub-contractors for the design, supply and erection of the super-structure including the roof and roof lights. H were specialists in 'system built' superstructures.

H completed the work in May 1967 and the respondents completed work under the main contract in August 1967. In February 1968 a major leak occurred in the roof of the factory, which was found by an arbitrator to be. due 85% to defective design of the roof lights, 12% to use of inferior materials and 3% to bad workmanship.

The trial judge held that:
(a) The respondents were liable for so much of the damage as was due to the use of inferior material by, and bad workmanship of, the sub-contractor; and
(b) The respondents were not liable for the loss which was due to the defective design of the roof lights.

The claimants appealed against (b), but the appeal was dismissed.

HENCHY J: In all cases of supply and installation by a sub-contractor I conceive the law to be that, unless the particular circumstances give reason for its exclusion, there is implied in the contract a term to the effect that the contractor will be liable to the employer for any loss or damage suffered by him as a result of the goods, materials or installations not being fit for the purpose for which they were supplied. The basis for this rule is that, while the contractor is thus made primarily liable, he will be able, under the sub-contract, to have recourse, by third-party procedure or otherwise, against the sub-contractor for the indemnity in respect of the contractor's liability to the employer. I do not think there is any dispute about the law as stated thus. The cleavage of opinion arises as to whether the circumstances of this case are such as to exclude an implied term in the contract making the contractor liable for the sub-contractor's default.

An implied term in a contract to the effect that goods or installations supplied under it will be suitable for their purpose cannot be held to exist unless it can be said to have come within the presumed intention of the parties: see *G.H. Myers & Co* v. *Brent Cross Service Co* (1934) which was approved by all the judges in the House of Lords in *Young & Marten Ltd* v. *McManus Childs Ltd* (1969).

As I say, the real question in the present case is whether the implied warranty of fitness is excluded in the particular circumstances. It is common ground – at least there is no appeal from the judge's decision to that effect – that Sisk are liable to Norta under an implied term in the contract in respect of the defective materials and the bad workmanship employed in erecting the roof. The judge went on to hold that there was no such implied term in regard to the defective design of the roof lights. . . . The modern cases (particularly the decisions of the House of Lords in *Young & Marten Ltd* v. *McManus Childs Ltd* (1969) and *Gloucestershire County Council* v. *Richardson* (1969)) show that the liability of a contractor under an implied term in the contract is approximated to the liability of a seller under section 14 of the Sale of Goods Act, 1893. Thus he may be liable, as Sisk are here, for bad workmanship done or inferior materials supplied by the sub-contractor. But if the subject matter of the employer's complaint is something which under its patent or trade name he has required to be supplied or installed under sub-contract, he cannot hold the contractor liable if, because of bad design, it turns out not to be suitable for a particular purpose. If he nominated it under its patent or trade name he has, in effect, accepted the design as being suitable for the particular purpose. If that be the position here, then the trial judge's reasoning for rejecting Norta's claim against Sisk as unanswerable.

However, I am inclined to think that the defective design of this roof did not simply render it unsuitable for a *particular* purpose. The arbitrator has held that 'the roof has now deteriorated to the extent that, in my opinion, it is not suitable for its purpose.' By that he meant, presumably, that the roof is not fit for the purpose or purposes for which a roof of that kind is commonly bought, in that it is not rain-proof. If that be so, then it might be held to be wanting in merchantable quality: *per* Lord Denning MR in *Cehave NV* v. *Bremer Handelsgesellschaft mbH* (1976). Without more findings than the arbitrator has set out in the case stated, one would not be justified in firmly reaching such a conclusion. But such a conclusion is not necessary for the purpose of deciding the case. Even if the defect in the design of the roof is such as to justify describing it as lacking merchantable quality, a warranty on the part of the contractor that it will not have such a defect should not be read into the contract unless it would be reasonable to do so.

(iii) PROGRESS AND COMPLETION

Normally, the contractor undertakes to complete the work in its entirety. He will normally be in breach of contract if he fails to complete, unless the contract has come to an end for some reason. (See Part III – Discharge of Contract, page 273)

If the contract is 'entire' the contractor who does not complete substantially will not be entitled to payment. (See Chapter 7(iv) – Payment). The practical effect of this rule is often modified by provision for payment against certificates, but the contractor who stops work without reason still does so at his peril.

Ibmac Ltd v. Marshall (Homes) Ltd

COURT OF APPEAL   (1968), 208 Est. Gaz. 852

The defendants, who were building a council housing estate, employed the plaintiffs to build a roadway to serve it. The plaintiffs quoted a figure of £1,494 but also supplied a bill of quantities with rates for the work. (The Court of Appeal thought it unnecessary to decide whether this was a fixed price or a re-measurement contract).

The plaintiffs found the work was more difficult than they had expected as the site lay at the bottom of a steep hill and there were serious difficulties with water. The plaintiffs abandoned the works and claimed payment for the work they had done.

Held: Their claim must fail.

LORD DENNING MR: After a few weeks, the difficulties which they did not expect arose. The water hazard was indeed serious, and they did not keep to their side of the contract. They did not get a six-ton roller to consolidate the earth of the roadway, and so forth. So the difficulties of the water were enhanced and eventually, in June, they left the work. They said it was all Marshalls' fault. But it was not. On June 9, Ibmac wrote, 'We shall not return to the above site until the problem of surface water has been dealt with' – taking the line, as they did then, that the water was not their responsibility but the fault of Marshalls: Marshalls had not given them a proper site on which to do the work. There was some attempt to get the matter in order for Ibmac to come back. There was no doubt that they had left the work, especially on the trial judge's findings of fact, which he made after a full hearing, that they were not justified in leaving the work. They abandoned it without just cause and excuse when they had about one-third of the work done. In point of law, on that footing they were not entitled to anything, because of the rule in *Sumpter* v. *Hedges* (1898), which was quite clear, that if one had an entire contract where one had to complete the work, if one abandoned it half or a third of the way through without just excuse one could not recover anything.

Note: In the absence of contractual provision, the contractor would only be required to complete by the agreed completion date or, if no date is expressly agreed, within a reasonable time. In practice however the contract will usually require the contractor, for example, 'to begin the works forthwith and regularly and diligently proceed with the same', etc. In practice, this obligation is important in relation to disputes concerning liquidated damages and extensions of time (see Chapter 13 – Liquidated damages and extension of time), and determination of the contractor's employment under express terms (see *J.M. Hill & Sons Ltd* v. *London Borough of Camden* (1980), page 299).

Where the contract requires the contractor 'regularly and diligently (to) proceed with' the works (JCT 80, clause 23.1., JCT 63, clause 21) whether this standard is achieved is probably to be judged according to the usage of the industry.

Hounslow Borough Council v. Twickenham Garden Developments Ltd

CHANCERY DIVISION   (1970), 7 BLR 81

MEGARRY J (discussing the phrase 'regularly and diligently' in JCT 63, clause 21): These are elusive words on which the dictionaries help little. The words convey a sense of activity, of orderly progress, and of industry and perseverance; but such language provides little help on the question of how much activity, progress and so on is to be expected. They are words used in a standard form of building contract and in those circumstances it may be that there is evidence that could be given, whether of usage among architects, builders and building owners or otherwise, that would be helpful in construing the words. At present, all that I can say is that I remain somewhat uncertain as to the concept enshrined in these words.

(iv) INDEMNITIES AND INSURANCE

The contract will normally contain provisions as to which of the parties is to insure against certain risks. These provisions may present serious difficulties of interpretation as illustrated by the next two cases.

Gold v. Patman & Fotheringham Ltd

COURT OF APPEAL   [1958] 2 All ER 497

The plaintiff engaged the defendants under a standard RIBA Form (1939 edition with quantities, 1952 revision). Clause 14 required the contractors to indemnify the employer in respect of injury to persons and property (subject to certain exceptions). Clause 15 required the contractors to 'effect . . . such insurances . . . as may be specifically required by the bills of quantities'. The risks required to be insured by the bills included 'insurance of adjoining properties against subsidence or collapse'.

A neighbouring owner brought a claim against the plaintiff in respect of subsidence caused by the building operations and it appeared that the defendants had taken out an insurance policy which covered their liability for subsidence but not that of the plaintiff. The plaintiff brought an action claiming that the defendants were in breach of their obligations under clause 15.

Held: The claim must fail.

> HODSON LJ: We must deal separately with the question who is to be insured, though it falls for consideration on the question of certainty; but, so far as the other matters are concerned, we agree with the learned judge that there is no uncertainty as to the risk to be covered by insurance. That which was to be insured against was the legal liability to adjoining owners arising out of the subsidence or collapse of the premises. It would be wholly wrong in our judgment to say, in the words of Lord Wright, that the language used was so obscure and so incapable of any definite or precise meaning that the court is unable to attribute to the parties any particular contractual intention.
>
> The only real difficulty which arises is as to the determination of the question who is to be covered by insurance, but the court ought not to be deterred by any difficulty of interpretation which can fairly be surmounted. Looking first at condition 14 of the contract dealing with (a) injury to persons, (b) injury to property, the contractors are bound to indemnify the building owner against loss in the circumstances therein set out. When one comes to condition 15, the contractors are thereby required to effect insurance or cause any sub-contractor to do so as may be specifically required by the bill of quantities. This condition prima facie refers to the indemnity required by condition 14, for it is introduced by the words 'without prejudice to his liability to indemnify the employer under clause 14 hereof', which indicates plainly that the indemnity is to stand independently of the insurance and that the contractors cannot claim to have fulfilled their obligation under the indemnity clause in respect of injury to persons or property by effecting a policy or policies of insurance or causing any sub-contractor to do so.
>
> These policies and their premium receipts must be produced as and when required by the architect. Loss or damage by fire is dealt with in clause 15(b)[A], and there, in contrast to the policies referred to in clause 15(a), the contractors must insure in the joint names of the building owner and themselves with a company or

companies approved by the architect, and the contractors are obliged to deposit (not only produce when required) the policies and premium receipts with the architect. It would thus appear that there is a contrast between policies in which the contractors are insuring themselves, no doubt for the additional security of the building owner, and cases where the contractors are insuring in the joint names of the owner and themselves. Thus, when the contract requires the building owner (described in the contract as the employer) to be insured by the contractors, it is expressly so provided. Further, it would appear appropriate in cases where the insurance is for the benefit of the building owner to require approval by his architect and deposit of the policy of insurance with him. Condition 25, which deals with fluctuations, refers also to the burden of insurance falling on the contractors under the heading of expenses, including the cost of employer's liability and third-party insurance, and lends support to the contention of the contractors that these are among the specific insurances contemplated by condition 15(a) of the contract. These considerations derived from the conditions in the main contract are persuasive to suggest that the insurances specifically required by the bill of quantities will be found to be insurances effected by the contractors for themselves and not for the building owner.

When one turns to the bill of quantities itself, nothing is found to lead to a contrary conclusion. In construing the obligation beginning with the words 'The contractor shall insure', one looks at the whole list of matters in respect of which they have to insure or make payments, and nothing is found in that list pointing to an obligation to insure anyone else but themselves. We cannot find anything in the initial phrase or in the context to lead one to understand that the obligation is to insure not the contractors alone but the building owner either in addition to the contractors or separately from him. The list following the initial phrase is not very illuminating and there is on any view some duplication. . . . All the items in the list would appear to be cases where the contractors must insure themselves, at least so far as they are intelligible, bearing in mind that 'Unemployment Insurance Acts' do not exist, and insurances against specific industrial diseases do not add anything to the words above, viz., 'National Insurance Acts and National Insurance (Industrial Injuries) Acts.' Damage by storm, etc., would be the contractors' risk for them to cover, for they would not be relieved of their obligation by such damage. . . .

We find no sufficient reason for thinking that, when one comes to the last item 'insurance of adjoining properties against subsidence or collapse', the obligation to insure is for the first time in the list an obligation to insure, not the contractors themselves, but the building owner. If there were to be a change of obligation, one would expect some words to be used to show that there is a change, and here there are none; nor would one expect to find such an obligation in this place, having regard to the words of condition 15(a), where reference is made to such insurances to be effected as may be specifically required by the bill of quantities, the policies to be produced when required and not only to be approved by and deposited with the architect. It is true that there is duplication, since the heading 'Employer's liability, common law, third party' no doubt produces the same result; but duplication, as we have said, appears elsewhere in the miscellany and the last item is only a specific identification of one kind of third-party risk.

Counsel for the contractors reinforced his argument by a contention that there would be grave difficulties in the way of a contractor knowing what to tender if he had to insure the building owner, for he would not survey the property before tendering, whereas in the case of insuring himself against the risk in question there is no practical difficulty, because the premium is based on the contractor's past record and the number of men he employs. Counsel sought in this connexion to rely on the oral evidence of an insurance expert, to show the practical difficulties involved. We see no ground for admitting evidence of this character to show the surrounding circumstances of this particular contract. No trade custom was sought to be proved and it is not contended that both parties were contracting in the light of some special insurance background which could be established by evidence. In any event, we should not have been deterred by the difficulty of effecting an insurance from finding that the obligation to insure the building owner had been placed on the contractors by the contract if the words used had produced that result. The evidence of [the expert] could, one would assume (although no argument was addressed to the court on this aspect of his evidence) be used to show uncertainty if there were a latent ambiguity in the use of the word 'insurance' in its context. We find nothing in his evidence which was read de bene esse to lead to the conclusion that the word 'insurance' as here used is uncertain in its meaning.

English Industrial Estates Corporation v. George Wimpey & Co. Ltd

COURT OF APPEAL

(1972), 7 BLR 122

LORD DENNING MR: This case arises out of a big fire at Hartlepool on 18 January 1970. Much of a new factory was gutted. The damage is estimated at £250,000. Who is to bear the loss?

The factory was occupied by Reeds Corrugated Cases Ltd, who make corrugated cardboard. The factory and its site are owned by a statutory corporation called English Industrial Estates Corporation. The corporation let it on lease to Reeds.

In 1969 Reeds wanted to extend the factory greatly. It was to be three times as large as before. The Corporation agreed to make the extension and to let the whole to Reeds on a new lease. But Reeds wanted to continue making their corrugated cardboard all the time whilst the extension work was being done. In order to make the cardboard they intended to install a great new machine. It was to be fed with reels of paper weighing over a ton each. The machine could eat up these reels at such a rate that Reeds needed storage space for hundreds of these big reels of paper.

On 17 December 1968, George Wimpey & Co Ltd, the big contractors, put in a tender. They offered to build the extension for £687,860. In making the tender they had before them the form of agreement, the RIBA conditions, and the bills of quantities. They agreed that, if the tender was accepted, they would enter into a formal contract on those terms. Their tender was accepted.

On 7 February 1969, the Corporation and Wimpeys duly entered into articles of agreement. It was on the standard RIBA form and incorporated the bills of quantities. The most important provision for present purposes was clause 20A(1) of the RIBA form. It provided that the contractors, Wimpeys, should insure the works against fire and other perils until they were practically completed. Wimpeys had a floating policy which covered this.

In 1969 Wimpeys started on the work. By January 1970 they had done a great deal of it. Reeds had installed their great new machine and started producing corrugated cardboard. They had stored over 1500 big reels of paper in the new reel warehouse. But the contractors had not finished their work when on 18 January 1970, there was this disastrous fire. Much of the new extension was gutted. The loss is said to be £250,000. On whom should this loss fall? The Corporation say that the works had not been completed: that it was the duty of Wimpeys to insure: and that the loss should fall on them or their insurance company. But Wimpeys (or rather, I expect, their insurers) said that the Corporation, through Reeds, had taken possession of several parts of the works: and that, under clause 16 of the RIBA form, the risk had passed from Wimpeys to the Corporation so far as those parts were concerned.

To decide the case, I must set out some parts of the RIBA form, and also of the bills of quantities.

2. THE RIBA FORM

The most material clauses were as follows:

'20A(1) The Contractor shall . . . insure against loss and damage by fire . . . all work executed and all unfixed materials and goods . . . and shall keep such work, material and goods so insured until Practical Completion of the Works, and until the employer shall authorise in writing the cancellation of such insurance.

16   If at any time . . . before Practical Completion of the Works, the Employer, with the consent of the Contractor, shall take possession of any part or parts of the same (any such part being hereinafter in this clause referred to as "the relevant part") then notwithstanding anything expressed or implied elsewhere in this Contract . . .

(a) Within seven days from the date on which the Employer shall have taken possession of the relevant part, the Architect shall issue a certificate stating his estimate of the approximate total value of the said part . . .

(d) The Contractor shall reduce the value insured under Clause 20(A) of these Conditions by the full value of the relevant part, and the said relevant part shall, as from the date on which the Employer shall have taken possession thereof, be at the sole risk of the Employer.'

It is plain that, in applying clause 16, the great question is: Had the employers, before the date of the fire, 18 January 1970, 'taken possession' of any part or parts of the works? If they had done so, that part was at the risk of the employers. If they had not done so, it was at the risk of the contractors. The contractors say that 'taking possession' is a question of fact which is answered here by the fact that the employers were using and occupying the factory at the time of the fire. The employers say that 'taking possession' only took place when the relevant part was handed over to them and accepted by them: and that never took place.

2. THE FACTS AS TO 'TAKING POSSESSION'

(i) *The Car Park*

The car park was ready for use by September 1969, but there was some work yet to be done, such as painting the white lines to show the spaces for the cars. The contractors asked the architect to inspect it prior to handover. The architect did inspect it. On 23 September 1969, he accepted the handover and issued a certificate on a printed form issued by the RIBA. It said:

> 'I certify that, subject to the making good of any outstanding items, or of any defects, shrinkages and other faults which appear during the defects liability period . . . a part or section of the works, namely, Car Park, the value of which I estimate to be £10,000 was *completed to my satisfaction* and taken into possession on 22 September 1969, and that in relation to the said part or section of the works, the said defects liablity period will end on 23 March 1970.
>
> I declare that one moiety of the retention moneys deducted under previous certificate in respect of the said section is to be released.'

The architect gave evidence that that form of certificate was in general use, and was the normal certificate which was used by architects for sectional completion. He said that, when issuing that certificate, he understood in principle that clause 16 required him to certify that a section or part of the works had been completed to his satisfaction and taken into possession.

Since this case started (and I expect because of it) the RIBA have issued an amended form in which the words 'completed to my satisfaction' have been struck out: but have left the rest of the form as it was.

That certificate as to the car park was the only certificate of the kind issued by the architect in respect of any part of the works. No such certificate was issued for any other part.

(ii) *The Board Machine House*

At the date of the fire – 18 January 1970 – the board machine house was virtually complete. The new 87-inch board machine had been installed and begun working. It was in production. 120 tons of boarding had been produced, about 80 per cent of which was suitable for marketing. There were some things still to be done by the contractors. For instance, a sliding fire door had yet to be fixed. The architect had inspected the floor but not the rest of the machine house.

On 12 December 1969, the architect had written to Wimpeys:

'We have inspected the floor of the Board Machine House and accepted this on behalf of our clients . . .

This letter is in no way a Practical Completion Certificate: this will be issued at the appropriate time for the whole of the Board Machine House.'

The position was, therefore, that the architect had not inspected the machine house itself, nor agreed to accept it.

(iii) *The Reel Warehouse*

The reel warehouse has three bays, namely, the eastern, central and western bays. At the date of the fire the whole was roofed and walled in, but the outside doors were not completed. In the central and western bays the sprinkler works and permanent lighting had been installed; but not in the eastern bay. There was some painting and glazing work to be done. The emergency exit doors had still to be fixed. Reeds had for some weeks used the warehouse for storing reels of paper. They had stored 1500 tons on the floors of the central and western bays, and 300 to 400 tons in the eastern bay. The floors had been inspected by the architect, but not the rest of the reel warehouse. On 8 December 1969, the architects had written to Wimpeys:

'We inspected the floor only of the central bay of the reel warehouse and accepted this on behalf of our client . . .

This letter is in no way a Practical Completion Certificate: this will be issued at the appropriate time for the whole of the reel warehouse.'

On 9 January 1970, the architects wrote a letter in the same terms as to the floor of the western bay of the reel warehouse.

The position was, therefore, that the central and western bays were approaching completion, save for the doors: but the eastern bay had still much to be done to it. The architects had not inspected the reel warehouse itself, nor agreed to accept it.

(iv) *Factory Extensions South and East*

At the time of the fire, these were well advanced, 90 reels were stored in the southern extension and 45 in the eastern extension. But work remained to be done. The architects had not inspected these extensions, nor accepted them. Nor had they written any letter about them.

(v) *The Respective Contentions*

The contractors, of course, relied greatly on the use which Reeds had made of the works. Reeds had installed a great new machine in the board machine room. They had stored nearly 2,000 tons of paper in the reel warehouse. This use by Reeds hampered Wimpeys very much in their work. So much so, that Wimpeys said that the use and occupation by Reeds amounted to 'taking possession' of those parts.

But the Corporation replied that the parties themselves did not so regard it. They said that there was no 'taking possession' until the relevant part or section had been handed over by the contractors to the employers. They referred especially to a letter of 22 December 1969, in which Wimpeys wrote to the architect:

'We write to inform you that sectional handovers of completed works are being delayed due to a number of impediments which are beyond our control. The major disruptions to our Contract are:
  (a) Tenants' reel movement and storage;
  (b) Excessive occupation of areas by plant erectors . . .
The effect of those factors has been to retard our performance and, in consequence, we apply for an extension of section completion dates . . .'

3. PROVISIONS 'C' and 'D'

One of the most important points in this case is whether the employers can rely on two special provisions (C and D) in the bills of quantities. These show that the parties contemplated that Reeds would install plant and equipment, and would occupy and use part of the works: but that the contractors were still to keep the works covered by insurance: and they could charge the increased cost of insurance to the employers. The two provisions were in these words:

' "C" *Employer may install equipment as the work proceeds*
As the work proceeds, the contractors shall permit the Employer or any other person authorised by him to place and install as much plant and equipment during the progress of the building as is possible before the completion of the various parts of the works, and such placing and installing shall not in any way alter the period fixed for completion of the works or the liability of the Contractor hereunder provided that:
  (a) in making arrangements for such placing and installation of plant and equipment, due regard shall be made to the requirements of the Contractor;
  (b) if by reason of such installation, the cost of any insurance which the Contractor is required to maintain under this contract shall be increased, then the amount of such increased cost shall be in addition to the contract sum.

"D" *Occupation of completed parts of the Works*
The Contractor shall allow the Employer or any person authorised by him to occupy and use any part or parts of the Works as soon as, in the opinion of the Architect or Employer, such part or parts can be occupied and used without impeding the progress of completion of the Works, and such occupation shall not in any way alter the period fixed for completion of the Works or the liability of the Contractor hereunder, provided that if, by any reason of such occupation, the cost of any insurance which the Contractor is required to maintain under this contract shall be increased, then the amount of such increased cost shall be an addition to the contract sum.'

Wimpeys contended that it was not permissible for the court to have regard to those special provisions 'C' and 'D'. They relied on clause 12(1) of the RIBA form, which says:

'. . . Nothing in the Contract Bills shall override, modify, or affect in any way whatsoever the application or interpretation of that which is contained in these Conditions.'

Mocatta J left to himself would have had regard to provisions 'C' and 'D', despite clause 12. But he felt bound by some observations by Lord Hodson in *Gold* v. *Patman & Fotheringham Ltd* (1958); *North-West Metropolitan Regional Hospital Board* v. *T.A. Bickerton & Son Ltd* (1970); just as he had in *M. J. Gleeson (Contractors) Ltd* v. *London Borough of Hillingdon* (1970). So, he thought it was best not to place any reliance on clauses C and D in interpreting clause 16.

4. TYPE VERSUS PRINT

I must say that I think that, in construing this contract, we should have regard to provisions C and D. They were carefully drafted and inserted in type in the bills of quantities. They were put in specially so as to enable the contractors to make their calculations. It was on the basis of these that the contractors made their tender and the employers accepted it. They were incorporated into the formal contract just as much as the conditions in the RIBA form. In contrast, conditions 12 and 16 were not specially inserted at all. They were two printed conditions in the middle of 23 pages of small print. It was in quite general terms. On settled principles they should have taken second place to the special insertion – see the passages collected by Scrutton LJ in *Sutro & Co* v. *Heilbut, Symons & Co* (1917). If and in so far as conditions 12 and 16 contain anything inconsistent with provisions C and D, they should be rejected, just as the printed form was in *Love and Stewart Ltd* v. *Rowtor S.S. Co Ltd* (1916).

5. THE CONSTRUCTION OF 'C' AND 'D'

In my opinion, however, the various provisions can all be reconciled in this way: The general printed condition 16 is designed to deal with the usual building contract in which the employer does not take possession of the works, or any particular part of the same, until it is practically complete. Whereas the special provisions C and D are designed to deal with this special contract in which it is contemplated that Reeds will be installing plant and equipment whilst the works are still in progress and will be using and occupying parts before they are fully completed. These special provisions apply to the situation which they expressly cover and should be given their proper effect without being restricted to condition 16.

Applying conditions C and D they show that the employers do not 'take possession' of a part of the works simply by installing plant and equipment in that part or by occupying and using that part. That is all the employers have done here, in regard to the board machine house, the reel warehouse and the factory extensions south and east. So the contractors remain liable to insure it and could charge the increased cost to the employers.

6. APART FROM 'C' AND 'D'

I would, however, not leave the matter there. I am quite prepared to consider clause 16 without placing any reliance on provisions 'C' and 'D'. In my opinion the words 'take possession' of a part of the works must be so interpreted as to give precision to the time of taking possession and accuracy in defining the part. I say this because of the consequences which follow on it. [We were given] a list of seven. They include these:

(i) That part is deemed to be practically complete and the defects liability period begins to run;

(ii) A moiety of the retention money, in respect of that part, becomes payable;

(iii) The architect is, within seven days, to issue a certificate of value of that part;

(iv) That part is at the sole risk of the employer and not of the contractor; and the contractor is to reduce the value insured, accordingly.

In order to achieve the requisite precision and accuracy, the parties themselves, as I read the evidence, evolved a suitable machinery to determine it. It was by way of a definite handing over of the part by the contractors to the employers. The practice was for the contractors to tell the architect that a part was ready for handover. The architect would then inspect it to see if the part was in a condition to be accepted. If it was, he would accept it on behalf of the employers. He would give a certificate which defined the part, and would give its value and date of taking possession, and went on to state the consequences. So the handover was precise and definite. I regard that conduct as admissible either as explaining an ambiguity or uncertainty in the contract as it originally stood, or, alternatively, as an agreed gloss or variation on it made subsequently to the contract. I realise that the original form of certificate issued by the architect was inaccurate in speaking of 'practical completion' of the part. But that does not alter the fact that it was the accepted means of defining the handover. It should be given effect, accordingly.

7. CONCLUSION

In my opinion the contractors, at the time of the fire, had not handed over to the employers the responsibility for the board machine house, the reel warehouse, or the factory extensions east and south. Reeds were using and occupying those places but, until actual handover, it was the responsibility of the contractors to insure them. The risk had not passed to the employers. So it remained with the contractors. They, or their insurers, must bear the loss.

Note: Edmund Davies and Stephenson LJJ delivered judgments reaching the same conclusion but differing as to the effect of clause 12. See p. 28.

In practice, the insurance needs of the parties will also be determined by the express contractual terms as to the transfer of risks. For instance, it is commonly provided that the works are at the contractor's risk until practical completion. The contractor therefore needs insurance cover against the possibility of the works being damaged by causes which are not his fault. The clause discussed in the following case has since been amended.

A.E. Farr Ltd v. The Admiralty

QUEEN'S BENCH DIVISION   [1953] 2 All ER 512

The plaintiffs contracted to construct a destroyer jetty at a naval base. Clause 26(2) of the contract provided:

'The works and all materials and things whatsoever including such as may have been provided by the Authority on the site in connection with and for the purposes of the contract shall stand at the risk and be in the sole charge of the contractor, and the contractor shall be responsible for, and with all possible speed make good, any loss or damage thereto arising from any cause whatsoever other than the accepted risks. . . .'.

After the plaintiffs had done a substantial part of the work, the jetty was damaged by a destroyer (owned by the employer) which collided with it. The plaintiffs claimed the cost of repairing the jetty.

Held: Because of clause 26(2) the plaintiffs were liable for the repairs to the jetty.

PARKER J: However one approaches this matter, one always gets back to what is the true construction of condition 26(2). It is argued forcibly by counsel for the plaintiffs that, if a party desires to be free of liability for negligence, he must say so in express words. That is undoubtedly, true, and at first sight, when one looks at the facts of this case, it does seem as if very clear words were needed to make contractors liable for damage caused by the negligent navigation of an Admiralty ship which damages part of the works they have constructed, but it all comes back to the meaning of the words in that clause, and, in particular, to what is meant by 'any cause whatsoever'. . . . It seems to me that the words in the present case are about as wide as they can be and that I must read 'any cause whatsoever' as if it included and expressly said 'including damage caused by the negligent navigation by an Admiralty servant of a ship'. If one reads the words in that way, there is no possibility of the plaintiffs recovering the sum expended under condition 7(2) of the contract, because, by condition 26(2), they are responsible for the damage and have to make it good, and, therefore, the expense which they have incurred in so doing is not an expense beyond that provided for in, or reasonably contemplated by, the contract.

It is, however, said that there is an implied term in the contract that the contractor should have uninterrupted and undisturbed possession of the site, and that the building owner must not come, either negligently or deliberately, and knock down the works, and, accordingly, 'any cause whatsover' must be read subject to that implied term. It seems to me that that is the wrong approach to the matter. Once one realises that condition 26(2) must be given the wide construction which, I think, it must be given, there is no room for any implied term, however it is worded, which, in effect, destroys the construction ordinarily given to those words. It was faintly argued by counsel for the plaintiffs that, even if he was wrong, it might be a possible view that the provision under condition 26(2) that the contractor was to make good was merely machinery, and that, having made good, it would be possible for him to obtain the sum that he has expended against the Admiralty for breach of contract or in negligence. It seems to me, however, that condition 26(2) is plain beyond words that, not merely as a matter of machinery has he got to carry out repairs, but that he is wholly responsible for the loss or damage. In my view, therefore, this claim, however hard it may be on the contractors, must fail.

An insurance policy taken out by virtue of the express insurance requirements of the contract may itself present problems of interpretation.

Manufacturers' Mutual Insurance Ltd v. Queensland Government Railways

HIGH COURT OF AUSTRALIA (1968), 42 ALJR 181

The insurance company issued a policy of insurance to the respondents covering all loss or damage arising in connection with the respondents' contract to supply and erect a railway bridge except 'loss or damage arising from faulty design'. The bridge was to replace one built in 1897 and swept away by a flood in March 1956. In February 1958, three concrete piers erected in the bed of the river as part of the supports for the bridge were brought down by a flood higher than any previously recorded. The arbitrator found that there was no negligence in the design of the piers, but that with the benefit of hindsight and further research it was possible to say that inadequate allowance had been made for the transverse forces which operate on piers in a stream.

Held: 'Faulty design' did not imply any element of negligence, and that the loss was due to 'faulty design'.

BARWICK CJ, McTIERNAN, KITTO AND MENZIES JJ: We have come to the conclusion that upon a proper construction of the relevant exclusion, the loss which occurred did arise from faulty design. Let it be accepted, as the arbitrator found, that the piers, as designed, failed to withstand the water force to which they were subjected because they were designed in accordance with engineering knowledge and practice which was deficient, rather than because the designer failed to take advantage of such professional knowledge as there was, nevertheless the loss was due to 'faulty design' and the arbitrator has done no more than explain how it happened that the design was faulty. We think it was an error to confine faulty design to 'the personal failure or non-compliance with standards which would be expected of designing engineers' on the part of the designing engineers responsible for the piers. To design something that won't work simply because at the time of its designing insufficient is known about the problems involved and their solution to achieve a successful outcome is a common enough instance of faulty design. The distinction which is relevant is that between 'faulty', i.e., defective, design and design free from defect. We have not found sufficient ground for reading the exclusion in this policy as not covering loss from faulty design when, as here, the piers fell because their design was defective although, according to the finding, not negligently so. The exclusion is not against loss from 'negligent designing'; it is against loss from 'faulty design', and the latter is more comprehensive than the former.

The performance of the contract is often secured by a performance bond, which is often in turn secured by a further guarantee. It is again a question of construction in what precise circumstances the bond and guarantee may be payable.

General Surety and Guarantee Co. Ltd v. Francis Parker Ltd

QUEEN'S BENCH DIVISION　　　　　　　　　　　　　　(1977), 6 BLR 18

DONALDSON J: These preliminary issues concern a performance guarantee and counter-guarantee in the building trade. The issue, which I am told is of general importance, is whether the terms of the counter-guarantee are such that the defendants are liable to make payment under it before the plaintiffs are in their turn liable to pay under the guarantee.

In September 1972 the Corporation of the City of Liverpool entered into a contract under which Francis Parker (Contracts) Ltd agreed to build 177 houses at a price of £1,162,326. The plaintiffs, who are a well-known insurance company, gave a performance guarantee to the Corporation limited to 10 per cent of the contract price of £116,232 60p. At the same time the plaintiffs took a guarantee from the defendants, who are the parent company of Francis Parker (Contracts) Ltd.

Disputes arose between the Corporation and the contractors, and on 25 April 1974, the Corporation gave notice purporting to determine the employment of the contractors. On 1 May 1974, the contractors in their turn purported to accept that notice as a wrongful repudiation of the contract. Whichever is right – and this still has to be determined – the employment of the contractors came to an end at that time and the Corporation thereafter contracted with different builders for the completion of the works.

On 22 August 1974, the City Solicitor wrote to the plaintiffs stating that the building contract had been terminated because the contractors had failed to proceed regularly and diligently with the works. The letter concluded:

'Your Company are sureties in respect of Phase II of this project and I write to give you notice that the City Council will eventually be submitting a claim under the Bond and will let you know in due course the amount involved.'

This letter was not a demand for payment, but rather notice that 'eventually' such a demand would be made. This fact created a problem for the plaintiffs. On no view of the true construction of the counter-guarantee could the defendants be under any liability before the Corporation had made a claim under the principal guarantee. But 'eventually' could be a very long time. The Corporation and the contractors were likely to be locked in mortal arbitral combat, in which rounds were likely to be measured in months, if not years, rather than minutes. When the Corporation 'eventually' made a claim on the plaintiffs and the plaintiffs were in a position to make a claim under the counter-guarantee, the defendants' financial status might be very different. This is no reflection upon the defendants. It is a fact of economic life. Meanwhile the plaintiffs wanted to be able to improve the security afforded to them by the counter-guarantee by obtaining a cash payment thereunder.

The plaintiffs thereupon set about the delicate task of stimulating the Corporation to put forward a claim, without giving the slightest indication that it would be met. In a letter dated 29 November 1974, the plaintiffs referred to their wish to prepare 'our year-end claims reserves', and continued:

> 'In these circumstances, would it now be advisable to let us have a formal claim for payment of the Bond amount, which could be accompanied, if you so desire, by a statement that you do not propose to take action unless and until you see fit to do so.'

The Corporation duly obliged by a letter dated 9 December 1974, in which the City Solicitor wrote:

> 'I refer to your letter of 27 November 1974, and now write to make a formal claim for payment of the Bond amount in respect of the above contract. I look forward to receiving your cheque in the sum of £116,232 in settlement at your earliest convenience.'

I rather doubt whether the City Solicitor thought that the plaintiffs' 'earliest convenience' would occur in the foreseeable future, but that is by the way.

The plaintiffs in their turn wrote to the defendants on 13 December 1974, as follows:

> 'I refer to the above Bond and regret to advise you that we have now received a formal claim for payment of the Bond from Liverpool Corporation as per the attached copy letter, dated 9 December 1974. We herewith make formal demand for payment of this amount from yourselves in accordance with Clause 1 of the counter indemnity given by yourselves to us, a copy of which is attached.'

I now turn to the wording of the guarantee. That given by the plaintiffs to the Corporation was in these terms:

'General Surety & Guarantee Co Ltd of 76, Cross Street, Manchester hereby guarantee to the Corporation the punctual, true and faithful performance and observance by the contractor of the covenant on his part contained in the said Agreement, and undertake to be responsible to the Corporation, their successors or assigns as sureties for the contractor for the payment by him of all sums of money, losses, damages, costs, charges and expenses that may become due or payable to the Corporation, their successors or assigns by or from the contractor by reason or in consequence of the default of the contractor in the performance or observance of his said covenant, but so nevertheless that the total amount to be demanded or recovered by the Corporation, their successors or assigns of or from us as sureties shall not exceed 10 per centum of the total prices payable under the said Agreement.'

The counter-guarantee in which the defendants were described as 'the Guarantor' and the plaintiffs as 'General Surety' was, so far as is material, in these terms:

'(1) The Guarantor hereby undertakes that the contractor will duly perform and observe all its obligations under the contracts and that if General Surety is called upon to make any payment or payments under any of the said Guarantee Bonds the Guarantor will forthwith on demand pay to General Surety the full amount of such payment or payments.

(2) The undertaking herein contained is irrevocable.

(3) General Surety shall not be required before enforcing the undertaking herein contained to take any step to recover from any other person whether a principal debtor or surety any sum due to General Surety or be required to enforce any of General Surety's rights in any order of priority and no time given to any other person whether a principal debtor or surety or any variation or release of any obligation of any other person or release of any other security shall in any way lessen or affect the liability of the Guarantor hereunder.

(4) General Surety shall be under no obligation to obtain the prior consent of the Guarantor or give notice to the Guarantor before or after entering into a Guarantee Bond in respect of the obligations of the contractor.'

It seems to me to be clear – and [counsel] for the plaintiffs did not seek to persuade me otherwise – that the plaintiffs are only liable to the Corporation to the extent that it can be proved that the contractor has made default and that sums of money, losses, damages, costs, charges and/or expenses have become due and payable by the contractor to the Corporation by reason or in consequence of that default. Whether or not the day will come when this can be proved, I know not. But it has not yet come and the liability of the plaintiffs to the Corporation has yet to arise. Indeed, whilst no criticism can be made of the Corporation's notice of intention to lodge a claim, the claim itself was clearly premature.

But what of the counter-guarantee? [Counsel] submits that the first rule of construction is that documents are to be taken to mean what they say, unless what they say is so repugnant to good sense that some other meaning must have been intended. In this case the defendants agreed that if the plaintiffs were called upon to make any payment under the guarantee bonds, the defendants would forthwith on demand pay them the full amount of such payment. The plaintiffs have been called upon to pay the sum of £116,232 60p and the defendants are accordingly liable to make a similar payment.

I accept that effect must be given to the counter-guarantee in accordance with the meaning of the words used by the parties. But what is that meaning? The words 'if the General Surety is called upon to make any payment or payments under any of the said Guarantee bonds' show clearly that actual payment by the plaintiffs is not a condition precedent to liability on the part of the defendants. But do they go further than this? The concept of being called upon to make a payment is ambiguous. It may mean that A is liable and ought to pay B. Or it may mean merely that a demand has been made that A pay B regardless of whether or not the demand is well-founded. I think that the former is the true meaning of the words in this particular contract.

My reasons for reaching this conclusion are as follows:

(i) If this is, as it says, a contract of guarantee, the background must be one in which there is a principal debtor, namely the plaintiffs as General Surety, and a secondary debtor, namely the defendants. On the plaintiffs' construction, there need be no principal debtor.

(ii) If this is a contract of indemnity, and the argument proceeded upon the basis that it was, the background is one of a legal relationship in which liability at law arises only when the promisee (the plaintiffs) has been damnified by payment and in equity when the promisee has incurred a liability. On the plaintiffs' construction, it is unnecessary that the plaintiffs should either have been damnified or be liable to be damnified in the future.

(iii) If the counter-guarantee is a hybrid, which is probably the case, these considerations overlap.

(iv) Whatever its category, a contract of guarantee or indemnity is to be strictly construed in favour of the promisor.

(v) The promissory words in clause 1 of the counter-guarantee are ancillary to the primary undertaking that the contractor will duly perform and observe all its obligations under the contracts. If the plaintiffs are right, these words are otiose, since liability can arise whether or not the contractor is in default. So too are many of the provisions of clause (3) which exonerate the plaintiffs from taking steps which they could only take if the contractor was in default.

(vi) The promissory words require there to be a demand 'under . . . the guarantee bonds'. A premature demand may purport to be made under the guarantee bond, but is not in law so made.

(vii) If the plaintiffs are right, the defendants are liable to pay them the sum of £116,232.60p, although it may eventually emerge that nothing is due from the contractors to the Corporation. I do not doubt that in such an event the money would be repayable, but it is not so certain that it would be repayable with interest. I find this situation very odd.

(viii) There is a simple and well-known way of achieving the result which the plaintiffs say that they have achieved. This is to provide that, as between the plaintiffs and the defendants, a demand by the Corporation shall be conclusive evidence of the plaintiffs' liability to the Corporation – see *Bache & Co (London) Ltd* v. *Banque Vernes et Commerciale de Paris SA* (1973). This course has not been adopted.

The argument before me has proceeded on the basis that either (a) there is an immediate liability on the part of the defendants to pay the plaintiffs the sum of £116,232.60 or (b) liability to make this payment is dependent upon the happening of any one of the events specified in the Order for the trial of the preliminary issues. No other position has been contended for. I therefore determine these issues by holding that:

'There can be no liability on the part of the defendants to pay the plaintiffs under the terms of the agreement made between the parties unless and until:

(a) the architect appointed under the contract between Liverpool Corporation and the defendants' subsidiary has properly certified that the Corporation has suffered loss and damage in excess of the sum which would otherwise have been due to the subsidiary under the said contract and/or

(b) any such certificate has been opened up, reviewed and if necessary revised by an arbitrator pursuant to the said contract and/or

(c) liability to the Corporation has been admitted by the subsidiary in the sum of £116,232.60 or greater sum and/or

(d) the Corporation has successfully sued or arbitrated with the subsidiary to judgment or to an award not subject to case stated for the aforesaid or a greater sum.'

. . . I have not gone into the details of this as to whether these are all conditions which trigger the guarantee, because that was not argued in front of me. I have simply taken it as a block.

Building contracts commonly provide in express terms that one of the parties shall indemnify the other against certain losses. Such a provision should not be read so as to impose on one party a duty to indemnify the other against the other's own negligence unless clear words to that effect are used. Indemnity clauses are strictly construed by the courts.

Walters v. Whessoe Ltd and Shell Refining Co. Ltd

COURT OF APPEAL (1960), 6 BLR 23

The plaintiff's husband was employed by W in operations on a site belonging to Shell. He was killed in an accident. Winn J held that both Whessoe and Shell were negligent, and apportioned liability 20% to Whessoe and 80% to Shell. Shell claimed to be entitled to indemnity from Whessoe by virtue of clause 3 of the contract between them which provided:

'*Injury and damage.* The contractors shall indemnify and hold Shell their servants and agents free and harmless against all claims arising out of the operations being undertaken by the contractor in pursuance of this contract or order or incidental thereto in respect of (a) personal injury, including death and industrial disease, sustained by any employee of the contractor or of a sub-contractor, (b) loss or damage to the property and personal injury, including death, to the person of any third party and (c) loss or damage to the property, equipment or tools of the contractor, a sub-contractor or any of their employees.'

Held: Shell's claim must be rejected.

DEVLIN LJ: It is now well established that if a person obtains an indemnity against the consequences of certain acts, the indemnity is not to be construed so as to include the consequences of his own negligence unless those consequences are covered either expressly or by necessary implication. They are covered by necessary implication if there is no other subject-matter upon which the indemnity could operate. Like most rules of construction, this one depends upon the presumed intention of the parties. It is thought to be so unlikely that one man would agree to indemnify another man for the consequence of that other's own negligence that he is presumed not to intend to do so unless it is done by express words or by necessary implications. The reason for the rule is illustrated by one of the earliest cases in which it was enforced – *Phillips* v. *Clark* (1857). In that case the question was whether a shipowner who had inserted in a bill of lading the stipulation, 'not accountable for leakage or breakage', was exempted from responsibility for a loss by leakage and breakage if it arose from his own negligence. In the course of the argument Cockburn CJ said: 'I think it can hardly be permitted to him to contend that he inserted the clause for the purpose of protecting himself against negligence . . . Can we in a court of justice put so absurd a construction upon language that is susceptible of another and a more rational construction.'

The law therefore presumes that a man will not readily be granted an indemnity against a loss caused by his own negligence. Such a loss is due to his own fault. No similar presumption can be made if he is made responsible for the negligence of others over whose acts he has no control. A responsibility of this sort may arise in law in the case of an employer and his workmen or an occupier and persons whom he invites to the premises. Similarly, a man may be responsible without negligence on his part for breach of statutory duty committed either by himself or by his servants or by someone for whom he was vicariously liable. In none of these cases is there any negligence on his part and there is therefore no reason to presume an intention to exclude them from the indemnity.

The indemnity which we have to construe does not contain any express words covering negligence. Its terms are wide enough to cover breaches of statutory duty and acts of independent contractors which may impose a liability on Shell without any fault on the part of themselves or their servants. It is not therefore necessary, in order to give the clause subject-matter, to construe it as covering negligence and so the indemnity does not cover the consequences of a negligent act committed by Shell's servants.

In two cases the court considered clause 14 of the then current RIBA contract (1939 edition) which provided:

'(a) *Injury to persons. The contractor shall be solely liable for and shall indemnify the employer in respect of any liability, loss, claim or proceedings whatsoever arising under any statute or at common law in respect of personal injury to or the death of any person whomsoever arising out of or in the course of or caused by the execution of the works, unless due to any act or neglect of the employer or of any person for whom the employer is responsible.'*

'(b) *Injury to property. Except for such loss or damage by fire as is at the risk of the employer under clause 15(b)(B) of these conditions the contractor shall be liable for and shall indemnify the employer against any loss, liability claim or proceedings in respect of any injury or damage whatsoever to any property real or personal in so far as such injury or damage arises out of or in the course of or by reason of the execution of the works, and provided always that the same is due to any negligence, omission or default of the contractor, his servants or agents or of any sub-contractor'.*

A.M.F. International Ltd v. Magnet Bowling Ltd

QUEEN'S BENCH DIVISION [1968] 2 All ER 789

MOCATTA J, having read clause 14, said: It is hard to imagine anyone being proud of the draftsmanship of this condition. Nevertheless, counsel for Magnet, whilst being freely critical, claimed that sub-paragraph (b) clearly covered him. Magnet have been held liable in respect of damage to A.M.F.'s property. Such damage arose out of or in the course of or by reason of the execution of 'the works' and was due to negligence, omission or default of Trenthams. The claim was prima facie, at any rate, unanswerable.

Counsel for Trenthams advanced a number of arguments to the contrary. First 'the works' in condition 14(b) could not include the work undertaken by A.M.F. under the permission granted by condition 23. This may be so, but the provisions in items 1/5/C, 1/6/E and 2/1/B of the bill of quantities, to which I have already referred were in my judgment part of 'the works'. It is clear on the facts as I have found them that had these provisions been complied with water should not have entered the building. For these reasons I reject this argument. Secondly, counsel for Trenthams said that condition 10 prevented Magnet placing any reliance on any of these items in the bill of quantities so as to entitle them to indemnity under condition 14(b). Condition 10 reads:

'The quality and quantity of the work included in the contract sum shall be deemed to be that which is set out in the bills of quantities mentioned in clause 2 of these conditions which bills unless otherwise expressly stated shall be deemed to have been prepared in accordance with the principles of the standard method of measurement of building works last before issued by the Royal Institution of Chartered Surveyors and the National Federation of Building Trades Employers, but save as aforesaid nothing contained in the said bills of quantities shall override, modify or affect in any way whatsoever the application or interpretation of that which is contained in these conditions.'

I accept the answer of counsel for Magnet to this point. The items in question constituted part of the quantity of the work included in the contract sum within the opening words of that condition. There is therefore no conflict, as counsel for Trenthams argued, between the last part of condition 10 and condition 14(b) to prevent Magnet from relying on the latter as they do.

Thirdly, counsel for Trenthams raised a point of law of general importance and some difficulty. Relying on my anticipated (and now actual) finding that the damage to A.M.F.'s property was partly caused by the negligence of Magnet, he submitted that on authority Magnet were unable to recover anything under the indemnity clause. He relied on *Rutter* v. *Palmer* (1922), *Joseph Travers & Sons Ltd* v. *Cooper* (1915), and similar well-known cases following on them, as establishing the general principle that an exception clause does not protect against negligence unless it expressly so provides on the true construction of the words used, or unless no content can be given to the words used unless they protect against negligence on the principle laid down in *Alderslade* v. *Hendon Laundry Ltd* (1945). These points are nowhere more clearly stated than in *Canada Steamship Lines Ltd* v. *Regem* (1952) in the well-known passage, which it is unnecessary for me to read. Condition 14(b) could be given content to protect Magnet against liability to third parties arising through no negligence on their part. Counsel for Trenthams instanced *Hughes* v. *Percival* (1883) as an example. In that case the building owner was held liable in tort for damage caused to his neighbour's house by bad work by his builder, notwithstanding that he had employed a competent architect and builder. Another example relied on was *Balfour* v. *Barty-King* (1957). On this point I think counsel for Trenthams was undoubtedly right. He further argued that these principles applied as much to claims by persons suing on indemnity clauses as to defendants seeking to escape liability under exception clauses. This undoubtedly seems to have been accepted by the Privy Council in *Canada Steamship Lines Ltd* v. *Regem* (1952). It seems to have been assumed, at least as a possibility, in *A.E. Farr Ltd* v. *The Admiralty* (1953) and *Westcott* v. *J.H. Jenner (Plasterers) Ltd and Bovis Ltd* (1962), though in both those cases the person relying on the indemnity succeeded by reason of the language used being held to have been sufficiently clear to have covered his own negligence.

The most relevant and powerful authority relied on by counsel for Trenthams was *Walters* v. *Whessoe Ltd and Shell Refining Co. Ltd* (1960). . .

Counsel for Magnet sought to distinguish the present case in a variety of ways. First of all he submitted that the words of condition 14(b) were wide enough to include Magnet's negligence and breach of contract. He pointed out that the word 'solely', present in sub-paragraph (a), is absent from sub-paragraph (b). As against this counsel for Trenthams suggested that the effect must be the same by reason of the words 'in so far as' in sub-paragraph (b). Further, counsel for Magnet urged that condition 14 was a sweeping-up clause and that sub-paragraph (b) significantly lacked the words at the end of sub-paragraph (a) 'unless due to any act or neglect of the employer', etc. I have already said that the phraseology of the condition as a whole leaves much to be desired; I do not think any real assistance is to be derived from comparing the language of the two sub-paragraphs. If a party relying on an indemnity is to escape the effect of the principle laid down in the *Whessoe* case the language he relies on must be clear. In my judgment the language used in 14(b) is not clear enough to take Magnet out of the principle.

Counsel for Magnet then argued that I must distinguish between causa causans and causa sine qua non, that in the *Whessoe* case Whessoe's negligence was only a causa sine qua non and that the causa causans or real or effective cause was the negligence of Shell. Here he equated Magnet with Whessoe and Trenthams with Shell. Moreover, the passages which I have read from the judgments of Sellers and Devlin LJJ, in dealing with the principle, did not mention the partial responsibility in negligence of Whessoe. Unless one interpreted the reasoning of the Court of Appeal in this way, counsel submitted there was an insoluble logical conflict between the *Whessoe* case and a line of authorities holding that a plaintiff can recover as damages for breach of contract the damages, flowing therefrom that he has had to pay a third party for liability in tort. He referred to the well-known line of cases starting with *Mowbray* v. *Merryweather* (1895), continuing with *Scott* v. *Foley Aikman & Co.* (1899) and *The Kate* (1935), and most recently exemplified by *Sims* v. *Foster Wheeler Ltd* (1966). In each of these cases the plaintiff succeeded in contract because his liability in tort to a third party was due to a breach of duty owed to that third party and not to the defendant. He further relied for his interpretation of the *Whessoe* case on what was said by Devlin J, in *Compania Naviera Maropan S.A.* v. *Bowaters Lloyd Pulp and Paper Mills Ltd* (1955).

I am unable to see how I can distinguish the reasoning and decision in the *Whessoe* case on the lines suggested. What fraction of responsibility greater than twenty per cent is required before the other party's negligence ceases to be the effective cause? The question has only to be asked to be seen to be unanswerable. If it can be answered, I can see no grounds for holding sixty per cent responsibility to be the effective cause and forty per cent only a sine qua non. Further I think that there is a real distinction in principle between the indemnity cases and the line of contract cases cited. In the latter the defendant is under a contractual obligation to provide reasonably safe tackle, or a berth or scaffolding, etc., or warrants that what he has provided is reasonably safe to use. In the indemnity cases the party giving the indemnity merely undertakes to pay the other party money in certain circumstances. If, as a matter of construction, the event that has occurred falls outside those circumstances, no liability to indemnify arises. For these reasons and considering myself bound by the *Whessoe* case, even if not strictly bound by the *Canada Steamship* case, I reject Magnet's claim to be indemnified under condition 14(b). In doing so, however, I base my decision on my earlier finding of Magnet's liability to A.M.F. in tort and express no view whether, had I only held them liable in contract, the claim under the indemnity clause would have failed.

City of Manchester v. Fram Gerrard Ltd

QUEEN'S BENCH DIVISION (1974), 6 BLR 70

KERR J: I must turn to clause 14(b) on which ultimately everything depends. It does not appear ever to have been considered judicially except by Mocatta J in *AMF International Ltd* v. *Magnet Bowling Ltd* (1968), and he then said: 'It is hard to imagine anyone being proud of the draftsmanship of this condition'. One important issue to be decided on the short-route argument in relation to [the sub-sub-contractors] in this case, i.e., whether or not the clause covers persons in [their] position and sub-sub-contractors generally, is unclear and certainly lends support to this view.

I deal first with three general matters which are of some relevance to the plaintiffs' contentions both in relation to the long-route and short-route arguments. First, clause 14(b) deals with injury to property, whereas clause 14(a) deals with injury to persons. I have not set out the latter because [counsel] rightly did not press any argument to the effect that one gets any real assistance on the construction of sub-clause (b) from the wording of sub-clause (a) and I note that Mocatta J took the same view in the *AMF* case. Secondly, I approach the construction of this clause on the basis that, being an indemnity clause, it should be construed strictly, analogously to an exemption clause; at any rate to the extent to which the defendants are thereby sought to be held liable for defaults etc. by persons other than themselves over whom they have no control. This is the position here, since it is not suggested by the plaintiffs that there was any negligence, omission or default on the part of the defendants in any respect whatever. This general approach to the construction of the clause is in my view established by authority. It was the approach adopted by the Privy Council in the leading case of *Canada Steamship Lines Ltd* v. *Regem* (1952), where the Board was dealing both with an exemptions clause and an indemnity clause, and approached the construction of both in the same way. This approach was followed by Mocatta J in the *AMF* case and in the decision of the Court of Appeal in *Walters* v. *Whessoe Ltd* (1960) there cited. More recently there is further direct authority that this is the correct approach to indemnity clauses in *Gillespie Brothers & Co Ltd* v. *Roy Bowles Transport Ltd* (1973), where the *Canada Steamship Lines* case was again followed, at any rate by the majority of the court, in the context of an indemnity clause, and where Buckley LJ expressly equated the approach to the construction of indemnity clauses with that to exemption clauses. I therefore proceed on the basis that if and in so far as the wording of clause 14(b) does not clearly have the effect of rendering the defendants liable for the defaults etc. of other persons, then the plaintiffs cannot successfully rely upon it.

. . .

The plaintiffs' final, and in my view most difficult, submission was that the words 'or of any sub-contractor' are wide enough to include sub-sub-contractors. They contend that [K] were sub-sub-contractors of the defendants. It could not, of course, avail the plaintiffs merely to rely on the fact that [K] were sub-contractors of Smith's, since they are not suing Smith's and since there is no indemnity clause in force between the plaintiffs and Smith's. [Counsel] said – in my view rightly so far as it goes – that any one working on a site otherwise than under some contract of employment is in general parlance often referred to as a 'sub-contractor'. But I have come to the conclusion that such a wide construction cannot properly be given to the indemnity clause in this contract. There are a number of reasons for this. First, as conceded by [counsel], the contract nowhere uses the words 'sub-contractor' in a context which requires it to be read as including sub-sub-contractors. It especially provides in clause 13(a) for the subletting of part of the works by the main contractor with the written consent of the architect, and in clause 21 for the appointment of nominated sub-contractors by the main contractor. Both these clauses therefore refer only to sub-contractors of the main contractor, and the contract nowhere deals, either expressly or by necessary implication, with further generations of sub-contractors. I therefore see no sufficient reason for attributing any wider meaning to the words 'sub-contractor' in the indemnity clause. Secondly, if it had been intended that the indemnity clause should have the wide meaning for which the plaintiffs contend, then it would have been easy to say so. Instead of referring to 'any sub-contractor' it would have been easy to say something like 'anyone executing any part of the works' or 'anyone working on the site'. Further, I do not think that anything is added to the plaintiffs' submissions, or provides any answer to the foregoing points, by relying on the main contractors' obligations in relation to insurance, to which I have already referred. [Counsel] said that the employer would wish to be covered by insurance, and therefore also by indemnity, in relation to anyone working on the site and not merely in relation to the first generation of sub-contractors. But in my view this is little more than an argument *ad misericordiam*. If this had been intended then it was incumbent upon the employer to cover it by clear language. The other side of the coin in relation to this argument is that it would lead to hardship on the main contractor. He would no doubt wish at least to have the opportunity – as he expressly has in relation to nominated sub-contractors – to cover his liability under the indemnity clause by sub-contracts of indemnity. But in practice this would be extremely difficult, to say the least, in relation to every contractor on the site with whom the main contractor has no contractual nexus, particularly when it is well known that many individual workmen nowadays require to be employed as independent contractors and not under contracts of employment.

I have accordingly come to the conclusion that the words 'or of any sub-contractor' in clause 14(b) are not wide enough to cover further generations of sub-contractors beyond those who are sub-contractors to the main contractor.

It may be possible in some circumstances to infer that one party has impliedly undertaken to indemnify the other.

Hadley v. Droitwich Construction Co. Ltd

COURT OF APPEAL   [1967] 3 All ER 911

The plaintiff was injured when a small mobile crane collapsed. The crane had been hired by its owner to a builder and both owner and builder were sued by the plaintiff. The trial judge held that both defendants were at fault, the owner because he had negligently handed the crane over with an incorrect adjustment of the clearance between its rear roller and the bottom of its base and the builder because he had negligently failed to have the machine serviced properly and to provide a competent driver for it. The builder claimed that he was entitled to be indemnified by the owner against his liability to the plaintiff.

Held: The builder was not entitled to an indemnity as the owner, in a letter confirming the oral agreement, had written 'we thank you for your promise to put a competent man in charge of this machine who will operate it carefully and carry out the servicing properly'.

HARMAN LJ: [The learned judge goes on, however, to acquit the employers of all liability.] I cannot follow him. He can only make that finding, as it seems to me, if what had passed was that the owners gave an assurance, on which the employers were entitled to rely, that the crane was all right as handed over and need not be further examined. So far from that being the position, the owners, although they did not send their book of words, so to speak, or any specific information about the alterations which they had made to the machine, did write a letter to say they were glad that the employers agreed to put an experienced man on this job and to see that the machine was properly serviced. Any warranty that the owners may have impliedly given as to fitness of the machine for its task was undoubtedly, as I see it, qualified by those two conditions. Neither condition was performed: the warranty disappears, as the conditions were broken. So far from being entitled to put the whole liability on the owners, I think that the employers cannot shift it from their shoulders in so far as it was due to their own breach of duty towards their employee.

Mowbray v. *Merryweather* (1895) was quite a different case. There was there an assurance, on which the plaintiff (who had paid the workman) was entitled to rely, that everything was in order, and he was absolved from his duty vis-à-vis the second defendant to examine the chain in question. Here, owing to the presence of the letter, the boot is on the other leg, so it seems to me.

In a building contract there are nearly always several parties (contractors, sub-contractors, sub-sub-contractors, suppliers, etc.) involved. It is common for insurance policies to be taken out in the names of some or all of these parties. The legal effect of this was considered in:

Petrofina (UK) Ltd v. Magnaload Ltd

QUEEN'S BENCH DIVISION [1983] 3 All ER 35

An accident occurred at an oil refinery at Killingholme where a major extension was being built. The refinery was owned by Lindsey Oil, the third plaintiffs, who operated it for the benefit of the first and second plaintiffs. The works were financed by the fourth plaintiff Omnium Leasing, a consortium of companies. The main contractors were Foster Wheeler, who had sub-contracted heavy lifting operations to Greenham (Plant Hire). Certain specialist lifting equipment was provided by the first defendants, Magnaload, and the second defendants, Mammoet Stoof. The plaintiffs alleged that the accident was due to the negligence of the first and second defendants. The proceedings involved a preliminary issue as to whether the first and second defendants were covered by a contractor's 'all risks' policy taken out by Foster Wheeler.

LLOYD J: Counsel for the defendants told me that it is a question of great importance for the construction industry generally; for policies such as the one I have to consider are in common use. I am sure he is right. The answer in the present case turns on the construction of this particular policy, and on general principles of law. The policy is a 'contractors' all risks insurance policy'. By section 1 it provides:

'The insurers will indemnify the insured against loss of or damage to the insured property whilst at the contract site from any cause not hereinafter excluded occurring during the period of insurance.'

'The insured' are defined in the schedule as –

'Omnium Leasing (Owner) and/or Lindsey Oil Refinery and/or Foster Wheeler Ltd. and/or Contractors and/or Sub-contractors.'

'The insured property' is defined as –

Item No. 1. The works and temporary works erected . . . in performance of the insured contract and the materials . . . for use in connection therewith belonging to the insured or for which they are responsible brought on to the contract site for the purpose of the said contract . . .

Item No. 2. Constructional plant comprising plant and equipment . . . if and insofar as not otherwise insured belonging to the insured and for which they are responsible brought on to the contract site for the purpose of the insured contract.

'The insured contract' is defined as: 'Construction erection and testing of an extension to the Lindsey Oil Refinery at South Humberside.'

By an indorsement dated 23 April 1978 the policy was extended to include –

'The erection operation and subsequent dismantlement of the following items of plant which are being used on the Greenham (Plant Hire) Ltd. sub-contract for the erection of eight large vessels: (1) . . . £600,000; (2) . . . £250,000; (3) Hydrajack system £2,150,000'

giving a total of £3,000,000. The minimum premium was increased by £9,000.

The policy also provides in section 3, third party liability cover in standard terms, whereby the insurers agree to indemnify the insured –

'Against all sums for which the insured shall become legally liable to pay as damages consequent upon (a) Accidental bodily injury to or illness or disease of any person (b) Accidental loss of or damage to property occurring . . . as a result of and solely due to the performance of the insured contract happening on or in the immediate vicinity of the contract site.'

There is an exception to the third party liability cover which excludes property forming the subject of the insured contract.

The first point taken by the insurers is that the defendants are not 'sub-contractors' within the meaning of the policy, and are therefore not insured at all. If they are right about that, then none of the other points arise. At an early stage of the argument counsel for the plaintiffs very properly accepted, though he never formally conceded, that the first defendants, Magnaload, must be sub-contractors; but he maintained the argument in relation to the second defendants, Mammoet.

The facts are that in the initial stages of the negotiations it was contemplated that Mommoet would enter into a contract with Greenhams for carrying out the specialist lifting operation with their hydrajack. At the end of June there was a change of direction. At a meeting which took place between Magnaload and Mammoet on 27 June 1978 it was agreed that, for administrative reasons, the contract would be in the name of Magnaload, not Mammoet. Mammoet would, however, remain responsible for the operation itself. The reasons for this change do not matter. The sub-contract price had already been agreed between Greenhams and Mammoet at £185,000. It was agreed that Mammoet would reimburse Magnaload out of that sum for any services which Magnaload performed or any costs which they incurred. On 13 July 1978 there was a meeting between Magnaload, Mammoet and Greenhams at which Greenhams agreed that the contract would be with Magnaload provided they received '100 per cent support' from Mammoet. On 24 July Greenhams sent a telex to Magnaload as follows:

'Please find the draft of the purchase order for Foster Wheeler job at Immingham. Could we have your comments as soon as possible. *"Order on Magnaload.* For supply of lifting equipment and services in order to carrry out the erection of heavy vessels at the Lindsey Oil Refinery expansion project, Killingholme, South Humberside. All as more fully described in the attached scope of work and conditions of order. For the lump sum price of £185,000." '

Then, under the heading scope of work and conditions of order, I will quote two paragraphs:

'Erection of vessels at Lindsey Oil Refinery expansion project, Killingholme, South Humberside. Main contractor: Foster Wheeler Ltd. Heavy lifting sub-contractor: Greenham Plant Hire Ltd. Specialist lifting contractor: Magnaload Ltd. . . . General conditions of contract in line with those which Greenham Plant Hire have accepted from Foster Wheeler are understood to apply to this order.'

That telex is signed by Greenham Plant Hire.

The same day a copy of that telex was sent on by Magnaload to Mammoet. Mammoet came back with certain comments, which were sent on to Greenhams. According to W, who gave evidence at the hearing, and who had taken an active part in the negotiations as commercial director of Magnaload, all three parties reached agreement on all points of substance.

The term 'sub-contractor' is defined in the head contract as meaning –

'Any person to whom the preparation of any design the supply of any plant or the execution of any part of the works is sub-contracted, irrespective of whether the contractor is in direct contract with such person.'

There is a similar definition in the sub-contract between Foster Wheeler and Greenhams.

On the facts which I have set out, there was clearly a contract between Greenhams and Magnaload, as, indeed, counsel for the plaintiffs accepted. Strictly speaking Magnaload were not sub-contractors but sub-sub-contractors. But I would hold that they were nevertheless sub-contractors within the definition contained in the contracts and within the ordinary meaning of the word 'sub-contractors' as contained in the policy. To my mind the word 'sub-contractors' in the context of the policy must include sub-sub-contractors as well as sub-contractors.

The next question relates to the property insured. Counsel for the defendants submits that a 'contractors' all risk' policy in this form is an insurance on property, namely the works and temporary works belonging to the insured, or for which the insured are responsible. Counsel for the plaintiffs accepts that it is an insurance on property, but submits that the property insured must be read distributively, in other words, each insured is only insured in respect of his own property, or property for which he is responsible. I do not accept that construction. It seems to me that on the ordinary meaning of the words which I have quoted, each of the named insured, including all the sub-contractors, are insured in respect of the whole of the contract works. There are no words of severance, if I may use that term in this connection, to require me to hold that each of the named insured is only insured in respect of his own property. Nor is there any business necessity to imply words of severance. On the contrary, as I shall mention later, business convenience, if not business necessity, would require me to reach the opposite conclusion. I would hold, as a matter of construction, that each of the named insured is insured in respect of the entire contract works, including property belonging to any other of the insured, or for which any other of the insured were responsible.

But then comes the question: what is the nature of the interest insured? Counsel for the plaintiffs submits that the only possible insurable interest which one insured could have in property belonging to another, is in respect of his potential liability for loss of or damage to that property. Accordingly section 1 of the present policy is, he submits, a composite insurance. It is an insurance on property so far as it relates to property belonging to any particular insured; it is a liability insurance so far as it relates to all other property comprised in 'contract works'.

Counsel for the defendants, on the other hand, submits that section 1 of the policy is a policy on property, pure and simple.

In my judgment the submission of counsel for the defendants is correct. A very similar question arose in *Hepburn* v. *A Tomlinson (Hauliers) Ltd* (1966). In that case a firm of carriers took out a policy of insurance of goods in transit, on a form known as 'Form J'. A quantity of cigarettes were stolen while the goods were still in transit, by reason of the negligence of the owners, Imperial Tobacco Co. It was common ground that the carriers were not liable to the owners. On the carriers making a claim under the policy, for the benefit of the owners, it was argued that the policy was a liability policy, and since the carriers were not liable to the owners, they could not recover from the insurers. The argument was rejected by Roskill J, and his judgment was upheld by the Court of Appeal and the House of Lords.

It appears that underwriters had always regarded an insurance on 'Form J' as being a liability policy, and not a policy on goods. But Lord Reid said that the language of the policy showed conclusively that the policy was a policy on goods. I can find no relevant distinction between the language of the present policy and the language in *Hepburn* v. *Tomlinson*. Section 1 of the present policy covers 'all risks of loss or damage to the insured property'. It excludes the cost of replacing or rectifying insured property which is defective in material or workmanship, and so on. There is nothing in section 1 of the policy which is appropriate to an insurance against liability, whereas everything is appropriate to a policy on property. Section 2 of the policy covers the contractual liability of the contractors to maintain the contract works during the maintenance period. Section 3 covers third party liability. The very fact that third party liability is covered in a separate section, which expressly excludes liability for damage to property forming part of the contract works, shows to my mind that s. 1 of the policy is an insurance on property, and not an insurance against liability.

There is a passage in Jessel MR's judgment in *North British and Mercantile Insurance Co* v. *London, Liverpool and Globe Insurance Co* (1877) which points in the other direction. For he refers to a wharfinger's insurance on goods held in trust, or for which they are responsible, as being in terms an insurance against liability. But that dictum (for it was no more) cannot now stand against the decision of the House of Lords in *Hepburn* v. *Tomlinson*.

That brings me to the central question in the case. In *Hepburn* v. *Tomlinson* it was held, indeed it was conceded, that if the policy was an insurance on goods, then the carriers could, as bailees, insure for their full value, holding the proceeds in trust for the owners. In the present case the defendants could not be regarded as being in any sense bailees of the property insured under the policy. Does that make any difference? Can the defendants recover the full value of the property insured, even though they are not bailees? It is here that one leaves the construction of the policy, and enters, hesitatingly, the realm of legal principle.

What are the reasons why it has been held ever since *Waters* v. *Monarch Fire and Life Assurance Co* (1856) that a bailee is entitled to insure and recover the full value of goods bailed? Do those reasons apply in the case of a sub-contractor?

One reason is historical: the bailee could always sue a wrongdoer in trover. If his possessory interest in the goods was sufficient to enable him to recover the full value of the goods in trover, why should he not be able to insure that interest?

Another reason was that, as bailee, he was 'responsible' for the goods. Responsibility is here used in a different sense from legal liability. A bailee might by contract exclude his legal liability for loss of or damage to the goods in particular circumstances, e.g. by fire. But he would still be 'responsible' for the goods in a more general sense, sufficient, at any rate, to entitle him to insure the full value.

It is clear that neither of these reasons apply in the case of a sub-contractor.

But there is a third reason which is frequently mentioned in connection with a bailee's right to insure the full value of the goods. From a commercial point of view it was always regarded as highly convenient. Thus in *Waters* v. *Monarch Fire and Life Assurance* (1856) itself Lord Campbell CJ said:

'What is meant in these policies by the words "goods in trust"? I think that means goods with which the assured were entrusted; not goods held in trust in the strict technical sense . . . but goods with which they were entrusted in the ordinary sense of the word. They were so entrusted with the goods deposited on their wharfs: I cannot doubt the policy was intended to protect such goods: and it would be very inconvenient if wharfingers could not protect such goods by a floating policy.'

In the case of a building or engineering contract, where numerous different sub-contractors may be engaged, there can be no doubt about the convenience from everybody's point of view, including, I would think, the insurers, of allowing the head contractor to take out a single policy covering the whole risk, that is to say covering all contractors and sub-contractors in respect of loss of or damage to the entire contract works. Otherwise each sub-contractor would be compelled to take out his own separate policy.

This would mean, at the very least, extra paperwork; at worst it could lead to overlapping claims and cross claims in the event of an accident. Furthermore, as W pointed out in the course of his evidence, the cost of insuring his liability might, in the case of a small sub-contractor, be uneconomic. The premium might be out of all proportion to the value of the sub-contract. If the sub-contractor had to insure his liability in respect of the entire works, he might well have to decline the contract.

For all these reasons I would hold that a head contractor ought to be able to insure the entire contract works in his own name and the name of all his sub-contractors, just like a bailee or mortgagee, and that a sub-contractor ought to be able to recover the whole of the loss insured, holding the excess over his own interest in trust for the others.

If that is the result which convenience dictates, is there anything which makes it illegal for a sub-contractor to insure the entire contract works in his own name? This was a question which was much discussed in the early cases on bailment. But it was never illegal at common law for a bailee to insure goods in excess of his interest. As for statute, the Marine Insurance Acts obviously do not apply. It is true that the Life Assurance Act 1774 by s. 3 prohibited an insured from recovering more than his interest on the happening of an insured event. But policies on goods were specifically excluded by s. 4 of the Act. Accordingly it was held that neither at common law nor by statute was there anything to prevent the bailee from insuring in excess of his interest.

What about a sub-contractor? 'Goods' in s. 4 of the 1774 Act has always been given a wide interpretation, and would clearly cover contract works until they became part of the realty. Whether the works would remain 'goods' thereafter, and if so for how long, may be more difficult. But it was not suggested that the insurers would have any defence here under the Life Assurance Act, so I say no more about it. I would hold that the position of a sub-contractor in relation to contract works as a whole is sufficiently similar to that of a bailee in relation to goods bailed to enable me to hold, by analogy, that he is entitled to insure the entire contract works, and in the event of a loss to recover the full value of those works in his own name.

For the reasons which I have mentioned, I would hold, both on principle and on the authority of the *Commonwealth Construction* case (1976), that a sub-contractor who is engaged on contract works may insure the entire contract works as well as his own property, and that the defendants in this case were each so insured.

That brings me to the next question: does the fact that the defendants are fully insured under the present policy defeat the insurer's right of subrogation? In the *Commonwealth Construction*

case, and in the American cases there referred to, it was assumed
that it followed automatically that the insurers could have no right
of subrogation. In the *Commonwealth Construction* case it was
described as being a 'basic principle'. In one of the American cases
it was said that the rule was too well established to require
citation. In none of the cases is there any discussion as to the
reason for the rule, except for a brief reference in the
Commonwealth Construction case (1976) to *Simpson & Co* v.
Thompson Burrell (1877) as follows:

'The starting point of that submission is the basic principle
that subrogation cannot be obtained against the insured himself.
The classic example is, of course, to be found in *Simpson & Co* v.
Thompson Burrell (1877). In the case of true joint insurance
there is, of course, no problem; the interests of the joint insured
are so inseparably connected that the several insureds are to be
considered as one with the obvious result that subrogation is
impossible. In the case of several insurance, if the different
interests are pervasive and if each relates to the entire property,
albeit from different angles, again there is no question that the
several insureds must be regarded as one and that no
subrogation is possible.'

Chapter 7

The Employer's Obligations

Some of the employer's obligations are imposed by the express terms of the building contract; others are imposed by the general law.

(i) POSSESSION OF SITE

(a) Time and extent of possession

It is an implied term in every building contract that the employer will give possession of the site to the contractor in sufficient time to enable him to complete his obligations by the contractual date.

Freemen & Son v. Hensler

COURT OF APPEAL   (1900), 64 JP 260

The plaintiff contracted with the defendant to demolish old houses and to erect new ones, the work to be completed within 6 months. The contract was dated 4 July. Under its terms all the brickwork for the new houses was to be built up simultaneously, no part being raised more than 5 feet higher than the remainder. At the defendant's request, the plaintiff agreed to a fortnight's delay from 4 July, but it was not until some weeks after the expiry of that time that possession of part of the site was given. It was nearly five months before the plaintiff got possession of the whole site. The plaintiff claimed damages for breach of contract.

Held: It was an implied term of the contract that the contractor would be given possession of the site immediately. The agreement between the parties had waived that obligation, and substituted a reasonable time. As possession had not been given within a reasonable time, the plaintiff was entitled to damages for the loss which he had sustained by reason of the delay.

COLLINS LJ: The contract clearly involves that the building owner shall be in a position to hand over the whole site to the builder immediately upon the making of the contract. There is an implied undertaking on the part of the building owner, who has contracted for the buildings to be placed by the [contractor] on his land, that he will hand over the land for the purpose of allowing the [contractor] to do that which he has bound himself to do.

The general principle is that in a project for new works, the contractor is entitled to exclusive possession of the entire site, but the degree of possession to be given depends on the circumstances of the case.

London Borough of Hounslow v. Twickenham Garden Developments Ltd

CHANCERY DIVISION   (1970) 7 BLR 81

The facts of this case are not material for present purposes, save that it arose out of a dispute between the parties to a contract in JCT 63 terms. That contract (and JCT 80) impose an express duty on the employer to give possession of the site to the contractor. It is couched in the following terms:

'On the Date of Possession possession of the site shall be given to the Contractor who shall thereupon begin the Works . . .': JCT 80, clause 23.1.

'The site' is nowhere specifically defined, and its extent is a question of fact to be determined in light of all the circumstances.

MEGARRY J: The contract necessarily requires the building owner to give the contractor such possession, occupation or use as is necessary to enable him to perform the contract, but whether in any given case the contractor in law has possession must, I think, depend at least as much on what is done as what the contract provides. . . .

The employer must give the contractor a sufficient degree of possession to permit the execution of the work unimpeded by others.

The Queen in Right of Canada v. Walter Cabott Construction Ltd

CANADIAN FEDERAL COURT OF APPEAL   (1975), 21 BLR 42

Cabott contracted with the Crown for the erection of a hatchery building. The contract was one of six contracts for the project as a whole. The work required under two of the later contracts interfered with Cabott's work because they encroached on the site. One of the later contracts was awarded to a third party, but Cabott successfully tendered for another to mitigate the effects of that contract on its work. Cabott claimed damages for breach of implied terms relating to possession.

Held: Cabott's claim succeeded. The Crown was in breach of contract by denying Cabott part of the site of the work. A clause of the contract providing that there were no implied terms was ineffective because it is fundamental to a construction contract that working space should be provided unimpeded by others.

URIE J: The learned trial judge was correct in finding that the appellant was in breach of its contract with the respondent by denying the respondent a portion of the site of the work which she was obliged to furnish to permit compliance with the contract.

The second breach found by the learned trial judge was that the appellant failed to observe an implied term that the respondent would have a sufficient degree of uninterrupted and exclusive possession of the site to permit it to carry out its work unimpeded and in the manner of its choice.

The appellant relied on general condition 6 for demonstrating the alleged error of the learned trial judge. General condition 6 reads as follows:

'6. No implied obligation of any kind by or on behalf of Her Majesty shall arise from anything in the contract, and the express covenants and agreements herein contained and made by Her Majesty are and shall be the only covenants and agreements upon which any rights against Her Majesty are to be founded; and, without limiting the generality of the foregoing, the contract supersedes all communications, negotiations and agreements, either written or oral, relating to the work and made prior to the date of the contract.'

In my opinion, it has no application in the case at bar because it is fundamental to a building contract that work space be provided unimpeded by others. The proposition of law is succinctly put by the learned author of Hudson's *Building and Engineering Contracts*, 10th ed. (1970), at p. 318, as follows:

'Since a sufficient degree of possession of the site is clearly a necessary pre-condition of the contractor's performance of his obligations, there must be an implied term that the site will be handed over to the contractor within a reasonable time of signing the contract (see, e.g. *Roberts* v. *Bury Commissioners* (1870) and, in most cases it is submitted, a sufficient degree of uninterrupted and exclusive possession to permit the contractor to carry out his work unimpeded and in the manner of his choice. This must particularly be so when a date for completion is specified in the contract documents.'

This statement of the law was adopted by Spence J, in the *Penvidic* case (1975). The learned trial judge was, therefore, in my opinion, clearly right when he found the second breach of contract.

The employer is not in breach of his obligation to give sufficient possession of the site to the contractor if he is wrongfully excluded from the site by a third person for whom the employer is not responsible in law and over whom he has no control.

Porter v. Tottenham Urban Council

COURT OF APPEAL [1915] 1 KB 1041

The plaintiff contracted to build a school for the council upon land belonging to them. The contract provided that he should be entitled to enter on the site immediately, and that he should complete the work by a specified date. The only access to the site was from a road, and as the soil of the council's land and of the adjoining road was soft, the contract provided that the plaintiff lay a temporary sleeper roadway from the road to the site for access, and subsequently provide a permanent pathway. The plaintiff began work but was forced to abandon it because of a threatened injunction from an adjoining owner, who claimed that the road was his property. The third party's claims were held to be unfounded. The plaintiff then resumed and completed the works. He claimed damages against the council in respect of the delay caused by the third party's action.

Held: The plaintiff's claim must fail. There was no implied warranty by the council against wrongful interference by third parties with the free access to the site.

(b) Site conditions

Under the general law there is no warranty by the employer that the site is fit for the works or that the contractor will be able to construct the building on the site to the employer's design.

Appleby v. Myers

COURT OF EXCHEQUER CHAMBER (1867), LR 2 CP 651

The plaintiff contracted to supply and erect machinery on the defendant's premises, payment to be on completion. Part of the work had been completed when a fire accidentally broke out on the defendant's premises, destroying both the premises and the works.

Held: Since neither party was at fault, both parties were excused further performance, but on the true interpretation of the contract the plaintiff was not entitled to recover any payment until the whole work had been completed.

BLACKBURN J: The whole question depends on the true construction of the contract between the two parties. We agree with the court below in thinking that it sufficiently appears that the work which the plaintiff agreed to perform could not be performed unless the defendant's premises continued in a fit state to enable the plaintiff to perform the work . . . [and] that if by any default on the part of the defendants these premises were rendered unfit to receive the work, the plaintiff would have had the option to sue the defendants for this default, or to treat the contract as rescinded, and sue on a quantum meruit. But we do not agree with them in thinking that there was an absolute promise or warranty by the defendant that the premises should at all events continue so fit.

Thorn v. London Corporation

HOUSE OF LORDS (1876), 1 App. Cas. 120

T contracted with the corporation to demolish an existing bridge and to construct a new one, in accordance with plans and a specification prepared by the corporation's engineer. The plans and specification proved defective in important respects and, as a result, T lost money and the bridge had to be built in a different way. The descriptions given were said to be believed to be correct but were not guaranteed. T brought an action claiming reimbursement for his loss of time and labour and alleged that there was an implied warranty that the bridge could be built inexpensively in accordance with the plans and specification.

Held: His claim must fail as no such warranty could be implied. T's obligation to complete the works covered substantial deviations from what had been planned.

Note: Lord Cairns LC also remarked that 'if the additional or varied work is so peculiar, so unexpected, and so different from what any person reckoned or calculated on, it may not be within the contract at all . . .'.

In the absence of any specific guarantee or definite representations by the employer as to the nature of the soil in which the works are to be executed, the contractor is not entitled to abandon the contract on discovering the nature of the soil.

Bottoms v. Lord Mayor of York

COURT OF EXCHEQUER   (1812), 2 HBC (4th edition) 208

B contracted to execute sewerage works for the defendants, intending to use poling boards for the excavations. The soil turned out to be unsuitable and necessitated extra works, which the engineer refused to authorise as a variation. B thereupon abandoned the works and sued for the value of the work done. Neither party had sunk boreholes, but before signing the contract the defendants had been advised that B's price was such that he was bound to lose money in the type of soil conditions expected.

Held: The plaintiff's claim must fail. He was not entitled to abandon the contract on discovering the nature of the soil or because the engineer declined to authorise extra payment.

Where the contract incorporates the rules of a method of measurement which makes provision for extra payment in respect of unexpected ground conditions, the contractor will be entitled to recover.

C. Bryant & Son Ltd v. Birmingham Hospital Saturday Fund

KING'S BENCH DIVISION [1938] 1 All ER 503

The contractor agreed to erect a convalescent home, the contract being in an earlier version of what is now the JCT form, clause 11 of which provided that 'the quality and quantity of the work included in the contract sum shall be deemed to be that which is set out in the bills of quantities, which bills, unless otherwise stated shall be deemed to have been prepared in accordance with the standard method of measurement last before issued by the Chartered Surveyors' Institution'. The Standard Method of Measurement required that, where practicable, the nature of the soil should be described and attention drawn to any existing trial holes, and that excavation in rock should be given separately. The bills referred the contractor to the drawings, a block plan and the site to satisfy himself as to the local conditions and the full extent and nature of the operations, but contained no separate item for the excavation of rock. The architect knew of the existence of rock on the site, but this was neither shown on any of the plans nor referred to in the Bills.

Held: The contractor was entitled to treat the excavation in rock as an extra, and to be paid the extra cost of excavation, plus a fair profit.

Note: The position would appear to be the same as regards contracts in JCT 63 and JCT 80 terms. JCT 80, clause 2.2.2.1. requires that the bills are to have been prepared in accordance with SMM 6, unless otherwise specifically stated, and claims for site conditions may be based on the principle laid down in this case.

A1 of the General Rules of SMM6 provides:

'The Standard Method of Measurement provides a uniform basis for measuring building works and embodies the essentials of good practice but more detailed information than is required by this document shall be given where necessary in order to define the precise nature and extent of the required work.'

D3.1(a) specifically requires the employer to provide information about the ground water level, the level when various excavations have been carried out and accurate details as to periodical changes if any.

By D3.1(c) the employer has to provide 'Particulars of underground services'.

But the most demanding provision is that contained in D3.2:

'If the above information is not available a description of the ground and strata which is to be assumed shall be stated.'

The terms of the tender documents may be such as to give rise to a collateral warranty or implied term that the ground conditions are as described.

Bacal Construction (Midlands) Ltd v. Northampton Development Corporation

COURT OF APPEAL   (1975), 8 BLR 88

As part of their tender, B submitted sub-structure designs and detailed priced bills of quantities for six selected blocks of dwellings and ancillary works in selected foundation conditions. These formed part of the contract documents. The foundation designs had been prepared on the assumption that the soil conditions were as indicated on the relevant borehole data provided by NDC, and the designs were adequate on that assumption. The NDC tender documents stated that the site was a mixture of Northamptonshire sand and upper lias clay. As work proceeded, tufa was discovered in several areas of the site. This necessitated re-design of the foundations in those areas and the carrying out of additional work. Bacal claimed that there had been a breach of an implied term or warranty by NDC in that the ground conditions should accord with the hypotheses upon which they had been instructed to design the foundations.

Held: On the proper interpretation of the contract, there was an implied term or warranty as claimed by B, and they were entitled to damages for its breach.

BUCKLEY LJ: Bacal have submitted that there are strong commercial reasons for implying such a term or warranty in the contract as they have suggested. First, before designing the foundations for any building it is essential to know the nature of the soil conditions. Secondly, where the contract is for a comprehensive development of the kind here in question, the contractor must know the soil conditions at the site of each projected block in order to be able to plan his timetable and to estimate his requirements for materials. These are matters which relate directly to the contract price. Thirdly, if the work is interrupted or delayed by unforeseen complications, the contractor is unlikely to be able to complete his contract in time. Clause 22 of the contract requires the contractor to pay or allow liquidated damages in such an event. The corporation have in fact retained a very substantial sum by way of liquidatd damages to which they claim to be entitled because Bacal did not complete the contract within 65 weeks. Clause 23 of the contract makes provision for extension of the contract time in certain circumstances, but the only provision of that clause which might possibly apply to the present case is sub-paragraph (c), which permits an extension of time if by reason of the architects' instructions issued under clause 1(2), clause 11(1) or clause 21(2) the completion of the works has been delayed beyond the due completion date. The delay resulting from the discovery of tufa was not occasioned by any instructions issued by the architect, and clause 1(2) cannot, in my opinion, apply. Clause 11(1) can only apply where the architect has issued instructions requiring a variation or he has sanctioned a variation made by the contractor otherwise than pursuant to such an instruction. That sub-clause cannot apply in the present case if upon the true construction of clause 1(4) we are not concerned with any variation. Clause 21(2) only applies if the architect has issued an instruction postponing any work to be executed under the contract, which is not the present case. Accordingly, in my opinion, sub-paragraph (e) of clause 23 can have no application in the present case. The learned judge accepted Bacal's argument and held that the corporation was liable on an implied term of warranty. I agree with him. . . .

Note: The contract discussed was substantially in JCT 63 form.

In an appropriate case the contractor may have a claim against the employer for misrepresentations about site and allied conditions made during pre-contractual negotiations.

Morrison-Knudsen International Co. Inc. v. Commonwealth of Australia

HIGH COURT OF AUSTRALIA   (1972), 13 BLR 114

This was an action by a contractor against the employer for negligence. The contractor claimed that basic information supplied by the employer at pre-tender stage 'as to the soil and its contents at the site of the proposed work was false, inaccurate and misleading . . . the clays at the site, contrary to that information, contained large quantities of cobbles'. The issue was a preliminary one, the court holding that the effect of the documents at issue could not finally be determined until all the relevant facts were established.

BARWICK CJ: The basic information in the site information document appears to have been the result of much highly technical effort on the part of [the employer]. It was information which the [contractors] had neither the time nor the opportunity to obtain for themselves. It might even be doubted whether they could be expected to obtain it by their own efforts as a potential or actual tenderer. But it was indispensable information if a judgment were to be formed as to the extent of the work to be done. . . .

A factual misrepresentation made during pre-contractual negotia-
tions by one party and relied on by the other may give rise to
liability under the Misrepresentation Act 1967. In an appropriate
case such liability could attach to inaccurate information about
ground conditions.

Howard Marine and Dredging Co. Ltd v. A. Ogden & Sons (Excavations) Ltd

COURT OF APPEAL   (1977), 9 BLR 34

O wished to hire barges for use in connection with excavation work. In the course of negotiations, H's manager misrepresented the barges' deadweight. He stated that the payload of each barge was 1600 tonnes: it was only 1055. The misstatement was based on his recollection of a figure given in Lloyd's Register, which was incorrect. He could have ascertained the correct figures from H's shipping documents.

Held: H was liable in damages for the misrepresentation under s. 2(1) of the Misrepresentation Act 1967.

BRIDGE LJ: The first question then is whether [Howard] would be liable in damages in respect of [their manager's] misrepresentation if it had been made fraudulently, that is to say, if he had known that it was untrue. An affirmative answer to that question is inescapable. The judge found in terms that what [the manager] said about the capacity of the barges was said with the object of getting the hire contract for the respondents, in other words with the intention that it should be acted on. This was clearly right. Equally clearly the misrepresentation was in fact acted on by [Ogden]. It follows, therefore, on the plain language of the statute that, although there was no allegation of fraud, [the manager] must be liable unless they proved that [Howard] had reasonable ground to believe what he said about the barges' capacity.
    It is unfortunate that the learned judge never directed his mind to the question whether [the manager] had any reasonable ground for his belief. The question he asked himself, in considering

liability under the Misrepresentation Act, 1967, was whether the innocent misrepresentation was negligent. He concluded that if [the manager] had given the inaccurate information in the course of the April telephone conversations he would have been negligent to do so but that in the circumstances obtaining at the Otley interview in July there was no negligence. I take it that he meant by this that on the earlier occasions the circumstances were such that he would have been under a duty to check the accuracy of his information, but on the later occasions he was exempt from any such duty. I appreciate the basis of this distinction, but it seems to me, with respect, quite irrelevant to any question of liability under the statute. If the representee proves a misrepresentation which, if fraudulent, would have sounded in damages, the onus passes immediately to the representor to prove that he had reasonable ground to believe the facts represented. In other words the liability of the representor does not depend upon his being under a duty of care the extent of which may vary according to the circumstances in which the representation is made. In the course of negotiations leading to a contract the statute imposes an absolute obligation not to state facts which the representor cannot prove he had reasonable ground to believe.

The substituted section 3 of the Misrepresentation Act 1967 restricts the employer's power to exclude liability for misrepresentation. Even at common law a contracting party cannot escape liability for fraudulent statements made by him or on his behalf.

S. Pearson & Son Ltd v. Dublin Corporation

HOUSE OF LORDS [1907] AC 351

The contract between the parties provided that 'the contractor is to satisfy himself as to . . . all . . . things so far as they may have any connection with the works in the contract, and to obtain his own information on all matters which can in any way influence his tender'. Plans prepared by the corporation's engineers showed a wall as existing when in fact it did not exist, and the contractor tendered on a lower price on this basis. The statement in the plans was found to be fraudulent. The contractors discovered the absence of the wall, which they had proposed to use in their works, after they had begun work. They had to revise their plans as a result and incurred considerable extra cost.

Held: The contractors were entitled to succeed.

LORD LOREBURN LC: It seems clear that no one can escape liability for his own fraudulent statements by inserting in a contract a clause that the other party shall not rely upon them. . . .[The corporation say] that, though a principal is liable for the fraudulent representation of his agent, yet that rule only applies where the representation has in fact been made by the agent. I cannot accept that contention. The principal and the agent are one, and it does not signify which of them made the incriminated statement or which of them possessed the guilty knowledge.

(ii) CO-OPERATION

It is an implied term of every building contract that the employer will do all that is reasonably necessary on his part to bring about completion of the contract.

'If A employs B for reward to do a piece of work for him which requires outlay and effort on B's part . . . generally speaking, where B is employed to do a piece of work which requires A's co-operation . . . it is implied that the necessary co-operation will be forthcoming': *per* Viscount Simon LC in *Luxor (Eastbourne)Ltd* v. *Cooper* [1941] 1 All ER 33.

'In general . . . a term is necessarily implied in any contract, the other terms of which do not repel the implication, that neither party shall prevent the other from performing it': *per* Lord Asquith in *Cory Ltd* v. *City of London Corporation* [1951] 2 All ER 85.

If the employer hinders or prevents the contractor from performing the contract, either himself or through his servants or agents, he is in breach of contract. The employer will thereby be disentitled from enforcing any liquidated damages clause if delay results, and the contractor may have a claim for damages against him.

Lawson v. Wallasey Local Board

COURT OF APPEAL (1882), 11 QBD 229

L contracted with the defendants to remove 10,000 cu. yds. of the bed of the River Mersey within a specific period. There was a temporary staging on the site of the work, and the contract provided that if it was not removed by the defendants in sufficient time to enable L to complete the work, he was to be entitled to such extension of time as the engineer might think reasonable. The contract provided for monthly payment against engineer's certificates up to 80 per cent of the value certified by the engineer, and the balance on completion. The work was not completed on time. L alleged that this was due to the non-removal of the staging.

Held: There was an implied term that there would be no unreasonable delay in the removal of the staging, and if L was in fact prevented from completing on time by reason of delay in so doing, he was entitled to damages.

(iii) INSTRUCTIONS

An employer's failure to issue plans, drawings and other information necessary for the execution of the works and at the proper time is a breach of contract by the employer, through the agency of the architect, unless the express terms of the contract state otherwise. In the absence of an express term, the employer's obligation is to issue instructions within a reasonable time. What is a reasonable time depends on the circumstances of the case and on the express terms of the contract. Under JCT terms the obligation is to provide drawings and details 'as and when from time to time may be necessary'. The implication is that the information will be issued from time to time as the contract progresses at a time which is reasonable in all the circumstances.

Neodox Ltd v. Borough of Swinton & Pendlebury

QUEEN'S BENCH DIVISION   (1958), 5 BLR 34

The dispute arose under a civil engineering form of contract, but it is submitted that the same principle is applicable to JCT terms.

DIPLOCK J: It is clear . . . that to give business efficacy to the contract, details and instructions necessary for the execution of the works must be given by the engineer from time to time in the course of the contract and must be given in a reasonable time. In giving such instructions, the engineer is acting as agent for his principals, the corporation, and if he fails to give such instructions within a reasonable time, the corporation are liable in damages for breach of contract.

What is a reasonable time does not depend solely upon the convenience and financial interests of the [contractors]. No doubt it is to their interest to have every detail cut and dried on the day the contract is signed, but the contract does not contemplate that. It contemplates further details and instructions being provided, and the engineer is to have a time to provide them which is reasonable having regard to the point of view of him and his staff and the point of view of the corporation, as well as the point of view of the contractors.

In determining what is a reasonable time as respects any particular details and instructions, factors which must obviously be borne in mind are such matters as the order in which the engineer has determined the works shall be carried out . . . , whether requests for particular details or instructions have been made by the contractors, whether the instructions relate to a variation of the contract which the engineer is entitled to make from time to time during the execution of the contract, or whether they relate to part of the original works, and also the time, including any extension of time, within which the contractors are contractually bound to complete the works.

In mentioning these matters, I want to make it perfectly clear that they are not intended to be exhaustive, or anything like it. What is a reasonable time is a question of fact having regard to all the circumstances of the case . . .

(iv) PAYMENT

Most modern building contracts provide for payment by instalments as the work proceeds, normally against an architect's certificate. Failure to make payment when due does not normally amount to a repudiation of the contract, nor entitle the contractor to repudiate the contract. The contractor's remedy is to sue on the certificate because when the certificate has been issued and falls due, a debt arises from the employer to the contractor, e.g., JCT 80, clause 30.1.1.1.

The express terms of the contract may confer additional rights on the contractor, e.g., JCT 80, clause 28.1.1., which makes the employer's failure to pay on any certificate a ground on which the contractor may determine his employment under the contract.

See also *C. M. Pillings & Co. Ltd* v. *Kent Investments Ltd* (1986), Appendix, p. 351.

(a) Lump sum contracts

A lump sum contract is one to complete the whole work for a lump sum.

In an ordinary lump sum contract, the contractor is entitled to payment if he achieves 'substantial completion' and he is then entitled to the contract price less a sum by way of set-off or counterclaim for any defects.

H. Dakin & Co. Ltd v. Lee

COURT OF APPEAL   [1916] 1 KB 566

Dakin contracted to carry out alteration works to a house in accordance with a specification for a lump sum of £264. The work was not done strictly in accordance with the specification. The employer refused any payment.

Held: Dakin was entitled to recover £264, subject to a deduction of the amount necessary to make the work correspond with that contracted to be done. The defects and omissions amounted only to a negligent performance of the contract, and not to its abandonment or a failure to complete.

Hoenig v. Isaacs

COURT OF APPEAL　　　　　　　　　　　　　　　　[1952] 2 All ER 176

The plaintiff was employed by the defendant to decorate and furnish a flat for the sum of £750, the terms of payment being 'net cash, as the work proceeds; and balance on completion'. The defendant paid £400 by instalments, but refused to pay the balance of £350 on the ground that some of the design and workmanship was defective.

Held: There had been a substantial performance of the contract and the defendant was liable for £750, less the cost of remedying the defects, which was assessed at £56.

DENNING LJ: This case raises the familiar question: Was entire performance a condition precedent to payment? That depends on the true construction of the contract. In this case the contract was made over a period of time and was partly oral and partly in writing, but I agree with the official referee that the essential terms were set down in the letter of 25 April 1950. It describes the work which was to be done and concludes with these words:

> 'The foregoing, complete, for the sum of £750 net. Terms of payment are net cash, as the work proceeds; and balance on completion'.

The question of law that was debated before us was whether the plaintiff was entitled in this action to sue for the £350 balance of the contract price as he had done. The defendant said that he was only entitled to sue on a quantum meruit. The defendant was anxious to insist on a quantum meruit, because he said that the contract price was unreasonably high. He wished, therefore, to reject that price altogether and simply to pay a reasonable price for all the work that was done. This would obviously mean an inquiry into the value of every item, including all the many items which were in compliance with the contract as well as the three which fell short of it. That is what the defendant wanted. The plaintiff resisted this course and refused to claim on a quantum meruit. He said that he was entitled to the balance of £350 less a deduction for the defects.

In determining this issue the first question is whether, on the true construction of the contract, entire performance was a condition precedent to payment. It was a lump sum contract, but that does not mean that entire performance was a condition precedent to payment. When a contract provides for a specific sum to be paid on completion of specified work, the courts lean against a construction of the contract which would deprive the contractor of any payment at all simply because there are some defects or omissions. The promise to complete the work is, therefore, construed as a term of the contract, but not as a condition. It is not every breach of that term which absolves the employer from his promise to pay the price, but only a breach which goes to the root of the contract, such as an abandonment of the work when it is only half done. Unless the breach does go to the root of the matter, the employer cannot resist payment of the price. He must pay it and bring a cross-claim for the defects and omissions, or, alternatively, set them up in diminution of the price. The measure is the amount which the work is worth less by reason of the defects and omissions, and is usually calculated by the cost of making them good. . . . It is, of course, always open to the parties by express words to make entire performance a condition precedent. A familiar instance is when the contract provides for progress payments to be made as the work proceeds, but for retention money to be held until completion. Then entire performance is usually a condition precedent to payment of the retention money, but not, of course, to the progress payments. The contractor is entitled to payment pro rata as the work proceeds, less a deduction for retention money. But he is not entitled to the retention money until the work is entirely finished, without defects or omissions. In the present case the contract provided for 'net cash, as the work proceeds; and balance on completion.' If the balance could be regarded as retention money, then it might well be that the contractor ought to have done all the work correctly, without defects or omissions, in order to be entitled to the balance. But I do not think the balance should be regarded as retention money. Retention money is usually only ten per cent, or fifteen per cent, whereas this balance was more than fifty per cent. I think this contract should be regarded as an ordinary lump sum contract. It was substantially performed. The contractor is entitled, therefore, to the contract price, less a deduction for the defects.

In considering whether there is substantial performance it is relevant to take into account both the nature of the defects and the proportion between the cost of rectifying them and the contract price.

Bolton v. Mahadeva

[1971] 2 All ER 1322

The plaintiff agreed to install a central heating system in the defendant's house. The contract price was a lump sum of £560. On completion, the defendant complained that the system was defective and refused payment. The defects were such that the central heating did not perform effectively. It gave off fumes and failed adequately to heat the house. The cost of remedying the defects was £174.

Held: A plaintiff was only entitled to recover on a lump-sum contract, subject to set-off in respect of defects, where he had substantially performed the contract. In considering this, it was relevant to take into account the nature of the defects and the proportion between the cost of remedying them and the contract price. As installed, the system failed to heat the house adequately. Taking into account the cost of remedying the defects, it was impossible to say that there had been substantial performance.

CAIRNS LJ: In considering whether there was substantial perform-ance I am of opinion that it is relevant to take into account both the nature of the defects and the proportion between the cost of rectifying them and the contract price. It would be wrong to say that the contractor is only entitled to payment if the defects are so trifling as to be covered by the de minimis rule.

The main matters that were complained of in this case were that, when the heating system was put on, fumes were given out which made some of the living rooms (to put it at the lowest) extremely uncomfortable and inconvenient to use; secondly, that by reason of there being insufficient radiators and insufficient insulation, the heating obtained by the central heating system was far below what it should have been. There was conflicting evidence about those matters. The judge came to the conclusion that, because of a defective flue, there were fumes which affected the condition of the air in the living rooms, and he further held that the amount of heat given out was such that, on the average, the house was less warm that it should have been with the heating system on, to the extent of 10 per cent. But, while that was the average over the house as a whole, the deficiency in warmth varied very much as between one room and another. The figures that were given in evidence and, insofar as we heard, were not contradicted, were such as to indicate that in some rooms the heat was less than it should have been by something between 26 and 30 per cent.

The learned judge, having made those findings, came to the conclusion that the defects were not sufficient in degree to enable him to hold that there was not substantial performance of the contract. . . .

For my part, I find it impossible to say that the judge was right in reaching the conclusion that in those circumstances the contract had been substantially performed. The contract was a contract to install a central heating system. If a central heating system when installed is such that it does not heat the house adequately and is such, further, that fumes are given out, so as to make living rooms uncomfortable, and if the putting right of those defects is not something which can be done by some slight amendment of the system, then I think that the contract is not substantially performed. . . .

The actual amounts of expenditure which the judge assessed as being necessary to cure those particular defects were £40 in each case. Taking those matters into account and the other matters making up the total of £174, I have reached the conclusion that the judge was wrong in saying that his contract had been substantially completed; and, on my view of the law, it follows that the plaintiff was not entitled to recover under that contract.

I have reached that conclusion without taking into account an argument that was pressed on us in this court by the defendant to the effect that it was a term of the contract that no payment should become due until the work had been completed to such an extent and in such a manner that he could properly sign a satisfaction note to be handed to an insurance company which was guaranteeing payment of the contract price. That contention must necessarily depend on the existence of some implied term to that effect, because there is nothing expressly in the contract about it. If the defendant wanted to rely on such an implied term, I think it was necessary for him to plead it, which he did not. It does not seem that any evidence was directed at the hearing to any such implied term, and no reference to the matter is made in the notice of appeal.

Where the contract provides that retention money is trust money, as by using the phrase 'the employer's interest in any amounts so retained shall be fiduciary as a trustee', the employer must, if called upon by the contractor, set aside and hold the retention money in a separate bank account, subject to the discretion of the court.

Rayack Construction Ltd v. Lampeter Meat Co. Ltd

CHANCERY DIVISION   (1979), 12 BLR 30

The plaintiffs contracted to build meat processing plants for the defendants, the contract being in JCT 63 form. There was a retention of 50 per cent to be retained during the whole of a 5 year defects liability period. The plaintiffs claimed that the defendants were obliged to pay the retention into a separate trust bank account.

Held: The retention clause imposed an obligation on the defendants to set aside the money retained into a separate trust account and, in the circumstances, a mandatory injunction would be granted to that effect.

(b) Other contracts

Questions of payment under other types of contract, such as 'measure and value' contracts and 'cost plus percentage' contracts depend on the terms of the contract in question and no general principle can be laid down.

The general rules relating to payment and related matters under building contracts were summarised as follows in

Holland Hannen & Cubitts (Northern) Ltd. v. Welsh Health Technical Services Organisation

QUEEN'S BENCH DIVISION (1981), 18 BLR 89

The facts of this case are not material for present purposes; it is further referred to on p. 211.

JUDGE JOHN NEWEY QC: It seems to me that the law is clear and can be shortly stated as follows:

(1) An entire contract is one in which what is described as 'complete performance' by one party is a condition precedent to the liability of the other party: *Cutter* v. *Powell* (1795), and *Munro* v. *Butt* (1858).

(2) Whether a contract is an entire one is a matter of construction; it depends upon what the parties agreed. A lump sum contract is not necessarily an entire contract. A contract providing for interim payments, for example, as work proceeds, but for retention money to be held until completion is usually entire as to the retention moneys, but not necessarily the interim payments: Denning LJ in *Hoenig* v. *Isaacs* (1952).

(3) The test of complete performance for the purposes of an entire contract is in fact 'substantial performance': *H. Dakin & Co. Ltd* v. *Lee* (1916) and *Hoenig* v. *Isaacs*.

(4) What is substantial is not to be determined on a comparison of cost of work done and work omitted or done badly: *Kiely & Sons Ltd* v. *Medcraft* (1965) and *Bolton* v. *Mahadeva* (1972).

(5) If a party abandons performance of the contract, he cannot recover payment for work which he has completed: *Sumpter* v. *Hedges* (1898).

(6) If a party has done something different from that which he contracted to perform, then, however valuable his work, he cannot claim to have performed substantially: *Forman & Co. Proprietary Ltd* v. *The Ship 'Liddesdale'* (1900).

(7) If a party is prevented from performing his contract by default of the other party, he is excused from performance and may recover damages: dicta by Blackburn J in *Appleby* v. *Myers* (1866); *Mackay* v. *Dick* (1880).

(8) Parties may agree that, in return for one party performing certain obligations, the other will pay him on a *quantum meruit*.

(9) A contract for payment of a *quantum meruit* may be made in the same way as any other type of contract, including conduct.

(10) A contract for a *quantum meruit* will not readily be inferred

from the actions of a landowner in using something which has become physically attached to his land: *Munro* v. *Butt* (1858).

(11) There may be circumstances in which, even though a special contract has not been performed there may arise a new or substituted contract; it is a matter of evidence: *Whitaker* v. *Dunn* (1886).

(c) *Quantum meruit*

Where the contractor does work under a contract and no price has been agreed he is entitled to be paid a reasonable sum for work and materials. There may also be a claim on a quantum meruit where there is a contract for specified work and, at the employer's request, the contractor does work outside the contract. Whether or not such a claim may be made is a question of interpreting the contract.

Sir Lindsay Parkinson & Co. Ltd v. Commissioners of Works

COURT OF APPEAL   [1950] 1 All ER 208

Contractors agreed to erect an ordnance factory for £3,505,834 by 30 January 1939. Under the contract the employers were entitled to require the contractors to execute additional work on a measured basis, and the Bills of Quantities stated that its value would be in the order of £500,000. Owing to delay in receipt of information from the employers, the contractors became entitled to extensions of time. The employers were anxious that work be completed by 30 January 1939, and so the parties entered into a deed of variation under which it was agreed that the contractors should adopt uneconomic working methods to achieve completion by the due date, at an estimated additional cost of £1 million, and that the contractors would be allowed a net profit of at least £150,000 but not more than £300,000 on the cost of the works, i.e., a net profit of 3 per cent and 6 per cent respectively on the total contract price of £5 million. As a result of extras ordered by the employer, the works executed eventually amounted to in excess of £6½ million.

Held: On the true interpretation of the deed of variation, it was not within the contemplation of the parties that there would be such large increases in the amount of work to be executed. The parties contemplated work to the approximate value of £5 million, and the maximum profit of £300,000 was agreed on that basis. Since the additional work had been performed by the contractors at the employer's request, the contractors were entitled to be paid a reasonable sum in respect of it.

The same principle may apply where work is done pursuant to a letter of intent: see British Steel Corporation v. Cleveland Bridge & Engineering Co. Ltd. (1982), pp. 3-8.

(d) Set-off

Most standard form contracts contain a provision entitling the employer (or, in a sub-contract, the contractor) to set-off by way of deduction from sums certified by the architect certain specific and limited sums. The exact effect of such a clause depends on its wording. The set-off clause may have the effect of excluding implied terms as to set-off. Thus, the JCT Standard Form of Nominated Sub-Contract, NSC/4 provides 'The rights of the parties . . . in respect of set-off are fully set out in [this sub-contract] and no other rights whatsoever shall be implied as terms of the sub-contract relating to set-off'.

JCT 80 contains no such express provision on this matter but it is clear that the so-called principle in *Dawnays Ltd v. F.G. Minter Ltd* (1971) is no longer good law. In effect, whether or not the equitable right of set-off is excluded depends on the terms of the contract.

Note: See *Tubeworkers Ltd v. Tilbury Construction Ltd* (1986), Appendix, p. 354.

Dawnays Ltd v. F.G. Minter Ltd

(1971), 1 BLR 16

Clause 13 of the 'green' form of nominated sub-contract provided: 'The contractor shall notwithstanding anything in this sub-contract be entitled to deduct from or set off against any money due from him to the sub-contractor (including any retention money) *any sum or sums which the sub-contractor is liable to pay* to the contractor under this sub-contract'. There was also a comprehensive arbitration clause. Main contractors refused to pay over to the sub-contractors moneys due to them in an interim certificate on the grounds that they were entitled to damages for delay caused by the plaintiffs.

Held: On the true construction of the contract, the only sums deductible under clause 13 are liquidated and ascertained sums which are established as admitted to be due.

Dawnays Ltd v. F.G. Minter Ltd was over-ruled by the following case. The basic principle is that the right of set-off persists unless, looking at the contract as a whole, it is excluded in express terms or by necessary implication.

Gilbert-Ash (Northern) Ltd v. Modern Engineering (Bristol) Ltd

(1973), 1 BLR 73

The sub-contract in this case was not in standard form, the set-off clause providing:

'If the sub-contractor fails to comply with any of the conditions of this sub-contract, the contractor reserves the right to suspend or withhold payment of any monies due or becoming due to the sub-contractor. The contractor also reserves the right to deduct from any payments certified as due to the sub-contractor and/or otherwise to recover the amount of any bona-fide contra accounts and/or other claims which he, the contractor, may have against the sub-contractor in connection with this or any other contract'.

Held: On the true construction of the contract, the main contractors were entitled to set-off claims in respect of alleged breaches of contract by the sub-contractors and that the common law and equitable right of set-off was not excluded by the terms of the contract.

VISCOUNT DILHORNE: My lords, the question raised in this appeal, one of great importance to the building trade, has been considered by the Court of Appeal in a number of cases in recent years. This is the first occasion on which it has come to this House. Shortly stated, it is whether, when under a contract in the RIBA form an architect has given an interim certificate, the employer is bound to pay to the contractor the amount certified without deduction or set-off save as permitted by the contract; and where the architect has told the contractor the amount forming part of the total certified attributable to work executed by a sub-contractor, the contractor is bound to pay that amount to the sub-contractor without deduction save as permitted by the sub-contract.

In his judgment in this case Lord Denning MR put the matter thus:

'When the main contractor has received the sums due to the sub-contractor – as certified or contained in the architect's certificate – the main contractor must pay those sums to the sub-contractor. He cannot hold them up so as to satisfy his cross-claims. Those must be dealt with separately in appropriate proceedings for the purpose. This is in accord with the needs of business. There must be a "cash flow" in the building trade. It is the very lifeblood of the enterprise. The sub-contractor has to expend money on steel work and labour. He is out of pocket. He probably has an overdraft at the bank. He cannot go on unless he is paid for what he does as he does it. The main contractor is in a like position. He has to pay his men and buy his materials. He has to pay the sub-contractors. He has to have cash from the employers, otherwise he will not be able to carry on. So, once the architect gives his certificates, they must be honoured all down the line. The employer must pay the main contractor, the main contractor must pay the sub-contractor, and so forth. Cross-claims must be settled later.'

Counsel for the respondents sought to support this passage with one qualification. He agreed that deductions might be made from the amount certified where authorised by the contract and that deductions authorised by the sub-contract might be made by the contractor from the sum which he had received from the employer attributable to work done by the sub-contractor and that the contractor was then only obliged to pay the balance.

Counsel for the appellants challenged this. He contended that neither the contract nor the sub-contract excluded the right of the employer to rely on his common law and equitable rights of counterclaim and set-off when sued by the contractor for the amounts certified and the right of the contractor to do so when sued by a sub-contractor for the amount contained in the certificate attributable to his work.

Which view is right must in every case depend on the terms of the contract or sub-contract in question. Changes are made from time to time in the form of the RIBA contracts and sub-contracts may take very different forms. Prior to 1971 and the decision of the Court of Appeal in *Dawnays Ltd* v. *F.G. Minter Ltd* (1971) it appears to have been generally thought that an employer was entitled to set off or counterclaim against the amount certified. In Halsbury's *Laws of England* it is said:

'the contractor is entitled to immediate payment thereof [the amount certified] subject to the terms of the contract and any right of the employer to any set-off or counterclaim (for example, for liquidated damages).'

Indeed, no scintilla of authority prior to the decision in the *Dawnays'* case can be found for the proposition that the amount certified by the architect in an interim certificate as the value of work done and consequently payable to the contractor is in a special position in that the employer cannot lawfully counterclaim and set off against the liability amounts due to him from the contractor. . . . A great deal has been said in *Dawnays'* case and the cases which followed it, as well as in this case, as to the importance of a 'cash flow' in the building industry. I cannot think that the building industry is unique in this respect. It is, of course, true that the contract makes provision for payments as the work proceeds, but, it is to be observed, a fact to which I feel insufficient attention has been paid, that the contractor is only entitled to be paid for work properly executed. He is not entitled to be paid on interim certificates for work which is defective. The architect should only value work executed properly, that is to say, to his reasonable satisfaction (clause 1); and no interim certificate is of itself conclusive evidence that the work was in accordance with contract (clause 30(8)).

Under clause 30(2) it is for the architect to deduct the retention money and the instalments previously paid and the balance is the amount to which the contractor is by clause 30(1) entitled to payment.

When one has regard to other provisions of the contract this does not mean payment without deduction. For instance, if the contractor has failed to comply with an instruction and the employer has employed others to do so, all costs incurred in doing so are recoverable from the contractor by the employer as a debt 'or may be deducted by him from any monies due or to become due to the contractor under this contract' (clause 2(1)).

Similarly if the contractor does not comply with the insurance requirements of the contract the employer may himself insure 'and may deduct a sum equivalent to the amount paid in respect of premiums from any monies due or to become due to the contractor' (clause 19) (see also clause 20(a)). The employer may also deduct under this contract £400 per week as liquidated and ascertained damages for non-completion by the due date (clause 22).

The existence of these provisions in the contract lends no support to the contention that the amount certified is of a special sacrosanct character which must be paid without deduction. If these deductions can be made, as they clearly can, from the amount certified, the amount which clause 30(1) says the contractor is entitled to be paid, why should it be inferred that the contract impliedly, for there is nothing express, excludes reliance by the employer on his common law and equitable rights to counterclaim and set-off for the amount certified. I see no ground for any such inference.

The following cases should also be studied.

Mottram Consultants Ltd v. Bernard Sunley & Sons Ltd

HOUSE OF LORDS   (1974), 2 BLR 28

There was a contract in special form, on a prime cost fixed fee basis. The relevant clause read:

'The Co-ordinator shall within seven days of receipt by him from the architect of any interim certificate pay to the contractor in local currency and/or sterling as required by the contractor any amount shown therein as due to the contractor less only:
(i) Retention money as hereinafter described
(ii) Any sum previously paid . . .'

The original printed form had contained a further sub-clause, which the parties had deleted:

'(iii) Any amount which the employer or the co-ordinator on his behalf shall be entitled to deduct or set off against any money due from him to the contractor (including any retention money) in virtue of any provisions of the contract or any breach thereof by the contractor'.

The question arose as to whether sums other than items (i) and (ii) could be deducted by way of set-off.

Held: The contract provided for only two permissible deductions and nothing else could be deducted from payments due to the contractor. In construing a printed contract (*per* Lord Cross) where parts of it are deleted, regard can be paid 'to what has been deleted as part of the surrounding circumstances in the light of which one must construe what they have chosen to leave in. The fact that they deleted (iii) shows that these parties directed their minds . . . to the question of deductions . . . and decided that no such deductions should be allowed'.

Redpath Dorman Long Ltd v. Tarmac Construction Ltd

COURT OF SESSION (OUTER HOUSE)   1982 SLT 442

Redpath were sub-contractors to Tarmac under the 'blue' form with an arbitration clause. The set-off clause provided:

> 'The contractor shall be entitled to set off against any money . . . otherwise due under this sub-contract the amount of any claim for loss and/or expense which has actually been incurred by the contractor by reason of any breach of, or failure to observe the provisions of, this sub-contract by the sub-contractor, provided,
>
> (a) the amount of such set-off has been quantified in detail and with reasonable accuracy by the contractor; and
> (b) the contractor has given to the sub-contractor notice in writing specifying his intention to set off the amount quantified . . . and the grounds on which such set-off is claimed to be made.'

Tarmac admitted that by November 1978, £290,204 had been certified by the architect as due to Redpath, but claimed a set-off against this sum of the same amount as damages for delay and non-performance. Tarmac quantified two items as having been incurred by 8 November 1978, namely '17 weeks' inefficient utilisation of labour and plant: £76,500,' and £3476 because 'efficient re-deployment of resources was impossible', a total of £79,976. Redpath sued for the full amount and Tarmac applied to have the action stayed.

Held: The terms of the clause did not allow any claim for loss or expense which the contractor might incur in future. Apart from the two items quantified and totalling £79,976, the rest of the claims related to possible future losses, and there was thus no dispute arising out of, or in connection with, the sub-contract as respects those claims because the sub-contract did not provide for set-off for future losses. Only quantified claims to set-off fell within clause 15(2). Redpath were entitled to judgment for £210,228 with interest at 11 per cent from 8 November until payment, but the action relating to the quantified sum of £79,976 would be stayed and the dispute referred to arbitration.

LORD ROSS: I am satisfied that except for the two items mentioned namely section 1A, item 4 (inefficient utilisation of labour and plant) and section 2 (control building) the claims do not fall within clause 15(2), as they do not bear to be claims 'for loss and/or expense which has actually been incurred by the Contractor'. With these two exceptions the claims bear to be claims for loss and/or expense which will be incurred in the future, and as such, they clearly do not fall within clause 15(2); accordingly so far as these claims are concerned no dispute or difference between the parties has arisen out of or in connection with the sub-contract.

Counsel for [Tarmac] founded on *Dunlop* v. *McGowans* (1980) and contended that loss was incurred as soon as breach of contract had taken place, and that loss is incurred once and for all even though some of the loss will follow in the future. I cannot agree. . . . The situation is different here where the sub-contract provides for set-off of the amount of any claim for loss and/or expense which has actually been incurred. *Redpath Dorman Long Ltd* v. *Cummins Engine Co Ltd* (1982), was a case where set-off was claimed at common law rather than under the provisions of the contract, and the case was accordingly different to the instant case.

[Here Redpath] are claiming payment of a sum which has been certified as due. The only defence put forward is a claim by [Tarmac] that they are entitled to set-off under clause 15(2). As already indicated, an examination of their claim makes it clear that, with two exceptions, their claim to set-off does not fall within clause 15(2). It follows that, apart from these two exceptions, there is no relevant defence to the action, and no dispute between the parties arising out of or in connection with the sub-contract which could go to arbitration. As regards the two exceptions, there is a dispute between the parties arising out of and in connection with the sub-contract, and that dispute must go to arbitration.

(v) MISCELLANEOUS

There are other obligations which may be imposed on the

employer by the general law in appropriate cases, some of a positive and others of a negative nature.

(a) Nomination of sub-contractors and specialists

Where the contract provides that the employer may nominate sub-contractors, a term may be implied that the specialist will be nominated in time for the contractor to incorporate the work into his programme. A term may also be implied that the nomination will be of a specialist willing to enter into a sub-contract which is consistent with the terms of the main contract.

Leslie v. Metropolitan Asylums District Managers

COURT OF APPEAL   (1901), 1 LGR 862

ROMER LJ: The [main contractors] clearly could not be compelled to enter into a [sub-]contract which did not adequately protect them. . . . If the terms of those [sub-]contracts prove insufficient properly to protect the [main contractors], that is their fault or misfortune and they cannot hold [the employers] liable in any way.

(b) Objection to nominees

Where the contract gives the main contractor the right to object to a proposed nominee, as does JCT 80, clause 35.4.1. (JCT 63, clause 27(a)) the contractor's remedy is to exercise that right of objection in either of the foregoing cases, and if he does not he cannot rely on any such implied term.

Percy Bilton Ltd v. Greater London Council

COURT OF APPEAL   (1981), 17 BLR 9

For the facts see pp. 249, 341 where extracts from the speeches in the House of Lords are given. The following statement is unaffected by the decision in the House of Lords.

> STEPHENSON LJ: [The plaintiff argued] that at the time of the application for re-nomination, the new sub-contractor's date for completion was later than the plaintiff's date for completion and that, since this would make it impossible for the plaintiff both to accept the new sub-contractor and to comply with the provision in their own contract as to the time for completion, therefore the time provision must go completely. . . .
>
> I do not accept that argument. The contractor, faced with a sub-contract with such a provision as to completion, would be entitled to refuse to accept that sub-contractor under clause 27. . . .

(c) Displacement of implied terms

Any implied term will be displaced to the extent that there is an express term dealing with the same matter.

(d) Form of sub-contracts

Where the main contract contemplates the use of nominated sub-contractors, there is no implied term that a particular form of sub-contract should be used, but the express terms of a particular form of contract, such as JCT 80, may raise this implication or make the use of a particular form mandatory.

James Longley & Co. Ltd v. Reigate & Banstead Borough Council

COURT OF APPEAL   (1982), 22 BLR 31

The facts of this case, which was decided under JCT 63 terms, are not relevant for present purposes.

WALLER LJ: [The employers] sought to support their case by referring to the sub-contract which is in a standard form and in which [they] submitted it was an implied term between the building owner and the main contractor that it should be used. In our opinion this case must be decided on the main contract alone, and no reference should be made to the sub-contract. . . . While the sub-contract refers to the main contract, there is nothing in the main contract which dictates the form of the sub-contract.

(e) Effect of failure

If a nominated sub-contractor or supplier fails during the continuance of the works, the employer is under a duty to re-nominate, and to pay the actual cost of the substituted sub-contractor's work. See: North-West Metropolitan Region Hospital Board *v.* T.A. Bickerton & Son Ltd (1970), *p. 334.*

(f) Default in supply of goods

Where goods are to be provided by or on behalf of the employer, a term may be implied that they will be supplied in time to enable the contractor to carry out the works expeditiously and in accordance with his planned progress, and the employer will be liable for his supplier's default.

Thomas Bates & Son Ltd v. Thurrock Borough Council

Thurrock employed Crux Limited (C) as management consultants in connection with a housing estate which they were developing. C were not paid fees, but it was agreed that only Crux components would be used in the houses to be erected, and that they were to be paid a price for the components so used. Thurrock contracted with Bates in JCT 63 terms for the construction of the houses. Crux components were detailed in the specification as 'to be supplied by C' and were not p.c. items. C were described as 'appointed suppliers'. The contract over-ran by 25 weeks and Bates claimed damages against Thurrock, attributing the delay and disruption to them through the shortcomings of C as their agents. They said that C failed to supply the components in due time and that the components were of inferior quality. Bates had no contract with C. The council settled the claim for £65,000 and claimed this amount from C. The contract between Thurrock and Bates provided that C's components 'shall be incorporated in the works and be delivered to the site in accordance with a programme to be agreed between the Supervising Officer, Bates, and C.

Held: C were liable. Having agreed to supply the components it was right to imply that they would live up to their programme so that the contract works could be kept up to date. As Thurrock had undertaken to supply the components they were liable to Bates on the basis of an implied term that the components would be supplied in time to avoid disruption and delay. Thurrock could pass that obligation over to C.

LORD DENNING MR: [It is clear from the main contract] that Thurrock were undertaking this obligation that Crux items should be incorporated in the works. Thurrock were undertaking that the obligation would be fulfilled. It is Thurrock's obligation and Thurrock's undertaking. The fact that payment was to be made by Bates to Crux is neither here nor there. . . . It is plain that Bates can sue Thurrock because Thurrock promised that the items shall be supplied. If A promises B that C will supply goods to B in a reasonable time and C does not supply them or supplies them late, then B can sue A for damages for C's default. Equally, Thurrock can pass over the obligation to Crux.

Note: In the High Court, Bates alleged four terms were to be implied in their contract with Thurrock:

(a) That the Supervising Officer and/or his agents Crux would furnish Bates with all necessary information in due time to enable Bates to carry out and complete the works expeditiously and in accordance with the contract.
(b) That Thurrock by their agents Crux would supply Crux components in accordance with the programme envisaged by the contract.
(c) That neither Thurrock, the Supervising Officer, nor Crux would hinder or prevent Bates from carrying out and completing the works expeditiously.
(d) That Crux components when supplied would be of good quality and fit for their intended purpose.

Thurrock admitted (a), (c) and (d); (b) was held to be an implied term.

JUDGE WILLIAM STABB QC: Commonsense seems to me to dictate that the scheme was designed to provide that Crux were to supply drawings and materials and that Thurrock were to be responsible for seeing that the contractor used those drawings and took those materials. In so doing Thurrock were ... assuming responsibility to the contractor for the Crux drawings and for the materials which they, Thurrock, were requiring the contractor to use and take, and, accordingly, would be liable to the contractor in the first instance for default or defect in the drawings or materials with a write-over against Crux who supplied them.

Chapter 8

Ownership and Vesting of Materials

(i) MATERIALS AND GOODS

A building contract is one for work and materials. Materials belonging to the contractor and brought on site remain his property, but when they are incorporated in the works they become the property of the employer, unless the contract provides otherwise.

Appleby v. Myers

COURT OF EXCHEQUER CHAMBER   (1867), LR 2 CP 651

The facts are set out on page 116.

BLACKBURN J: Materials worked by one into the property of another become part of that property. This is equally true whether it be fixed or moveable property. Bricks built into a wall become part of the house; thread stitched into a coat which is under repair, or planks and nails and pitch worked into a ship under repair, become part of the coat or ship, and, therefore, generally, and in the absence of something to show a contrary intention, the bricklayer, or tailor, or shipwright, is to be paid for the work and materials he has done and provided. . . .

The express terms of the contract may alter the position as between employer and contractor, but terms in the contract between a supplier and the contractor cannot affect the employer's position since there is no contractual relationship between the supplier and the employer. As soon as the on-site goods and materials are incorporated in the works, they become the property of the employer, even if supplied to the contractor under a contract containing a retention of title clause.

Reynolds v. Ashby

HOUSE OF LORDS   [1904] AC 406

The appellant sold machines to a factory tenant on hire-purchase terms. The machines were fixed to concrete beds on the factory floor by means of nuts and bolts, and could be removed without damaging the building. The tenant mortgaged the factory, together with its fixtures and fittings, to Ashby, who entered into possession. The tenant defaulted under the hire-purchase agreement, and the appellant claimed the machines or their value from Ashby.

Held: The machines were fixtures and part of the building and Ashby was entitled to retain them as against the appellant.

LORD LINDLEY: I pass now to consider whether, in this case, the fact that the mortgagor Holdway had not acquired the ownership of the machines by paying for them entitles the appellant to recover their value from the respondents. The title to chattels may clearly be lost by being affixed to real property by a person who is not the owner of the chattels. . . . Holdway agreed to buy the machines; but the appellant knew what he wanted them for, and that, before paying for them, he intended to put them up in his factory and to use them. The appellant knew that the factory was mortgaged, and he ran the risk of the machines being claimed as fixtures. In effect, Holdway was authorised by the appellant to convert the chattels into fixtures, subject to the right of the appellant to enter and retake them if he did not pay them. But, apart from this knowledge and authority, the result would be the same, although not so obvious. After the machines were fixed, and before the appellant claimed them, the second mortgagee took possession; the appellant's right to enter and remove the machines, resting as it did on his contract with Holdway, ceased to be exercisable. The appellant has no greater rights against the respondents than he had against the second mortgagee, whose rights the respondents have acquired. The machines had ceased to be chattels belonging to the appellant; they were not chattels wrongfully detained by the respondents. The machines had become fixtures which the appellant was not entitled to remove from the possession of the mortgagees.

Hobson v. Gorringe

COURT OF APPEAL   [1897] 1 Ch 182

H sold K an 11 horse power gas engine on hire purchase terms. The contract provided that if K defaulted on the instalment payments, H was at liberty to remove and repossess the engine, K having no claim for any money paid. It was also provided that the contract should not operate as a contract of sale, but only as an agreement for the hire of the engine, unless and until K had paid all the instalments.

K installed the engine in his saw-mill. It was fixed by screws and bolts to a concrete base laid in the soil. K defaulted on the instalments and mortgaged his saw-mill and fixtures to a third party, G, who took possession of the premises together with the engine.

Held: On the true interpretation of the agreement, coupled with the annexation of the engine to the soil, the engine became a fixture, subject as between H and K to H's right to unfix it and retake possession if the hirer defaulted. This right imposed no legal obligation on the third party and, therefore, the owner's remedy was by action against the hirer personally. The action against G failed.

In re Morrison, James & Taylor, Ltd

COURT OF APPEAL   [1914] 1 Ch 50

By an agreement made between the applicants and the company, the applicants agreed to erect and install a sprinkler system in the company's premises. The sprinkler was to be paid for by instalments, on terms that the system would remain the property of the applicants until the entire price was paid. Subsequently, the company issued debentures and thereafter a receiver was appointed on behalf of the debenture holders. The applicants were not paid the price of the system and claimed to be entitled to enter the company's premises and remove it.

Held: The applicants were entitled to remove the installation, notwithstanding the appointment of the receiver.

The position is different in the case of unfixed goods and materials, whether on or off site. In that case, even if the employer has paid for the goods and materials, unless the contractor is their owner he cannot pass title to the employer. A vesting clause in the building contract is only effective to transfer property if the contractor has good title himself.

Dawber Williamson Roofing Ltd v. Humberside County Council

QUEEN'S BENCH DIVISION   (1979), 14 BLR 70

The council contracted in JCT 63 form with a main contractor for the erection of a school. The main contractor sub-contracted the roofing work to the plaintiffs on the 'blue' form of sub-contract which provided by clause 1(1) that 'The sub-contractor shall be deemed to have knowledge of all the provisions of the main contract . . .'. The 'blue' form did not expressly provide when property in unfixed goods and materials was to pass to the main contractor, although clause 14(1) of JCT 63 provided that, subject to certain conditions, where the value of any unfixed goods or materials 'has been included in an interim certificate under which the contractor has received payment, such materials and goods shall become the property of the employer'. The plaintiffs delivered 16 tons of slates to the site, and their value was included in an interim certificate issued under the main contract. The main contractor did not pay the plaintiffs and subsequently became insolvent. The council refused to return the slates, maintaining that they were their property. The plaintiffs sued for the return of the slates and damages, or the value of the slates.

Held: The plaintiffs were entitled to succeed as title in the slates had not passed to the main contractor. Clause 1(1) of the sub-contract did not have the effect of making main contract clause 14(1) a term of the sub-contract.

MAIS J: The plaintiffs submit that their contract with the main contractor was a 'supply and fix' contract. They were not selling the slates to the main contractor. However, there was no privity of contract between the plaintiffs and the defendants, and my attention was drawn to Hudson's *Building Contracts*, 10th edition, at page 742, and the following pages and authorities.

The defendants drew attention to clause 1(1) of the sub-contract between the plaintiffs and the main contractor. By this clause the plaintiffs are deemed to have notice of all the provisions of the main contract. Whilst this is so, that does not, in my judgment, mean that there is any privity of contract between the plaintiffs and the defendants. [Counsel for the plaintiffs], however, submitted that material brought on to the site on a 'supply of material and fix' contract remained the property of the sub-contractor until it was fixed. He referred me to *Tripp* v. *Armitage* (1839) . . . I would refer also to the case of *Seeth* v. *Moore* (1886), where Lord Watson said:

> 'Materials provided by the builder and portions of the fabric, whether wholly or partially finished, although intended to be used in the execution of the contract, cannot be regarded as appropriated to the contract or as 'sold' unless they have been affixed to or in a reasonable sense made part of the corpus . . .'

and unless there be any good reason to the contrary, this appears to me to conclude the case against the defendants.

The defendants submit, however, and contend that they are entitled to the property in the slates, and they rely, first of all, on clause 14(1) of the main contract, whereby:

'any unfixed materials and goods delivered to, and placed on, or into the works, shall not be removed except for use upon the works, unless consent in handwriting be given, and when the value of the goods has been included in any interim certificate under which the contractor has received payment, such material and goods shall become the defendants' property.'

In my judgment, this presupposes there is privity between the defendants and the sub-contractor, which there is not in the present instance, or the main contractor has good title to the material and goods. If the title has passed to the main contractor from the sub-contractor, then this clause has force.

Secondly, as I indicated earlier, the defendants at one time relied upon s.25(2) of the Sale of Goods Act 1893, but [counsel for the defendants] rightly says that he can no longer rely upon that clause. In my judgment, there being no privity of contract between the plaintiffs and the defendants, the defendants are not entitled to claim property in the slates by reason of any clause in a contract they had with the main contractor.

Note: JCT 80 and JCT Nominated Sub-contracts NSC/4 and 4a have been amended in an attempt to overcome this decision.

Where goods are supplied to the main contractor under a sub-contract, and the main contractor becomes insolvent before paying, the sub-contractor or supplier has no right to claim against the employer.

Pritchett & Gold and Electrical Power Storage Co. v. Currie

COURT OF APPEAL [1916] 2 Ch 515

Currie contracted with the second defendant for the supply of an electrical installation, which included a storage battery. The second defendant entered into a sub-contract with the plaintiffs for the supply and erection of the battery. The battery was delivered, and shortly afterwards the second defendant became insolvent. The plaintiffs claimed the price of the battery from Currie.

Held: The claim must fail.

Note: Although at first sight the sub-contract appears to be one for 'work and materials' it was in effect a contract for the sale of goods.

PICKFORD LJ: The plaintiffs sued the defendant, Mrs. Currie, for the value of materials which they had supplied to the defendant company for the purpose of their being erected by the defendant company on Mrs. Currie's premises as an electric battery as part of a much larger contract for the whole electrical installation, amounting to nearly £1,850. The master made an order that without prejudice to any questions between the plaintiffs and the defendant company Mrs. Currie should lodge in court a certain sum of money, a portion of the balance of the contract price alleged to be payable by her to the defendant company, and it was ordered that on such lodgment all proceedings should be stayed against Mrs. Currie and that the plaintiffs should within fourteen days deliver a statement of their claim to the defendant company who were within fourteen days thereafter to deliver their defence and be at liberty therein to set up any counter-claim. The defendant company had been added on the application of Mrs. Currie as defendants. The result of that was that a certain sum of money payable by Mrs. Currie in respect of the main contract was paid into court, and the plaintiffs were to deliver a claim by which they were to establish their right to that sum of money as against the defendant company. . . .

To establish that right, counsel for the plaintiffs, concedes, quite rightly, that he must establish that he has a lien upon or a right to follow that particular sum. He also concedes that, unless such a right is given to him by the authority of *Bellamy* v. *Davey* (1891), he cannot make out such a right because it is the assertion by a sub-contractor of a lien on the whole of the money payable by the principal debtor to the head contractor under the contract. He admits that, unless that is given to him by the authority of that case, it is contrary to the general rules of law. I do not think that it is in the least necessary for me to consider whether *Bellamy* v. *Davey* is right or is wrong. I think that it is enough to say that, assuming that the decision in that case was absolutely right, it is no authority for the proposal for which the plaintiffs contend in the present case. That was a case in which the contractor and the sub-contractor were concerned only with the article that was supplied by the sub-contractor. No doubt the contractor had charged a larger price to the building owner than the sub-contractor charged to him. But that and that only was the subject of the contract. It may be – I do not say whether it is or is not so – that in that case the sub-contractor may have a lien upon the amount payable to the contractor for the amount due to him, and that only the surplus went to the contractor, but it seems to me that that does not in any way establish that where there is a sum of money payable to a contractor, not in respect of one article only, but in respect of the whole contract which includes many things other than the subject-matter of the sub-contract, the sub-contractor has a lien upon a large sum which is not connected with him except indirectly for the payment of his debt. It seems to me that *Bellamy* v. *Davey* is no authority for such a proposition as that. The plaintiffs have to establish their right as against the defendant company. The defendant company are entitled to have this money paid to them by Mrs. Currie unless the plaintiffs have established that they have a lien upon it. The plaintiffs have not established any such lien. Therefore, it seems to me that they have not established that they are entitled to the money as against the defendant company or anybody else. I think that the action fails on that ground.

(ii) VESTING CLAUSES

The majority of building contracts contain a clause vesting in the employer the property in unfixed materials on site. Such a clause sometimes extends to the contractor's plant during the currency of the contract. JCT 80, clause 16, is a typical vesting clause. Whether a vesting clause is effective to transfer property to the employer without qualification depends on its terms.

Bennett & White (Calgary) Ltd v. Municipal District of Sugar City No. 5

PRIVY COUNCIL [1951] AC 786

A construction contract between the appellants and the Canadian government provided that 'all machinery, tools, plant, materials . . . and things whatsoever provided by the contractors . . . for the works . . . shall from the time of their being so provided become and until the final completion of the said work shall be the property of His Majesty for the purposes of the said work'. The appellants provided plant and equipment, including motor vehicles, on site. The respondent local authority assessed all the plant to tax, and the appellants contended that His Majesty was the owner of the property and that they were not liable.

Held: The Crown was at all material times the 'owner' of the plant, materials and equipment in question, but in fact there was a re-delivery of possession to the contractors for the purpose of carrying out the works and, therefore, they were 'in legal possession' of the property for the purposes of the taxing statute.

Where a vesting clause provides for certification of the amount payable to the contractor upon delivery of materials, property passes when the certificate is issued.

Banbury & Cheltenham Railway Company v. Daniel

CHANCERY DIVISION (1884) 54 LJ Ch 265

A contract between the parties for the construction of a railway provided that the engineer should certify, on a monthly basis, the amount payable to the contractor in respect of 'the value of materials delivered', and that such certificates were payable by the employer 7 days after presentation.

Held: The property in the materials delivered passed to the company upon their being certified for by the engineer.

Note: Supply contracts

Contracts between a supplier and contractor for the supply of goods are sales of goods and are governed by the Sale of Goods Act 1979. Under s.17, ownership of goods sold or contracted to be sold passes to the buyer when the parties intend. If the contract of sale is not explicit, the question must be determined having regard to the terms of the contract, the conduct of the parties and the circumstances of the transaction. Section 18 of the Act sets out certain presumptions as to the intention of the parties.

Note: Vesting on insolvency

Some building contracts provide that, in the event of the contractor's insolvency, the employer shall be entitled to seize materials and plant on site and vest them in himself for the purpose of completing the unfinished work. Normally, such a clause will be good against the contractor's liquidator (or trustee in bankruptcy if the contractor is not a limited company). Such a provision will be invalid against other third parties, such as the owner of plant which is hired to the contractor.

See further Chapter 17.

Chapter 9

Acceptance and Defects

*The building owner is not prevented from complaining of defects in
the works by the fact that he has paid the price in full, even where he
has so paid to settle an action by the builder for the price.*

Davis v. Hedges

COURT OF QUEEN'S BENCH   (1871), LR 6 QB 687

HANNEN J: The plaint was for damages for the non-performance and
improper performance of certain works which the plaintiff had
employed the defendant to execute.

The defence set up was, that the defendant had sued the plaintiff
for the price of the work now alleged to have been improperly done,
and that the plaintiff had settled by paying the whole amount then
sued for; and as the plaintiff might have given the non-
performance and defective performance now complained of in
evidence, in reduction of damages, the plaintiff was precluded from
bringing a cross action for that which he might have availed
himself of as a defence to the former suit.

The county court judge decided in favour of the defendant,
holding that all the damages now sued for might have been used in
reduction of damages in the former action, and that for anything
which might have been so used no cross action could be
maintained.

We are of opinion that the decision of the learned judge was
erroneous. . . .

It appears, from . . . *Mondel* v. *Steel* (1841), that the present
practice of allowing the defence of the inferiority of the thing done
to that contracted for to be applied in reduction of damages was

introduced (on the same principle as the Statutes of Set-off were passed) for the benefit of defendants. It would greatly diminish the benefit, and in some cases altogether neutralise it, if the defendant was not allowed an option in the matter. The hypothesis is, that the plaintiff suing for the price is in default. The conditions on which he can bring his action are usually simple and immediate. The warranted chattel has been delivered, or the work contracted for has been done, and the right to bring an action for the price, unless there is some stipulation to the contrary, arises. On the other hand, the extent to which the breach of warranty or breach of contract may afford a defence is usually uncertain; it may take some time to ascertain to what amount the value of the article or work is diminished by the plaintiff's default. It is unreasonable, therefore, that he should be able to fix the time at which the money value of his default shall be ascertained. In many cases the extent to which the value of works may be diminished by defect in their execution may be altogether incapable of discovery until some time after the day of payment has arrived. Surely the right to redress for the diminution of value, when discovered, ought not to depend on the accident whether the contracting party in the wrong had or had not issued a writ for the price.

Another inconvenience which would result from holding that the inferiority of the thing done to that contracted for *must*, if an action be commenced, be used by way of defence, is, that instead of furthering the object for which this defence was permitted, namely, the prevention of circuity of action, it would in many cases tend to complicate and increase litigation. The cases are, perhaps, rare in which the consequences of defective performance of work are limited to the depreciation of the value of the work done; they usually involve consequential damage by reason of the necessity of repairing the defective work; and for this the case of *Mondel* v. *Steel* decides a separate action must be brought. Parke B., there says: 'All the plaintiff could by law be allowed in diminution of damages on the former trial was a deduction from the agreed price, according to the difference at the time of delivery between the ship as she was, and what she ought to have been according to the contract; but all claim for damages beyond that *on account of the subsequent necessity for more extensive repairs* could not have been allowed in the former action, and may now be recovered.' So that two litigations instead of one would be frequently necessary, with the additional inconveniences that it would be very difficult to discriminate in each action between that which ought to be allowed in diminution of price, and that which should be compensated for as special damage.

Where one party fails substantially to perform the contract, the other party may be entitled to terminate it, but is not obliged to do so. In some circumstances, the conduct of that other party may amount to an acceptance of the work done, which will give rise to an obligation to pay for it.

Tannenbaum and Downsview Meadows Ltd v. Wright-Winston Ltd

ONTARIO COURT OF APPEAL   (1965), 49 DLR (2d) 386

The defendant was the builder of a housing project and was required under the terms of his mortgage to construct a twelve inch trunk sewer from the project to a sewage treatment plant some 3,000 ft. away. The approximate cost was $65,000 ($38,000 for the sewer and the balance for a pumping station). The plaintiffs owned land nearby and agreed to build a 36 inch trunk sewer and pumping station to serve both properties in return for the defendant contributing $55,000 to the cost. The plaintiff built the sewer at a cost of $110,000 but, in breach of contract, did not build the pumping station, apparently because of a new plan under which the local authority was to extend the trunk sewer to a new and larger treatment plant. The defendant decided that it was too expensive to build a pumping station as a temporary measure and instead built a 'pumping main' at a cost of about $18,000 and cut into the sewer constructed by the plaintiffs.

Held: Although the plaintiffs' failure to build the pumping station was a breach of contract of a kind which entitled the defendants to terminate the contract, they had elected not to and by using the sewer they became bound to pay the agreed price, less the cost of building the 'pumping main'.

McLENNAN JA: I turn now to the main ground of appeal which is that the plaintiffs contracted to build the works for a lump sum, that is, it was an entire contract which the plaintiffs abandoned for their own purposes and are not entitled to claim under it on the principle stated in *Sumpter* v. *Hedges* (1898) and *Bradley* v. *Horner* (1957). A subsidiary proposition is that there is no evidence upon which to fix any amount on a *quantum meruit*. With that subsidiary proposition I agree and the plaintiffs' case must stand or fall on a claim under the contract.

In *Sumpter* v. *Hedges*, the plaintiff, a builder, agreed to erect buildings on the defendant's land for a lump sum. After he had done part of the work he abandoned the work for lack of funds. The defendant completed the buildings. In *Bradley* v. *Horner*, the plaintiff, a painter, agreed to paint the interior of a house for a lump sum but abandoned the work before it was half finished. In both cases the actions were dismissed because of the rule that where there is a contract to do work for a lump sum the price cannot be recovered until the work is completed. Other cases such as *Dakin & Co. Ltd* v. *Lee* (1916), and *Hoenig* v. *Isaacs* (1952), illustrate the modification of the rule. In cases where there is substantial performance of the contract, a defendant who gets the benefit of substantial performance is liable on the contract for the agreed price less the cost of making good the omissions.

There was no substantial performance in this case. The construction of a pumping station was a term of the contract, performance of which was essential to its completion because the whole purpose of the contract would be defeated if the pumping station were not erected. In *Sumpter* v. *Hedges* and cases of a like nature, where there was no substantial performance, the argument was advanced that because the defendant took possession and received the benefit of the plaintiff's work he ought to pay for it on a *quantum meruit* but as the Judges in *Sumpter* v. *Hedges* pointed out the defendant was always in possession and had no option as to whether he would take the benefit of the work or not.

There is, of course, a vital distinction between the facts in *Sumpter* v. *Hedges* and in *Bradley* v. *Horner* and this case. The works which the plaintiffs agreed to construct were not on the defendant's land and the contract was in the nature of a joint venture in that the completed works were for the mutual use and advantage of both parties and the defendant accepted the advantage and used the works. In *Sumpter* v. *Hedges*, A.L. Smith LJ, in relation to the facts of the case said, 'the plaintiff having abandoned the contract, what was the owner to do? He cannot keep buildings in an unfinished state on his lands'. The answer to the same question put about the defendant in this case is I think quite a different one. The works were not being constructed on the defendant's land, and the defendant did not have to accept the benefit of the works constructed by the plaintiffs but could have constructed its own sewer on the easements it held when it became aware of the plaintiffs' breach.

Counsel for the defendant laid great stress on the conclusion of the learned trial judge that the plaintiffs had abandoned the contract, as if abandonment discharged the contract and put an end to the obligations on both sides. It may well be that the inference made by the learned trial judge that the plaintiffs abandoned the contract is warranted but whether it was an abandonment or a refusal or neglect to perform an essential term is largely, I think, a question of semantics. I would prefer to put it this way. Because the plaintiffs did not perform an essential term of the contract the defendant had a right to terminate it and put an end to its obligation but until it did the contract was not discharged nor was the defendant's obligation discharged.

The authorities to which I refer for this proposition are recent decisions of the House of Lords: *Heyman* v. *Darwins Ltd* (1942), and *White and Carter (Councils) Ltd* v. *McGregor* (1962). The rule laid down in those cases is that where one party repudiates a contract by making it clear to the other party that he will not perform his obligations the other party has a choice. He may accept the repudiation and, if he chooses, sue for damages for breach or he may disregard the repudiation and the contract remains in full force. Repudiation by one party does not terminate the contract. Termination requires repudiation by one party and acceptance of the repudiation by the other.

The effect of that rule on this case is this. When the defendant became aware that the plaintiffs were not going to build a pumping station it could have notified the plaintiffs that it was treating their breach as discharging the contract and its obligation to pay, constructed its own sewer and pumping station and brought an action for damages for the difference between its cost and the contract price of $55,000. It will be recalled that the evidence disclosed the approximate cost to the defendant of the 12″ sewer and pumping station was $64,000. The defendant did not adopt this course. It took the plaintiffs over 8 weeks to construct the sewer – from 16 July to 13 September. The defendant knew of the plaintiffs' breach at the end of the first 2 weeks of that period. It did nothing while the plaintiffs were completing the most costly item and after that part of the works had been finished by the plaintiffs the defendant completed the works, used what the plaintiffs had constructed and accepted the benefits thereof. I am therefore of the opinion that the learned trial judge was right in holding that the defendant was liable on the contract. This result is consistent with the dicta of Denning LJ in *Hoenig* v. *Isaacs* (1952). While that was a case of substantial performance of a contract to decorate and furnish a flat Denning LJ held that, even if entire performance was a condition precedent to payment, by using the defective furniture the defendant waived the condition and must pay the contract price subject to a deduction to remedy the breach.

Chapter 10

Approval and Certificates

(i) GENERAL

Unless the terms of the contract specify a set form of certificate, all that is required is that 'it must clearly appear that the document relied upon is the physical effect of a certifying process'. One should, therefore, have some regard to the factors of 'form', 'substance' and 'intent' of which Devlin J spoke in . . . *Minster Trust Ltd* v. *Traps Tractors Ltd* (1954). . . The document should be 'the expression in a definite form of the exercise of the . . . opinion . . . of the . . . architect . . . in relation to some matter provided for by the terms of the contract': *per* Edmund Davies LJ in *Token Construction Co. Ltd* v. *Charlton Estates Ltd* (1973).

Most modern building contracts contain provisions that the work is to be executed to the satisfaction or reasonable satisfaction of the architect. When so certifying, the architect must act independently as between employer and contractor. Whether or not the work is to the 'reasonable satisfaction' of the architect is to be judged objectively.

Minster Trust Ltd v. Traps Tractors Ltd

QUEEN'S BENCH DIVISION   [1954] 3 All ER 136

This case involved a contract for the sale of re-conditioned machines, under which the sellers were to provide a 'Hunt Engineering Certificate that [the machines] have been fully re-conditioned to [Hunt's] satisfaction.' The following abstract from the judgment deals with certifiers generally.

DEVLIN J: There is no general rule of law prohibiting the influencing of certifiers. Apart from fraud, a duty not to influence can only be imposed by an implication arising from contract. Such an implication must, in accordance with settled principles, be both reasonable and necessary, and the contract must be examined to see what it yields in this respect. There is, after all, nothing to prevent a party from requiring that work shall be done to his own satisfaction. He might then choose to act on the recommendation of an agent. If an agent is named in the contract, it may be plain that he is to function only as the alter ego of his master, and then his master can tell him what to do: see *Ranger* v. *Great Western Ry. Co.* (1854) *per* Lord Cranworth LC and *per* Lord Brougham. Whether it be the act of the master or the servant, there may be a question (again depending on the implication to be drawn from the contract) whether the dissatisfaction must be reasonable, or whether it can be capricious or unreasonable so long as it is conceived in good faith. The tendency in modern cases seems to be to require the dissatisfaction to be reasonable.

What has to be ascertained in each case is whether the agent is or is not intended to function independently of the principal. The mere use of the word 'certificate' is not decisive. Satisfaction does not necessarily alter its character because it is expressed in the form of a certificate. The main test appears to be whether the certificate is intended to embody a decision that is final and binding on the parties. If it is, then it is in effect an award, and it has the attributes of its arbitral character. It cannot be attacked on the ground that it is unreasonable, as the opinion of a party or the certificate of one who is merely an agent probably can. On the other hand, it must be made independently, for independence is the essence of the arbitral function. If two parties agree to appoint an arbitrator between them, it would be, I think, implied in the contract in order to give it business efficacy (it is not now in practice necessary to consider the point because of the control over the arbitrator which the Arbitration Acts give to the court) that neither side would seek to interfere with his independence. If a party to a contract is permitted to appoint his agent to act as arbitrator in respect of certain matters under the contract, a similar term must be implied; but it is modified by the fact that a man who has to act as arbitrator in respect of some matters, and as servant or agent in respect of others, cannot remain as detached as a pure arbitrator should be.

There is another distinction between certifiers. If work under a contract has to be completed to the satisfaction of a certifier, it may mean that his duty is merely to see that the requirements of the contract are met, or it may mean that he is entitled to impose a standard of his own. It may be that his standard is that to which the parties submit and that it constitutes the only provision in the contract about quality; or it may be that his standard is an added protection, so that performance under the contract must satisfy both the contract requirements and the certifier. . . . So far as I am aware, there is no case in which a certifier who has to certify only according to his own standards has been held to be an arbitrator. It is not, however, possible under this head to draw a hard and fast line between the two categories. There must always be some room for the discretion of the certifier, however tightly the contract may be drawn. Even if he has to certify only whether an express term has been complied with, there must be room for the exercise of his judgment. *Cammell Laird & Co. Ltd* v. *Manganese Bronze & Brass Co. Ltd* (1934), is a case in which it was held that the satisfaction of a third party operated only within the requirements of the contract; that is to say, it was not permissible for the third party to put on the seller obligations as to quality outside the contract. Lord Wright said:

> 'It is true that the clause cannot be intended to deal with matters outside the scope of the respondents' obligations under the contract, or to enlarge these obligations except in so far as these are enlarged because there is super-added the judgment of a third party'.

Allied to this, there is a distinction between a certificate that is given only for the purposes of a particular contract and one that, although it may be called for by a particular contract, is not particularly related to that contract. The certifier may not then be an agent of either party. That is so in the case of a Lloyd's certificate or of a certificate by a public analyst, or others that may be obtainable on a fee. It has now become quite common to include among documents which have to be tendered against payment a certificate of inspection or of quality. Certificates of this sort are addressed to all the world or to all who may be concerned. If the phrase is not used as more than a label, they might be called certificates in rem, as compared with certificates in personam, which deal only with particular contracts and are addressed only to particular parties. The former carry the same meaning to all who read them. The latter may have to be interpreted in the light of particular contractual requirements or of information known only to the addressees. A certificate in rem certifies a standard of quality extraneous to the contract. It may be the certifier's own standard or it may be taken from some public or independent source. A certificate in personam may be based on the certifier's own standards or on standards prescribed by the contract. It may be a certificate of quality or a certificate that a contract has been carried out.

Finally, there is a distinction between a certifier whose function has been completed before the contract is entered into and one who has a function to perform under the contract. In the latter case there is room for an implied undertaking that he will not be improperly influenced. In the former case there can be no room for an undertaking that relates to the future. If anything is to be implied, it must be a warranty about the past. There may be grounds for implying a warranty that the certifier has not been improperly influenced, but, so far as I know, the point has never been decided. All the leading cases on the topic, such as *Hickman & Co.* v. *Roberts* (1913) and *Panamena Europea Navigacion (Compania Limitada)* v. *Frederick Leyland & Co. Ltd* (1947) deal with certifiers whose duties arose out of the contract. It seems clear that there cannot be the same necessity for implying a warranty in respect of the past as there is for implying an undertaking for the future. If a building owner makes the obligation of payment conditional on his architect's certificate of quality, he must not, for example, instruct his architect not to be content with less than three coats of paint, for he has impliedly undertaken that he will leave his architect free to judge independently, either by reference to the contract requirements or to his own standards of quality, whichever the case may be, how many coats of paint are required. But if he sells a completed house and agrees to supply with it his architect's certificate of quality, does he thereby warrant that he has not told his architect to be content with two coats of paint? It may be argued that there is no necessity for an implied warranty of that sort, and that the buyer is sufficiently protected by the fact that the seller would, in his own interests, have been concerned to get the best quality he could.

Panamena Europa Navigation Compania Limitada v. Frederick Leyland & Co. Ltd

HOUSE OF LORDS [1947] AC 428

Leyland had contracted to repair Panamena's ship, payment to be made after the issue of a certificate by Panamena's surveyor 'that the work has been satisfactorily carried out': clause 7. The surveyor contended that these words meant that his function was not confined to certifying the actual quality of the works done, but also covered the amount and value of the materials and labour used. He refused to grant a certificate unless he was supplied with information about these points, to which he considered himself entitled. Leyland had done everything under the contract which they were bound to do. Panamena supported their surveyor's view.

Held: The contention of the surveyor was wrong and, in the circumstances, Leyland were entitled to recover the amount due to them without the surveyor's certificate.

Note: The undermentioned extract is taken from the judgment of the Court of Appeal: see (1943) 76 Ll. L. Rep. 113.

SCOTT LJ: The problem calls for answers to two questions: (1) What is the scope of the function assigned to the certifying third party by the agreement of both the contracting parties? (2) What, if any, is the contractual undertaking, express or implied, of the party whose servant, or agent, or nominee such third party expert is? ... As I have already indicated, the submission of the appellants, made both here and below, is that the function of the surveyor, as defined by the contract was not merely to consider the quality of the work carried out by the respondents and then to express his satisfaction or dissatisfaction therewith, but to investigate the value of that work in relation to the amounts charged by the respondents ... The respondents, on the other hand, contend that the appellants' surveyor was concerned only with quality, ... The proper interpretation of the sentence in clause 7 which defines the surveyor's function assists materially in answering my second question, namely, is there any and what implied undertaking by the shipowner towards the repairer that the surveyor shall keep within the scope of the function assigned to him by the contract? That question of construction seems to me a fundamental one in the present contract; and I cannot help thinking that it is one which would have helped greatly towards the decision of many cases of building and similar contracts which have been before the courts, and have been decided without conscious or express reference to it, as an aid to adjudication. ...

The first thing seems to me to be to define the legal relationship between [the surveyor] and the appellants, for it is primarily on the right solution of this problem of construction that the legal consequences of the absence of [the surveyor's] certificate in the light of the actual facts of the case must depend. To shut out common complications which do not arise in the present case, it is important to note first that the surveyor appointed by the shipowners under the present contract, [the surveyor] (and he was obviously so appointed with the approval of the repairers), is not either an arbitrator, or a quasi arbitrator (a phrase which has no clear meaning and is best avoided), but only an expert whose opinion in regard to the quality of the work done, the contractor has to satisfy, as a condition precedent to the owners' undertaking to pay; secondly, that there is no arbitration clause to which the validity of the surveyor's action or inaction can be submitted by the aggrieved party; and thirdly, that he is put in the position of a person whose duty it is to act with scrupulous independence of both parties.

Where the contract provides for the issue of certificates by the architect, it is an implied term that the employer will do nothing to prevent the independent exercise of the architect's powers as certifier. If he does so, the need for a certificate is dispensed with and the contractor can sue without a certificate.

Hickman & Co. v. Roberts

HOUSE OF LORDS   [1913] AC 229

The employer instructed the architect not to issue a certificate until he had received the contractor's account for extras. The architect wrote to the contractor saying: 'Had you better not call to see my clients, because in face of their instructions to me I cannot issue a certificate whatever my own private opinion in the matter'.

Held: The employer had improperly influenced the architect. The contractor was entitled to payment even in the absence of a certificate.

It is a further implied term of the contract that the employer will do all that he reasonably can to ensure that the architect properly performs his duties as certifier.

A Building Contract Casebook

Perini Corporation v. Commonwealth of Australia

SUPREME COURT OF NEW SOUTH WALES   (1969), 12 BLR 82

A building contract contained a provision that the Director of Works, a government employee, could extend the time for completion 'if he thinks the cause sufficient . . . for such period as he shall think adequate'. The contractors applied to the Director of Works for extensions of time, but many of these applications were refused because of departmental policy.

Held: As certifier, the Director of Works had duties imposed on him under the contract. He had a discretion as to whether or not to grant an extension of time, and was entitled to rely on other people to supply him with information on which to exercise his discretion. He was bound to give his decision within a reasonable time and was not entitled to defer it. Departmental policy could properly be taken into account, but he should not regard it as controlling him. Terms were implied into the contract that the employer would not interfere with the Director's duties as certifier and that the employer would ensure that the Director did his duty as certifier.

MACFARLAN J: I must . . . express my opinion upon an important argument submitted by counsel for the defendant. This argument was that, not only was there not any authority justifying the implication of a term in an agreement of this type but that in fact the authorities were to the contrary. The argument was principally founded upon *Hickman & Co* v. *Roberts* (1913), and *Panamena Europea Navigation Compania Limitada* v. *Frederick Leyland & Co Ltd* (1947). It was argued that these authorities decided that if there had been a wrongful, in the sense of unauthorized, exercise of the powers by a certifier with the knowledge of the employer of the certifier, the employer being the other party to the contract pursuant to which the certifier was appointed, the only right of the contractor was that he was entitled to disregard the provisions of the agreement with respect to time and either to sue for the price or resist a claim for liquidated damages by way of penalty: *Dixon* v. *South Australian Railways Commissioner* (1923). While it is, in my opinion, the law that a contractor is entitled to disregard the provisions of the agreement with respect to time ... it does not follow, nor has it been decided that if the contractor has otherwise suffered damage he is not entitled to sue upon an implied term. ... I was referred to the decision of the Court of Appeal in *Panamena.* ... In that case the certifier adopted a wrong understanding of his functions and his employer, which was the other contracting party, adopted and encouraged the understanding of the certifier. In the result the certifier did not issue his certificate and the claim of the plaintiff was for the contracted price, or in the alternative, for damages in the same amount. The claim of the plaintiff was based in the alternative upon the implication of a term in the agreement that the defendant, building owner, would not encourage or influence the certifier wrongly to withhold the issue of his certificate. The defence was that a certificate had not issued and that in its absence and in accordance with the provisions of the agreement the plaintiff was not entitled to recover. ... Scott, LJ, said that the problem for decision called for answers to two questions which he stated as follows:

'(1) What is the scope of the function assigned to the certifying third party by the agreement of both the contracting parties?
(2) What, if any is the contractual undertaking, express or implied, of the party whose servant, or agent, or nominee such third party expert is?'

The learned Lord Justice also said:

'Let us consider the position. The repairers were entitled to rely on the surveyor doing what the contract said he was to do, that is, keeping to matters of quality only. It seems to me plain if the shipowners had known that he was departing from his proper function under the contract, it would have been their duty to stop him and tell him what the function was for which the contract provided ... It obviously was not the contractual duty of the repairers to bring him to book. It is equally obvious that they would count on his carrying out his proper function. In those circumstances I think the Court ought to imply an undertaking by the owners that in the event of its becoming known to them that their surveyor was departing from the function which both parties had agreed he was to perform, they would call him to book, and tell him what his real function was. This seems to me an implication exactly on the lines of all the authorities on implied terms from *The Moorcock* (1889) down to *Luxor (Eastbourne) Ltd* v. *Cooper* (1941).

At the same page, Scott LJ said:

'In the result, I am of opinion that they were under a contractual duty to keep their surveyor straight on the scope of what I metaphorically called his "jurisdiction" by which I do not mean that he was in any sense an arbitrator, but only that as an expert entrusted with the duty of impartiality within a certain sphere he had to form his opinion with judicial independence within that sphere. It follows from my premises that in failing to inform Dr Telfer that he was going outside and away from the limits of his function, they broke their implied undertaking.'

Luxmoore LJ agreed with the judgment of Scott LJ and also with that of Goddard LJ. Goddard LJ as I read his judgment . . . agreed with the construction of the contract stated by Scott LJ, and did not propose to add to it. Accordingly it is apparent that all members of the Court held the same opinion on the construction of the contract. However, Goddard, LJ then dealt with another question, namely, whether the defendant was entitled to rely on the absence of a certificate and he held, applying the principles of *Hickman & Co* v. *Roberts* (1913) that the defendant was not so entitled. Both these grounds were stated as grounds for dismissing the appeal in the Court of Appeal. It is significant, I think, that although the claim of the plaintiff was stated in the alternative to be for damages for breach of the implied term the damages claimed were the same amount as was claimed firstly for the contract price. It was accordingly sufficient in order that the case might be decided in favour of the plaintiff. In that sense *Panamena Europea Navigation Compania Limitada* v. *Frederick Leyland & Co Ltd* (1947) is different from the present case where the plaintiff, having been paid its contract price claims other damages for breach of the implied term which all members of the Court of Appeal in *Panamena* held to have existed.

When the case came to the House of Lords . . . the appeal was dismissed and it is true . . . that the House of Lords did not deal in any way with the plaintiff's claim based upon the existence of an implied term. . . . The defendant submitted that this was because the House of Lords was of the opinion that there was not any implied term in the contract and he relied upon the approval given by Lord Thankerton, with whose speech the other noble and learned Lords concurred, to the judgment of Goddard LJ, in the Court of Appeal. However, in my opinion the argument of the plaintiff is correct about the true extent of the decision of the House of Lords in *Panamena*. [From] the report it appears that specific questions were submitted to their Lordships in the course of argument and that the argument of counsel and the subsequent decision of the House was confined to those points. The fourth point only is material and that is stated in the following words:

'Are the respondents entitled to recover the amount claimed in
the action – and not merely such amount – as may be held to be
a reasonable price – without producing a certificate of the
appellant's surveyor in pursuance of clause 7 of the contract?'

The House of Lords answered this question in the affirmative and
dismissed the appeal. In my opinion in so treating the appeal the
House of Lords directed its attention to the alternative, albeit the
first alternative ground, upon which the plaintiffs in the action
rested their claim. The noble and learned Lords expressed their
opinions on that point and in my opinion it is correct to say they
approved the judgment of Goddard LJ. . . .

I find some comfort in the conclusion I have reached about
Panamena from the circumstance that Lord Devlin was counsel in
Panamena, both at the trial and in the two appeals. That learned
judge referred to *Panamena* in his judgment in *Minster Trust Ltd*
v. *Traps Tractors Ltd* (1954), and it would be surprising in the
circumstance if, having referred to *Panamena*, without making any
such criticism as a judge of first instance may be entitled to make
of a decision of the House of Lords, he should then have added in
the course of his judgment the passages which I have cited about
the implication of a term in such a contract. In my opinion Lord
Devlin must have held the same opinion of the effect of the
Panamena judgments as I have expressed.

I will now consider the affirmative aspect of the term
which . . . the plaintiff argued must be implied in this agreement.
Fundamentally he argued it is essential to consider this aspect
against the background that the Director of Works was at all times
a servant of the [government]; that the position he occupied as
certifier was part of the machinery set up by the agreement of the
[government] and the plaintiff for certifying applications for
extension of time. As such a servant of the [government] the
Director of Works was obliged by law to obey all lawful orders of
[his employer]. . . . If, for example, the [government] were to say to
the Director: 'You are ordered to act in a particular manner', and if
that order constituted a breach of the contractual mandate
conferred on him by clause 35, the Director would none the less be
still obliged to act in compliance with the [government] order.
Against this background I will assume that the certifier, with the
knowledge of both parties to the agreement, acted in breach of his
obligation. Could the [government] in those circumstances say: 'I
will not do anything to insure that he carries out these duties'? In
my opinion an application of the test stated in *The Moorcock*
(1889), would require the answer that the [government] was
contractually bound to order him to carry out his duties. In my
opinion the plaintiff and the defendant, being the parties bound by

his agreement, are bound to do all co-operative acts necessary to bring about the contractual result. In the case of the defendant this is an obligation to require the Director to act in accordance with his mandate if the defendant is aware that he is proposing to act beyond it. In *Mackay* v. *Dick* (1881) Lord Blackburn said:

> 'I think I may safely say as a general rule that where in a written contract it appears that both parties have agreed that something shall be done which cannot effectually be done unless both concur in doing it, the construction of the contract is that each agrees to do all that is necessary to be done on his part for the carrying out of that thing, though there may be no express words to that effect.'
>
> . . .

I am accordingly of the opinion that a term must be implied in the present agreement binding the defendant to insure that the Director of Works, its servant, performs his duties under clause 35 in accordance with this mandate.

(ii) NATURE AND EFFECT OF CERTIFICATES

The effect of a certificate depends on the wording used in the contract.

East Ham Borough Council v. Bernard Sunley & Sons Ltd

HOUSE OF LORDS   [1965] 3 All ER 619

Sunley erected a school for the council under RIBA terms. Clause 24 provided:

> '. . . (f) Upon expiration of the defects liability period . . . the architect shall . . . issue a final certificate and such final certificate . . . save as regards all defects and insufficiencies in the works or materials which a reasonable examination would not have disclosed, shall be conclusive evidence as to the sufficiency of the works and materials. (g) Save as aforesaid no certificate of the architect shall of itself be conclusive evidence

that any works or materials to which it relates are in accordance with this contract.'

The arbitration provision (clause 27) provided:

'. . . the arbitrator shall have power . . . to open up, review or revise any certificate . . . in the same manner as if no such certificate . . . had been given.'

Some two years after the architect had issued his final certificate, stone panels fixed to the exterior walls fell off owing to defective fixing by Sunley and were repaired by the employer, who sought to recover the cost from Sunley. One of the matters in dispute was whether the arbitrator had power to reopen the architect's final certificate.

Held: Clause 24(f) and (g) placed the architect's final certificate in a special position for a special purpose and therefore the wide general words of clause 27 could not be read literally because so to do would deprive clause 24(f) and (g) of any effect or operation. In order to give effect to both clauses, clause 27 must be read as being subject to clause 24(f) and (g). Accordingly the final certificate of the architect was conclusive and could not be reopened by the arbitrator save in the exceptional circumstances specified.

VISCOUNT DILHORNE: How are these two apparently conflicting provisions to be reconciled? It is unfortunate that a contract, in a form sanctioned by the Royal Institute of British Architects and presumably widely used, should contain such apparent inconsistencies.

I cannot see any way of interpreting this contract so as to give full effect to the words in clause 24(f) and in clause 27.

The appellants contended that the provision in clause 27 must be read as overriding and qualifying clause 24(f). Their argument in effect was that clause 24(f) must be read as if it provided that the certificate should be conclusive evidence save and except when there was an arbitration. The respondents, on the other hand, contend that the provision in clause 27 must be read as subject to clause 24(f), with the result that the arbitrator, despite the wording of clause 27, was bound to treat the certificate as conclusive evidence of sufficiency.

The appellants suggested that the conclusive effect of the final certificate was to be interpreted solely as terminating the contractor's liability under clause 12 to recall to remedy defects. I do not consider this suggestion well founded. The contractor's liability to recall ended at the end of the defects liabilty period and the termination of that liability was in no way dependent on the issue of the final certificate.

They further contended that to treat the certificate as conclusive evidence in litigation but not in an arbitration was not unreasonable as the parties might have been ready to agree to investigation of the sufficiency of the work before an expert arbitrator but not willing for that to take place in litigation. I do not myself regard it as right to base the interpretation of his contract on any such assumption as to the intention of the parties. Further, it appears to be thought in some quarters that, if special powers are given to an arbitrator, they devolve on the court should there be litigation. I do not regard the decision in *Robins* v. *Goddard* (1905) as establishing or, indeed, supporting such a proposition. In that case, as I understand it, the Court of Appeal held that the arbitration clause, which gave power to an arbitrator to open up, review and revise a certificate, showed beyond doubt that the certificates in that case were not conclusive and, the certificates not being conclusive, the court was not obliged to treat them as if they were.

If, however, it be the case that in litigation the court has such powers as the contract may give to an arbitrator, then if the appellants are right, the final certificate is not to be treated as concusive evidence of sufficiency either in litigation or in an arbitration.

The appellants sought also to rely upon the words 'Provided always' at the commencement of clause 27. They contended that these words showed that the rest of the contract was subject to clause 27. I do not read the contract this way. These words form part of the sentence entitling either party to go to arbitration and, in my view, make it clear that the whole contract was subject to arbitration. They do not form part of the final sentence of the clause and I do not think it is to be inferred from the inclusion of these words at the beginning of the first sentence in the clause that it was intended that the powers given to the arbitrator in the final sentence should override the conclusive effect of the final certificate.

If the final sentence of clause 27 had been intended to make it possible for the arbitrator to treat the final certificate as not conclusive on the very matters on which clause 24(f) said it was to be conclusive, one would have expected this to have been made clear beyond doubt. As the interim certificates were not made conclusive by clause 24(a) and (b), it is not possible to say with any certainty why it was necessary to say that the arbitrator had power to determine any matter as if no certificate had been given. Perhaps the explanation may be that these words in clause 27 were retained in the form of contract sanctioned by the RIBA lest, after the decision in *Robins* v. *Goddard*, their omission might lead to it being argued that the certificates were conclusive.

To read clause 24(f) as subject to the final sentence of clause 27 would, in my view and as the respondents contended, rob the final certificate almost entirely of its conclusive effect. To treat clause 27 as applying to all the certificates, interim and final, but not as giving power to the architect to ignore the conclusive effect of the final certificate, in my opinion does less violence to the language of the contract.

On this question I agree with the Court of Appeal. I think the answer to it is in the affirmative. In my opinion, the specific provision as to the conclusiveness of the final certificate prevails over the generality of the words in the final sentence of clause 27.

In *Windsor Rural District Council* v. *Otterway & Try Ltd* (1954) Devlin J came to the contrary conclusion when considering the same question on the same form of contract. For the reasons I have stated I am unable to agree with his conclusion.

Note: After the *Sunley* case, and until July 1976, JCT 63 clause 30(7) declared that the final certificate was, subject to certain provisos about the commencement of proceedings or the appointment of an arbitrator, declared to be 'conclusive evidence in any proceedings that the works have been properly carried out and completed in accordance with the terms of this contract' and subject to three exceptions set out.

P. & M. Kaye Ltd v. Hosier & Dickinson Ltd

HOUSE OF LORDS   [1972] 1 All ER 121

Contractors agreed to build a warehouse and offices for the employers on JCT 63 terms. Work on the warehouse was substantially completed in June 1967 although, with the contractors' consent, the employers had taken possession in April 1967. The architect issued interim certificates in April and July and the employers paid sums on account, leaving a balance of £14,861 unpaid. The employers complained that the warehouse floor was faulty. The contractors relaid the floor and then, on still not being paid, commenced proceedings. The employer did not seek to refer the dispute to arbitration, but sought leave to defend the proceedings by alleging that the flooring was still faulty and counterclaimed some £13,500. The action was referred to the Official Referee but then postponed. The contractor alleged that he tried to repair any further defects, but about 12 months later wrote to the architect requesting the final certficate, claiming that it was due since the works could not now be other than properly carried out.

The architect issued the final certificate showing a balance due to the contractor of £2,360. Three days after its issue the employer asked the contractor to concur in the appointment of an arbitrator. The contractor refused, alleging that it was now too late, and relying on clause 30(7) which then provided as follows:

'Unless a written request to concur in the appointment of an arbitrator shall have been given . . . by either party before the final certificate has been issued . . . the said Certificate shall be conclusive evidence in any proceedings arising out of this contract . . . that the works have been properly carried out and completed in accordance with the terms of this contract. . . .'.

The contractors issued a second writ claiming the amount of the final certificate.

Held: The words in clause 30(7) that the final certificate shall be 'conclusive evidence in any proceedings arising out of this contract' prevented any further legal action, including the legal proceedings started long before the certificate was issued.

Note: Lord Diplock gave a strong dissenting judgment, which is quoted below. Clause 30(7) was revised by the JCT as a result of this case. By virtue of JCT 80, clause 30.9, the conclusiveness of the final certificate is now limited.

LORD MORRIS OF BORTH-Y-GEST: Of the issues which are open for decision, the most difficult, in my opinion, is whether the words 'conclusive evidence in any proceedings arising out of this Contract' should be limited to proceedings commenced after the date of the final certificate or whether they also cover proceedings previously begun. As a matter of language it can hardly be doubted which alternative is to be preferred: to accept the former involves writing in a limitation which is not there. But there are more substantial issues involved, which relate, broadly, to the inter-action between an architect's certificate and the power and duty of the courts to decide disputes. Can the parties, it may be asked, by a contractual stipulation, exclude the courts from their constitu-tional responsibility? If one considers the closely analogous field of arbitration, one may contrast the willingness of the courts to stay court proceedings when there has been a submission to arbitration, while still retaining ultimate control, with their unwillingness, once court proceedings have started, to allow the question before the court to be decided by subsequent arbitration – see *Doleman & Sons* v. *Ossett Corporation* (1912), not cited before the Court of Appeal but referred to by Cairns LJ in his judgment. The present case resembles, it is said, the situation in that case.

My Lords, I am impressed, as was Lord Denning MR, with this argument, but on the whole I have come to a conclusion against it. To describe clause 30(7), as sought to be invoked here, as an ouster of the court's jurisdiction, seems to me to beg the question and in fact to misdescribe the effect attributed to it. The court pro-ceedings, as can be seen from the pleadings, raise the question whether the work done and the materials used were such as should have been done and used under the contract. An essential question must be, what standard is to be set? If the proper standard to which that work and those materials ought to conform were one to be fixed by the court (for example, by reference to what is reasonable), I could see good reasons for not allowing this matter to be decided conclusively outside the court by another person. But that is not what the contract provides. Throughout the contract, the architect is the person to pronounce on these matters. This appears right from the inception: in clause 1(1) where it is said that the works are to be carried out and completed (it may be noted that this language is repeated in clause 30(7)) to the reasonable satisfaction of the architect; and as regards materials it is for the architect to be satisfied that they are in accordance with the contract bills. This being so, what objection can there be to a clause which provides that, as regards these matters, as to which it is the architect's standard that is relevant, the architect's final certificate is to be conclusive evidence? The court has to find the facts. It has to do so in accordance with the contract. The clause provides a means – the means – of establishing the facts. The court retains ultimate control in seeing that the architect acts properly and honestly and in accordance with the contract. The method of proof, chosen by the parties, is legitimate, and by its terms binding, and I can see no reason for denying the contractual effect of the evidence to proceedings previously commenced, or, as I would prefer to put it, for reading in a limitation to subsequent proceedings. Allowing to *Doleman & Sons* v. *Ossett Corporation* its full force (and I note that four very eminent judges were equally divided even there) the present case is altogether different.

LORD DIPLOCK (dissenting): Clause 30(7) nevertheless provides that the final certificate shall be 'conclusive evidence' not only of a matter to which it does refer, i.e. the adjustment of the contract sum, but also a matter to which it does not refer at all, i.e. 'that the Works have been properly carried out and completed in accordance with the terms of this Contract.' In their natural meaning, it seems to me that these words are dealing not primarily with the activities of the contractor but with the state of the works as a result of the activities of the contractor, and are dealing with their state at the time of issue of the certificate. They mean no more than that the works are in accordance with the contract drawings and contract bills, subject to any variations authorised by the architect under clause 11, and that the workmanship and materials are of the quality required by the contract. They do not mean that at no time previously to the issue of the final certificate were there defects in the works which required remedying and had been remedied, for sub-clause (6) expressly contemplates that the final certificate must be issued notwithstanding this; and the parties should not lightly be held to have intended the final certificate to be conclusive evidence of the truth of anything that the certifier knew to be a lie.

This being, as I think, the natural meaning of the words directly under consideration, I return to see whether it requires to be qualified because of the remaining provisions of sub-clause (7). It is to be observed that the final certificate is to have no evidential value as to any of the matters referred to in the latter part of the paragraph unless –

'a written request to concur in the appointment of an arbitrator shall have been given under clause 35 of these Conditions by either party before the Final Certificate has been issued or by the Contractor within 14 days after such issue . . .'

This is obviously an elliptical phrase. 'A written request to concur in the appointment of an arbitrator' forms no part of the procedure in referring a dispute to arbitration under clause 35 if the parties agree on an arbitrator themselves. It would be quite irrational to make the evidential effect of a final certificate dependent on whether or not in an arbitration pending at the time of the issue of the final certificate the arbitrator had been appointed as a result of a written notice to concur instead of by agreement without such notice. Both parties, indeed, concede that it is a necessary implication that an agreement to refer a dispute to arbitration by an agreed arbitrator must have been intended to have the same consequences on the evidential effect of the final certificate as a written notice to concur in the appointment of the arbitrator. Furthermore, the phrase says nothing expressly about the nature of the dispute which must have been referred to arbitration. Yet disputes can arise under the contract which have nothing to do with the state of the works at any stage before or after the end of the construction period or the defects liability period. For instance, there may be a dispute arising under clause 7 which deals with the right of the employer to indemnify for infringements of patents, or clause 19 which deals with the obligation of the contractor to insure. It would be quite irrational for the parties to make the evidential effect of a final certificate dependent on the existence or non-existence of a pending arbitration in a dispute which had nothing to do with any of the matters of which the final certificate was to be conclusive evidence. It seems to me to be also a necessary implication that a prior reference to arbitration was intended to deprive the final certificate of its conclusive character only in those cases where the dispute referred to arbitration was one concerning the subject matter of which the final certificate was made conclusive evidence.

My Lords, if clause 30(7) is so construed, the system of certification provided for in the contract culminating in the final certificate makes business sense. The primary obligation of the contractor is to carry out and complete the work to the reasonable satisfaction of the architect. Prima facie the standard to be attained is a subjective one set by the architect, subject only to the qualification that it must be reasonable.

In the course of the progress of the works during the construction period until practical completion the architect signifies his satisfaction by the issue of interim certificates. By issuing his certificate of practical completion he signifies his satisfaction with the state of the works at the end of the construction period; but this is subject to any latent defects which may become apparent to him during the defects liability period. If none becomes apparent, he signifies his satisfaction with the state of the works at the end of the defects liability period by refraining from issuing a schedule of defects within the next 14 days. If latent defects do become apparent during the defects liability period and are either the subject of a schedule of defects or of written instructions by the architect to make good, he signifies his satisfaction that they have been made good by issuing a certificate to that effect under clause 15(4).

But the contract also provides that the parties may dispute the reasonableness of the architect's satisfaction as to any of these matters. The only method of doing so provided by the contract is by arbitration under clause 35. It is expressly provided by clause 30(8) that none of these previous certificates of the architect shall be conclusive evidence as to the condition of the works, materials or goods to which it relates and, by clause 35(3) that any of them may be treated by the arbitrator as if it had never been issued.

My Lords, in a contract of this character, there is good business sense in agreeing on a time limit, shorter than that laid down by general statues of limitation, within which any claims must be brought either for breaches of contract generally or alternatively, for breaches of contract of a particular kind. If it was intended to do the former it would be very simple to state this in an express provision. There are many precedents to be found in business contracts, but there is no such provision in the contract. Even clause 35 itself imposes no time limit within which disputes must be referred to arbitration.

Disputes under the contract may be of two kinds: disputes, on the one hand, as to whether the architect acted reasonably in signifying his satisfaction of the way in which the works had been carried out or completed or in ordering adjustments to be made to the contract sum, and disputes, on the other hand, as to matters in respect of which the architect has no functions to perform. In a contract by which the parties have agreed to accept 'the reasonable satisfaction of the Architect' as the prima facie criterion as to whether particular obligations under the contract have been performed, there is good business sense in distinguishing breaches of these obligations from other obligations arising under the contract, and in providing a time limit after which neither party can dispute the reasonableness of any satisfaction he has expressed as to the performance of any such obligation. In my opinion, that is what is done by clause 30(6) and (7).

The date of issue of the final certificate is so fixed by clause 30(6) that the parties will have already had an opportunity of referring to arbitration under clause 35 any dispute about the reasonableness of the satisfaction signified by the architect as to the condition of the works at the time that any previous certificate was issued by him, or about the reasonableness of any decision by him as to any adjustments to the contract sum which is necessary in accordance with the terms of the contract. If either party has availed himself of this opportunity, the final certificate has no evidential effect. But if neither party has done so before the issue of the final certificate (or in the case of the contractor within 14 days thereafter), both parties are deprived of any further right to dispute that the architect's satisfaction was not reasonable or that the obligation of which performance to his reasonable satisfaction was the criterion had not been performed; or that the adjustment to the contract sum was not properly made.

This, in my judgment, is the effect of the provision in clause 30(7) which makes the final certificate conclusive evidence that the works have been properly carried out and completed in all respects in accordance with the conditions of this contract. I can see no grounds in language or in reason for limiting its evidential effect to proceedings started after the date of the certificate. It is relevant evidence only if a fact to be proved in the proceedings is that the works have been properly carried out and completed to the reasonable satisfaction of the architect. There is no principle of law which prevents parties to litigation from agreeing in advance what shall be treated as conclusive evidence of a relevant fact – at any rate if such agreement is not a mere sham to induce the court to decide a hypothetical question on a hypothesis of fact known by both parties to be false; *Kerr* v. *John Mottram Ltd* (1940).

The way in which clause 30(7) works in practice is well-illustrated by the two actions which have been consolidated and are the subject of the instant appeal. The first action was brought during the defects liability period for the part of the works of which possession had been taken under the provisions for sectional completion to be found in clause 16. Two claims were advanced by the employers. The first was as to the liability of the contractors for consequential damage resulting from latent defects which had appeared during the defects liability period in respect of which remedial measures had already been undertaken by the contractors on the instructions of the architect. Under the contract the architect had the duty of satisfying himself whether there were defects but he had no functions in respect of any claim for consequential damage. His opinion whether such a claim was justified or not was irrelevant. The second claim was for unliquidated damages representing the cost of further remedial measures [which] had already been undertaken by the contractors to make the works comply with the terms of the contract. As I have pointed out earlier, this kind of claim was premature, even if it were well-founded, since under the contract the contractor had the right, as well as the duty, to undertake any further remedial works himself in accordance with the instructions, and to the reasonable satisfaction of, the architect. By the date at which the reply and defence to counterclaim had been delivered, the final certificate had been issued under clause 30(6). Accordingly, the architect must have been satisfied either that no further remedial measures were needed at the date of the first action to bring the works into conformity with the terms of the contract at the end of the defects liability period, or that such further measures as were needed had been satisfactorily completed. Of the fact that his satisfaction was reasonable, the final certificate was conclusive evidence. By the time of the hearing of the preliminary issue the employers had lost their opportunity to dispute this and, accordingly, of proving their claim, which had been prematurely advanced, for the cost of any remedial measures. The final certificate, however, was irrelevant to the claim for consequential damage. It could have no effect on it.

In the second action, brought after the date of the final certificate, identical claims were made in the defence and counterclaim, except that the date at which it claimed that further remedial measures were still necessary to make the works conform with the requirements of the contract was after the expiry of the period during which the contractor would have had the right, as well as the duty, to execute them himself. But the same principle applies. The final certificate is conclusive evidence that all necessary remedial works had been executed by the time of its issue. It, too, is irrelevant to the claim for consequential damage in respect of defects which had been found during the defects liability period.

There is no requirement that the architect should observe the rules of natural justice before issuing a certificate, unless the express terms of the contract so require.

Hounslow Borough Council v. Twickenham Garden Developments Ltd

CHANCERY DIVISION   (1970), 7 BLR 81

For the facts see p. 304.

MEGARRY J: It seems to me that an architect under a building contract has to discharge a great number of functions, both great and small, which call for the exercise of his skilled professional judgment. He must throughout retain his independence in exercising that judgment; but provided he does this, I do not think that, unless the contract so provides, he need go further and observe the rules of natural justice, giving due notice of all complaints and affording both parties a hearing. His position as an expert and the wide range of matters that he has to decide point against any such requirement; and an attempt to divide the trivial from the important, with natural justice applying only to the latter, would be of almost insuperable difficulty. It is the position of independence and skill that affords the parties the proper safeguards, and not the imposition of rules requiring something in the nature of a hearing. For the rules of natural justice to apply, there must . . . be something in the nature of a judicial situation; and this is not the case.

My view is fortified by two decisions, not cited in argument, which seem to me to be adverse to . . . the contractor's submissions. First, there is *Panamena Europea Navigacion (Compania Limitada)* v. *Frederick Leyland & Co Ltd* (1947). In this, Lord Thankerton, with whom the other Lords of Appeal agreed, indicated the true basis of the decision in *Hickman & Co* v. *Roberts* (1913), the limits of which, he said, had been misapprehended. The decision was based on the fact that the architect:

'. . . had referred to the owners for instructions and had accepted and acted on their instructions in reference to the matters submitted to him as arbitrator, regardless of his own opinion.'

The *Panamena* case concerned a contract for the repair of a ship under the supervision of the shipowner's surveyor, and it is clear from Lord Thankerton's speech that the House of Lords did not regard certain discussions between the shipowner and the surveyor, and certain expressions of opinion by the surveyor about what he would later have to certify, as being improper conduct.

Secondly, there is the New Zealand decision in *A.C. Hatrick (NZ) Ltd* v. *Nelson Carlton Construction Co Ltd* (1964). That concerned a certificate given by an engineer under a building contract, certifying that in his opinion the contractor had 'failed to make such progress with the works' as the engineer deemed sufficient to ensure their completion within the period specified in the contract. In a long and careful reserved judgment, Richmond J considered . . . the applicability of the rules of natural justice. The observations of Channell J in *Page* v. *Llandaff & Dinas Powis Rural District Council* (1901) to the effect that if the architect hears one side, he must also hear the other, were duly cited, but in the end the judge rejected the submission. The architect must be impartial and independent; but unless the contract so requires, expressly or by implication (and Richmond J stressed the importance of the terms of the contract), the architect is not required to hold anything in the nature of a hearing.

Interference or obstruction with the issue of a certificate means actual intermeddling rather than mere negligence or omission.

R.B. Burden Ltd v. Swansea Corporation

HOUSE OF LORDS   [1957] 3 All ER 243

The facts of this case are set out at page 289.

LORD TUCKER: I think, without attempting an exhaustive enumeration of the acts of the employer which can amount to obstruction or interference, that the clause is designed to meet such conduct of the employer as refusing to allow the architect to go on to the site for the purpose of giving his certificate, or directing the architect as to the amount for which he is to give his certificate or as to the decision which he should arrive at on some matter within the sphere of his independent duty. I do not think that negligence or errors or omissions by someone who, at the request, or with the consent, of the architect is appointed to assist him in arriving at the correct figure to insert in his certificate can amount to interference. Interference, to my mind, connotes intermeddling with something which is not one's business, rather than acting negligently in the performance of some duty properly undertaken. Nor do I think that the conduct found against [the quantity surveyor] obstructed the issue of a certificate; it may have resulted in the issue of a certificate for a smaller sum than that which was due, but that is, in my view, a matter for arbitration under clause 27 and not a ground for repudiation under clause 20. I think the use by the learned official referee of the words 'the sort of certificate' to which the appellants were entitled shows the difference between the interpretation which he put on these words and that which I have endeavoured to suggest as correct.

As a result of clause 30.1.1.1. of JCT 80 (and clause 30(1) of JCT 63) an architect's certificate is a condition precedent to payment to the contractor.

Dunlop & Ranken Ltd v. Hendall Steel Structures Ltd

QUEEN'S BENCH DIVISION [1957] 3 All ER 344

In proceedings by judgment creditors against sub-contractors nominated under the 1939 edition of the (then) RIBA form, the question was whether a sum payable by the main contractor to the nominated sub-contractor was a debt owing or accruing.

Held: There was no debt due to the sub-contractor until the architect's certificate had been issued.

LORD GODDARD CJ: A contractor who has all the expense of the materials and labour wants money from time to time, and it is perfectly clear that, until the architect has given a certificate, the contractor has no right at all to receive any sum of money from his employer by what I may call a drawing on account. Until the contractor can produce to the building owner a certificate from the architect, the contractor cannot get anything.

A certificate which is good on its face and within the authority of the architect is 'in accordance with these conditions' for the purposes of JCT 63, clause 35(2) and JCT 80, Article 5.2.

Killby & Gayford Ltd v. Selincourt Ltd

COURT OF APPEAL   (1973), 3 BLR 104

The plaintiff contractors sought payment of some £15,000 shown as due to them on an interim certificate. The defendants disputed the architect's authority to issue the certificate, and, while admitting that extra work had been ordered and carried out, also disputed its amount.

Held: Since the certificate had been properly issued under the terms of the contract, it must be honoured.

LORD DENNING MR: In the first place, the sixth interim certificate was given in accordance with the conditions of the contract. It must be honoured. Condition 30(1) says that 'the contractor shall be entitled to payment within 14 days from the issue of that certificate'.

In the second place, there is no suggestion that the work was done badly, or that there were any defects or delays. So there is nothing in the nature of a set-off or counterclaim. . . .

In the third place, Selincourt Ltd have produced no evidence to show that the variations or extras were not properly ordered. The architect was their agent. They should ask him about it. They cannot call upon the contractors to produce the written orders. The contractors can rely on the certificate without more ado.

If there were any ground for saying that the certificate was 'not in accordance with these conditions', that is a matter which under condition 35(2) could be the subject of an interim arbitration before the works were completed. . . . So long as a certificate is good on the face of it and is within the authority given by the contract, then it is in accordance with the conditions. It must be honoured. I do not think it is open to the employers of the contractors to challenge an interim certificate by saying that it is too much, or too little, or includes this or omits that, or that the extras were not sanctioned in writing. Such matters must be left till after the practical completion of the work. Then they will all be considered in the final arbitration at the end of the contract. In my opinion there is no ground whatever for saying that this interim certificate is not in accordance with the conditions of the contract. It should be honoured.

Under JCT terms, once the architect has issued the final certificate he is functus officio unless notice of arbitration has been given.

H. Fairweather Ltd v. Asden Securities Ltd

QUEEN'S BENCH DIVISION (1979), 12 BLR 40

The contract between the parties was in JCT 63 form, the date for completion being 5 March 1974. Thirty-one weeks' extensions of time were granted by the architect giving a revised date for completion of 10 October 1975. The date of practical completion was certified to be 1 December 1975. The architect issued a final certificate under clause 30(6) on 28 July 1977. On 26 March 1978 the architect issued a certificate under clause 22 to the effect that the works ought reasonably to have been completed by 10 October 1975. The defendants withheld £13,300 from sums otherwise certified in favour of the contractors, relying on the certificate of 26 March 1978.

Held: The contractors were entitled to recover £13,300, together with interest. The balance certified to be due in the final certificate was the amount due, and the subsequent purported certificate was invalid.

JUDGE STABB QC: If the architect were permitted to issue a certificate under clause 22 after the final certificate, an air of uncertainty would be introduced in the sense that neither party would know when the account could be regarded as finalised. I think that once the architect has issued his final certificate, and provided no notice of arbitration is given by either party within the permitted time, and provided that none of the three specific exceptions apply, the architect should be regarded as *functus officio*, and he is precluded thereby from issuing thereafter any valid certificate under clause 22 of the conditions of the contract.

Note: See also *Oram Builders Ltd* v. *M. J. Pemberton and C. Pemberton* (1985), 2 ConLR 94, Appendix, p. 357.

There is no express or implied term of JCT contracts which gives the quantity surveyor power or authority to decide liability. He has power under the express terms of the contract to decide quantum only.

John Laing Construction Ltd v. County & District Properties Ltd

QUEEN'S BENCH DIVISION   24 November 1982

Laing was main contractor under a JCT 63 (July 1969 revision) form for the construction of a shopping development at Chatham. The work began in 1971 and ended in 1976. Disputes arose between the parties about various matters, including the amount of fluctuations payable, under clause 31(d).

Held: The quantity surveyor has no power to agree liability as to an amount payable to or allowable by the contractor.

WEBSTER J: [The provisions of JCT 63] together with the implication that [the quantity surveyor] will probably have prepared the bills of quantity referred to in the recital to the articles of agreement, lead me to the conclusion that, save for the fact that if he prepares the bills he will have specified the quality of the works, his functions and his authority under the contract are confined to measuring and quantifying and that the contract gives him authority, at least in certain instances, to decide quantum but that it does not in any instance give him authority to determine any liability, or liability to make any payment or allowance. . . .

(iii) RECOVERY WITHOUT CERTIFICATE

Where the contractor can show that the architect (or other certifier) was disqualified from issuing a valid certificate, he can sue for moneys otherwise due to him.

Page v. Llandaff and Dinas Powis Rural District Council

QUEEN'S BENCH DIVISION   (1901), 2 HBC 316

P was employed by the council to construct certain works. The contract provided that 'the decision of the surveyor with respect to the value, amount, state and condition of any part of the works (etc.) shall be final and without appeal'. The surveyor duly gave his final certificate and P alleged that it 'was not honestly made or given in the exercise of, or in reliance upon, his own judgment, but was made and given by reason of the interference of, and in obedience to, the directions and orders of' the council. The council had, in fact, instructed their surveyor not to apply day-work rates for part of the works but that he was to use estimated quantities and apply a measured rate.

Held: This amounted to interference with the surveyor's functions. Accordingly, the final certificate was not conclusive and binding on the contractor.

Chapter 11

Variations

Most building contracts empower the architect to issue 'Variation Orders' in certain (often widely defined) circumstances. However, it is clear that the work which forms part of the contract cannot be taken away from the contractor and given to others, even where the variation clause is couched in wide terms.

Carr v. J.A. Berriman Pty Ltd

HIGH COURT OF AUSTRALIA   (1953), 27 ALJR 273

The defendant contracted to erect a factory for the plaintiff. The contract provided for the free issue of steel by the plaintiff and its fabrication by the defendant to the architect's instructions. The architect subsequently wrote to the defendant, stating that the plaintiff had awarded the contract for the fabrication of the structural steel to another company.

Held: Even if the architect's letter were treated as a variation order, it was outside the powers conferred by the variation clause and therefore amounted to a breach of contract entitling the defendant to terminate the contract.

FULLAGAR J: The relevant part of clause 1 of the conditions . . . is contained in the words: 'The Architect may in his absolute discretion and from time to time issue . . . written instructions or written directions . . . in regard to the . . . omission . . . of any work. ... The Builder shall forthwith comply with all Architect's Instructions.' Clause 1 is part of a printed form, and the powers conferred upon the architect extend to the giving of directions on a great variety of matters in addition to the 'omission of any work.' The clause is a common and useful clause, the obvious purpose of which – so far as it is relevant to the present case – is to enable the architect to direct additions to, or substitutions in, or omissions from, the building as planned, which may turn out, in his opinion, to be desirable in the course of the performance of the contract. The words quoted from it would authorize the architect (doubtless within certain limits, which were discussed in *R*. v. *Peto* (1826)) to direct that particular items of work included in the plans and specifications shall not be carried out. But they do not, in my opinion, authorize him to say that particular items so included shall be carried out not by the builder with whom the contract is made but by some other builder or contractor. The words used do not, in their natural meaning, extend so far, and a power in the architect to hand over at will any part of the contract to another contractor would be a most unreasonable power, which very clear words would be required to confer.

Commissioner for Main Roads v. Reed & Stuart Pty Ltd

HIGH COURT OF AUSTRALIA   (1974), 12 BLR 55

Under a contract for road-works, contractors were to move and spread top-soil as part of the works. The specification provided that if there was insufficient top-soil on site, 'the Engineer may direct the Contractor in writing to obtain top-soil' elsewhere, payment to be made at a prescribed rate. Instead of invoking this provision, the Engineer instructed a third party to bring more top-soil onto the site.

Held: This was a breach of contract on the employer's part. Although the contract entitled the employer to omit work from the contract, it did not entitle him to take away part of the contract work so that it might be given to another contractor.

STEPHEN J: The interaction of two factors, the shortage of on-site top-soil and the underestimation of the total amount of material required to carry out all necessary top-soiling, necessarily produced, in the case of a contract such as the present, an acute conflict between the interests of the respective parties to it. The underestimation meant that the contractor had, for the same lump sum reward, to perform considerably more top-soiling than was contemplated when tendering; but if, because of the shortage of on-site top-soil, it could invoke the fourth paragraph of clause B3.03 the entire financial position would alter, it would then be paid at what was apparently a profitable rate, for all imported top-soil needed to make good that shortage and this regardless of the initial underestimation.

When the shortfall of top-soil manifested itself the contractor sought to, but the commissioner refused to, invoke these provisions of the contract which were designed to deal with that eventuality; instead the commissioner adopted a quite different course. The commissioner's Engineer decided that, rather than incur the rate of £3 per cubic yard, he would instead, by the exercise of what he regarded as powers available to him under the contract, arrange for the work of importing top-soil onto the site to be done by a third party, no doubt at cheaper rates.

Were he legally entitled to do so it would, I think, run counter to a concept basic to the contract, namely that the contractor, as successful tenderer, should have the opportunity of performing the whole of the contract work. By the contract the contractor had covenanted that for the bulk sum of almost £5 million it would perform the works and supply all the materials shown in the other contract documents. That this included the placing of all top-soil called for by the contract drawings is clear from those drawings, from the definition of 'Works' in the general conditions of contract and from the concluding words of clause A1.22 of the specification, which expressly includes in the contract work the placing of top-soil as shown in the contract drawings. Consistently with this the second paragraph of clause B3.03 provides that top-soil shall be placed by the contractor 'on batters, medians and at other locations shown on the Drawings or as determined by the Engineer. . . .' Then the third paragraph of clause B3.03 gives meaning to the figure of 49,700 cubic yards of top-soil appearing in the schedule to the tender by describing it as the estimated quantity of top-soil required in place within the limits of the completed Works. Finally, to meet the contingency of a shortfall in on-site top-soil, there appears the fourth paragraph of clause B3.03, which I have already set out in full.

The engineer sought to rely upon two provisions of the contract to attain his purpose. He regarded the fourth paragraph of clause B3.03 as conferring upon him, or rather, in his words, upon 'the

Department', an option and initially told the contractor that in exercise of that option separate arrangements would be made for the supply of the shortfall of top-soil. He at the same time told the contractor that the Department would 'relieve your company of the responsibility of undertaking any further part of the work required under clause B3.03 of the specification' and that a deduction from the lump sum price would be made, at the rate of 15s. per cubic yard, in respect of the difference between the quantity of on-site top-soil already spread and the quantity, as estimated at time of tender, of 49,700 cubic yards; he relied upon clause 18 of the general conditions of contract for this action.

Clause 18 is a common enough provision to be found in engineering contracts and permits of the omission from time to time by the proprietor of portions of the contract works. What it clearly enough does not permit is the taking away of portions of the contract work from the contractor so that the proprietor may have it performed by some other contractor: *Carr* v. *J.A. Berriman Pty Ltd* (1953). Yet this was what the engineer sought to do in the present case in relation to spreading of top-soil.

The commissioner was, in my view, in breach of his contractual obligations under the contract. First there was the failure of the engineer to direct the contractor to obtain additional top-soil from outside the site once it was decided that the contract work of spreading top-soil to the extent shown in the contract drawings should proceed despite the shortfall of on-site top-soil. Secondly, there was the closely allied act of taking away from the contractor the balance of top-soil placement work, using imported top-soil, and, in consequence, the deduction from the lump sum price of an amount calculated by reference to the uncompleted portion of the originally estimated cubic yardage of top-soil required to be placed on site.

The shortfall having become apparent and it having been decided that all the originally contemplated top-soil spreading should nevertheless be carried out, the engineer, by failing to give a direction under the fourth paragraph of clause B3.03, rendered it impossible for the contractor to perform its contractual obligations; without such a direction it was confined to the use of on-site top-soil of which there was insufficient. The resultant situation is not dissimilar to that which arose on the facts in *Carr* v. *J.A. Berriman Pty Ltd*, where there had been a failure to give possession of the building site to the contractor, thus constituting a breach of contract, and see *Freeman & Son* v. *Hensler* (1900), where the members of the Court of Appeal, in separate judgments, each regarded as a term necessarily to be implied into a building contract the giving of possession of the site to the contractor. As Lord Atkin said, in a quite different context, in *Southern*

Foundries (1926) Ltd v. *Shirlaw* (1940), there is 'a positive rule of the law of contract that conduct of either promiser or promisee which can be said to amount to himself "of his own motion" bringing about the impossibility of performance is in itself a breach'.

It is often difficult to decide whether a departure from what the parties expected to be done under the contract is a variation. It must depend principally on careful construction of the contract, since it may be that the contractor has agreed to do all that is necessary to finish the job.

Sharpe v. San Paulo Brazilian Railway Co.

COURT OF APPEAL IN CHANCERY   (1873), 8 Ch App 597

Contractors agreed to construct a railway. The contract provided that the company should not, under any circumstances, be liable to pay to the contractors a greater sum than £1,745,000, and the contractors bound themselves to execute and provide not only all the work and materials specified, but also such other works and materials as in the judgment of the employer's engineer were necessarily or reasonably implied in and by, or inferred from, the specification, plans and sections. The contract also provided that the engineer's certificate of the ultimate balance of the account should be 'final and conclusive'.

After completion of the works, the contractors claimed to be entitled to payment in respect of extra works executed under the engineer's direction, alleging that they had done so upon the faith of his promise that they would be paid.

Held: (i) The contractors were bound by the contract to complete the work for a specified sum.

(ii) The engineer had no power or authority to alter the terms of the contract and that his promise to pay, if made, was unenforceable.

(iii) This was so even if the amount of work was understated in the specification.

(iv) In the absence of fraud, the engineer's certificate was conclusive between the parties.

SIR WILLIAM JAMES LJ: The first contract was that the line should be completed for a fixed sum. But the plaintiffs say they are, upon several heads, entitled to a great deal more than that sum. The first head is that the earthworks were insufficiently calculated, that the engineer had made out that the earthworks were two million and odd cubic yards, whereas they turned out to be four million and odd cubic yards. But that is precisely the thing which they took the chance of. They were to judge for themselves. There was no fraud; it is not alleged that [the engineer] had wilfully made miscalculations for the purpose of deceiving them; and if so, that would be the personal fraud of [the engineer] himself. But he made the calculations apparently to the best of his ability, and calculated that the earthworks would be of a certain amount. The plaintiffs says it is quite clear that this was a miscalculation. But that was a thing the contractors ought to have looked at for themselves. If they did not rely on [the engineer's] experience and skill as an engineer, they ought to have looked at the consequences and made out their own calculations. I think probably they did look at it, and satisfy themselves to some extent. I do not know whether they calculated it over again, or whether they did not take 5s. a cubic yard, which seems a large sum for earthworks, for the purpose of covering them in all possible contingencies in that respect. But that is one of the things which, in my mind, was clearly intended to be governed by the contract, the company virtually saying, 'Whether the earthwork is more or whether it is less, that is the sum we are to pay.'

Then there was a considerable item as to the inclines . . . but every statement in the bill, it seems to me, puts the plaintiffs completely out of Court as to that. The bill says that the original specification was not sufficient to make a complete railway, and that it became obvious that something more would be required to be done in order to make the line. But their business, and what they had contracted to do for a lump sum, was to make the line from terminus to terminus complete, and both these items seem to me to be on the face of them entirely included in the contract. They are not in any sense of the words extra works.

Then it is alleged that the engineer, finding out that this involved more expense than he had calculated upon, promised that he would make other alterations in the line, making a corresponding diminution so as to save the contractors from loss on account of that mistake. And then in the vaguest possible way it is said that all these promises of the engineer were known to and ratified by the company. I am of opinion you cannot in that way alter a contract under seal to do works for a particular sum of money. The plaintiffs cannot say that the company is to give more because the engineer found he had made a mistake and promised he would give more, and the company verbally, or in some vague way, ratified that promise. To my mind it was a perfectly *nudum pactum*. It is a totally distinct thing from a claim to payment for actual extra works not included in the contract.

Another situation which often arises is that the contract is of a 'measure and value' kind. Here the stated quantities are often no more than working figures and the agreement is to do so much of the work as may be required at the agreed rate.

Arcos Industries Pty Ltd v. The Electricity Commission of New South Wales

NEW SOUTH WALES COURT OF APPEAL   (1973), 12 BLR 65

The plaintiffs contracted to undertake extensive building work for the defendants at a power station on a 'schedule of rates' contract. When the work was in fact done, it was discovered that the amounts of earthworks and concrete works actually to be done fell short of the estimates in the schedule by more than 10 per cent. The plaintiffs sought a declaration that this was an omission which could not be made without their approval.

Held: The plaintiffs' claim must fail.

JACOBS P: Incorporated in the contract is a large schedule of rates covering the work in all its aspects. In the schedule there are at the beginning certain general items, then items related to earth works followed by items related to drainage, cable conduits, concrete work of many kinds, steel work, metal work, demolition, brick work, roof and wall sheeting, carpentry and joinery, plastering and so on through many different subject matters of construction. In respect of all of these items there are carried across the schedule various columns expressing first the unit of measurement of quantity, then the quantity of such units, then the rate per unit and lastly a multiplication of the quantity by the rate. This last column is described as the price. The price is added up in a summary of prices at the end of the schedule and comes to $6,797,512.38. The tender addressed to the Electricity Commission stated that Arcos offered to execute the works described for the prices and at the rates stated in the documents. It then noted that the total amount of the tender calculated in accordance with the above-mentioned prices and rates and the estimated quantities (if any) stated in such documents is $6,797,512.38.

So far, then, there can, in my opinion, be no doubt of the intention of the parties. The quantities stated in the schedule were estimated quantities only. The contractor was to be paid only for work actually done, and the price which it was to receive was to be the product of the actual quantities multiplied by the relevant rate per unit of quantity. If more were required to support this construction of the intention of the parties, it is to be found in a number of places. Particularly in the specification in its introductory section the attention of the tenderer and contractor is drawn to the likelihood of a variation between the levels and dimensions shown on the drawings provided as a basis for tendering and the actual levels and dimensions which will be required in execution of the works. . . .

The clause deals with variations of the work by way of addition thereto or omission therefrom. Such a variation is described by inference as an extra and in appropriate cases where there is authority of the engineer as an authorised extra. Then, with what to me appears to be a reference back to paragraph 1 of the clause, there is a provision in the nature of a proviso that in certain cases the approval of the contractor is required. There is to be no variation by way of addition or omission exceeding ten per cent of the contract price. The clause does not at this point refer to 'variation' because it proceeds at once to deal with additions and omissions; but such additions or omissions have already in the clause been described as things which 'vary the work' and 'extras', if they so vary the work. Then it is necessary to determine a basis of valuation so that the percentage of ten per cent or fifteen per cent may be applied to it. In the case of a lump sum contract there is no difficulty. But in the case of a schedule of rates contract a base must be specified. Therefore 'contract price' is given a very special meaning in paragraph 3 because, as the parties know, there is no fixed contract price in such a contract. So, the contract price is defined in such a case as the total value of the work; not the total estimated value, but the total value. Before it is possible to determine the total value of the work it is necessary to determine what the work is and this to me is the crucial point. The reason for having a schedule of rates contract is that the extent of the work cannot at the outset be predicated so that a contract price may be firmly stated. The nature of the work is certain, but its extent is not. The work must be completed, but the extent of it depends on information not available at the contract stage. So for the special purposes of clause 11 there must be a contract price, even in the case of a schedule of rates contract. However, that price is not the price stated in the contract as the estimated value; it is to be the true value of the work of the nature agreed to be done at the schedule of rates. This work is to be defined irrespective of any extras, that is to say, of variations in the nature of the work to be done. But variations in the levels and dimensions of the work are integral to a schedule of rates contract. The 'total value of the work' can never be determined until the extent of the work has been determined. Until then there is only a total estimated value, and that is not what the contract refers to.

I am reinforced in this view by the fact that a contrary view would seem to run counter to the whole basis of a schedule of rates contract. Its basis is that only actual work will be paid for at the rates in the schedule. Quantities as a result of the uncertainty of level or dimensions cannot initially be ascertained. If it were intended that the basis clearly expressed in clause 2(a) should be substantially modified, as it would be on the suggested interpretation of clause 11, one would expect to find such a modification of the apparent contractual basis otherwise than in a clause which on any view is primarily directed towards true variations in the works, a separate subject matter.

Grinaker Construction (Transvaal) Pty Ltd v. Transvaal Provincial Administration

SUPREME COURT OF SOUTH AFRICA   (1981), 20 BLR 30

The plaintiffs contracted to build a road for the defendants on a schedule contract. The plaintiffs argued that increases or decreases of the quantities listed in the schedule of quantities amounted to a variation.

Held: This argument should be rejected.

VILJOEN JA: I do not agree with this argument. I agree with counsel that sub-clause (4) should be construed in the light of the subclauses which precede it. When as construed clause 49(4) does not, in my view, include the so-called automatic increases and decreases. The basic flaw in counsel's argument is that it equates a variation of the quantity of the works as envisaged in sub-clause (1) with an increase or decrease of the quantities in the schedule of quantities. His entire further argument on the linguistic construction to be put upon clause 49 precede from this fallacious premise. In the normal course of the execution of the work in accordance with the drawings and specifications, even though the quantities involved may differ from those appearing in the schedule of quantities, there is no cause for the engineer to intervene and to make any variation of the form, quality or quantity of the works or part thereof. From the conditions of the contract quoted above it is abundantly clear that the schedule of quantities contains 'approximate' quantities only and that the final contract price is to be computed according to the actual quantities measured. Since the quantities therein specified are expressly stated to be 'approximate', the contract itself envisages that the final computation, after due completion of the works, may result in an increase or decrease in the quantities stated in the schedule of quantities.

The definition of 'Schedule of Quantities' emphasises the fact that the quantities entered therein are approximate quantities of work. This is repeated frequently throughout the contract. In my view it simply means that the quantities are an estimate by the engineer and the fact that the contractor is required to satisfy himself as to the conditions likely to influence the work is an indication that the employer does not warrant that the engineer's estimates are correct or even nearly correct. It stands to reason that in a contract of this nature no accurate or nearly accurate measurements in respect of certain items of work can be made unless intensive and extensive exploratory operations are embarked upon, which may be too costly from a practical point of view.

Where a departure from the specified work has been authorised by or on behalf of the employer, this may amount to a variation and entitle the contractor to payment even though the contractor is in breach of contract.

Simplex Concrete Piles Ltd v. Borough of St Pancras

QUEEN'S BENCH DIVISION   (1958), 14 BLR 80

S contracted with the defendants for pile-driving work. The contract incorporated the then current RIBA conditions, which authorised the architect to issue variation instructions and added that if compliance therewith 'involves the contractor in loss or expense beyond that provided for in . . . this contract, then, unless such instructions were issued by reason of some breach of this contract by the contractor, the amount . . . shall be added to the Contract Sum'. Another clause provided that S were to satisfy the engineer as to the stability and adequacy of the piles at all stages of construction and 'shall be liable to make good at his own cost any failure or inadequacy of the work . . . due to faulty design, materials or construction irrespective of any approvals that may have been given'.

As work proceeded, tests showed that the piles were unlikely to be able to carry the working load specified. S suggested to the architect that they should employ a third-party to install bored piles at extra cost and wrote to him seeking his 'instructions and views as to the extra cost which will be involved'. The architect replied that the council were 'prepared to accept your proposal that piles supporting Block "A" should be of the bored type in accordance with the quotations submitted by' the third party. S proceeded to use bored piles and claimed the extra cost involved on the basis that the architect's letter was a variation instruction. S conceded that, but for the work sanctioned by the letter, they would have been in breach of contract.

Held: The architect's letter constituted a variation instruction and S were entitled to payment.

EDMUND DAVIES J: [It was] submitted that, unless the letter amounted to a variation, the plaintiffs cannot succeed, whatever the architect may have thought about the matter. In my judgment, it is not immaterial that the architect intended to authorise a variation, thought he was authorising variation, and thought further that, as the employers were being rendered liable, he must check on the extra expense involved, all of which factors I find were present in [the architect's] mind when he wrote his letter.

On these findings, the one question that remains is: Was it a variation which, under contract, the architect was empowered to authorise? If it was, then the liability of the defendants would appear to be clear.

From clause 1 of the general conditions it is clear that the architect's instruction does not necessarily involve any variation at all. [Counsel] submitted that a variation must involve either (1) the doing of something extra which (in the absence of instruction) the contractor is not required to do, or (2) the omission of something which he would otherwise be obliged to do, or (3) a combination of both (1) and (2), but that the 30 July letter merely gave the contractor precise instructions as to the mode of doing that which he had already bound himself to do by clause 3(a) of the general conditions and preliminaries. He relied strongly upon clause 7 thereof, requiring the contractor:

'to make good at his own cost any failure or inadequacy . . . due to faulty design, materials or construction irrespective of any approvals that may have been given,'

and I accept that the piles driven by the plaintiffs were gravely inadequate by reason of 'faulty design' or 'construction' or a combination thereof, and I have already said that in my view the plaintiff company would have had no defence to an action for breach of contract when on 16 July they made it clear that their original design of 164 driven piles had proved in the circumstances to be an impracticable and inadequate proposition. On any view of the case, they were accordingly very fortunate in finding so amenable and co-operative an architect. . . .

[Counsel] is clearly right in his submission that an architect (whatever he himself may think) has in law no power to instruct or authorise as an extra something which the contractor is already bound to perform, but is that the correct way to describe what this particular architect did? I think not. In my judgment, it is an oversimplification of the case to say that, following upon the 30 July letter, the plaintiffs did no more than they were contractually bound to do and that, accordingly, they cannot claim extra payment. If they failed to carry out the Section 'A' works, their liability was in damages, which might as events turned out be substantial or small, according as to what alternative arrangements the defendants might have been able to make. What the contract did not oblige them to do was to get the work done by subcontractors in a wholly different way at considerable extra expense.

By clause 1 of the general conditions:

'The architect may in his absolute discretion . . . issue written instructions . . . in regard to (a) The variation or modification of the design quality or quantity of the works or the addition or omission or substitution of any work.'

Accordingly it is not right to say that a variation must involve an addition to or omission from the contracted works. The contractor's detailed specifications (schedule I) here are for 32ft long, 18-inch diameter Simplex Cast Insitu driven piles, and such specifications constitute or describe in part the 'design' or 'quality' of the works. I regard as immaterial the fact that such 'design' emanated from the contractor rather than from the employer, just as I do the fact that the first suggestion of bored piles also came from the contractor. I agree with [counsel for the plaintiffs], that the architect's letter of 30 July contained an instruction involving a variation in the design or quality (or both design and quality) of the works which the plaintiffs were being instructed to perform, and I have already indicated my view that he did so in circumstances in which he was accepting on the employers' behalf that they would be responsible for the extra cost involved. Such an action fell, in my judgment, within the 'absolute discretion' vested in him by clause 1 and was motivated by his great desire 'to get the job moving', as he put it, and regardless of the legal position of the plaintiffs under their contract. It was an action which led to the plaintiffs doing something different from that which they were obliged to do under their contract, and it was an action which involved the defendants in responsibility for the extra expense which it entailed.

In some cases the parties disagree as to exactly what the contract requires. The architect may insist that the contractor carries out the contract according to what he regards as its strict meaning. Supposing the court later holds that the contractor's view as to the meaning of the contract was correct, does the architect's insistence amount to a variation order? Somewhat different views have been taken on this question.

Molloy v. Liebe

PRIVY COUNCIL   (1910), 102 LT 616

A contract provided that no works beyond those included in the contract would be paid for without a written order from the employer and architect. While the works were in progress the employer required certain work to be done, which he insisted was within the contract. The contractor insisted that the work was not within the contract and must be paid for.

Held: It was open to the arbitrator, if he concluded that the works were not in fact included in the contract, to hold that the employer had impliedly promised to pay for them.

Peter Kiewit Sons' Company of Canada Ltd v. Eakins Construction Ltd

SUPREME COURT OF CANADA   (1960), 22 DLR (2d) 465

The appellant was engaged under a contract with the local authority for work on the approach road for a bridge across Vancouver Harbour. The appellant sub-contracted the piling work to the respondent. During the course of the piling work the authority's engineer insisted that many of the piles be driven to a greater depth than the respondent considered was required by the sub-contract. The respondent complained to the appellant that the work was not required under the contract but was told that he must comply with the instructions of the engineer. Both the appellant and the engineer on the behalf of the authority made it clear that they would not authorise extra payment.

The respondent claimed extra payment both from the authority and the appellant. Its action against the authority was dismissed by the trial judge and this judgment was affirmed by the British Columbia Court of Appeal and no appeal was made to the Supreme Court. The action against the appellant also failed at first instance but an appeal was allowed by the British Columbia Court of Appeal. The trial judge held that the engineer was entitled to act as he did, but the Court of Appeal held that the respondent's view of the contract's meaning was correct.

Held (without deciding): that on the correct interpretation of the contract the appeal succeeded and the respondent had no claim to extra payment. The majority took the view that if the engineer's instructions were not justified, the respondent's correct course was to refuse to carry them out without a variation order. Cartwright J, dissenting, took the view that this was an unrealistic course of action to propose.

CARTWRIGHT J (dissenting): Howell, with the fullest knowledge that the respondent was taking the position that it was being called on to do work entirely outside its contract and would expect and demand to be paid for it (a position which, in my opinion, both in fact and in law it was justified in taking) persisted in ordering that work to be done. In these circumstances the law implied an obligation on the part of the appellant to pay for that work of the performance of which it has had the benefit. I find some difficulty in basing the appellant's liability on an implied contract when the evidence shows that the respondent was repeatedly pressing the appellant to agree that it would pay for the work which it was doing and which did not fall within the terms of the sub-contract, and the appellant instead of so agreeing was making only 'nebulous statements' to the effect that the respondent ought to be paid or that 'there was something coming to' the respondent. I prefer to use the terminology which has the authority of Lord Mansfield and Lord Wright . . . and to say that the appellant having received the benefits of the performance by the respondent of the work which the latter did at the insistence of the former, the law imposes upon the appellant an obligation to pay the fair value of the work performed.

It is said that the respondent (who held what turns out to be the right view as to the meaning of the sub-contract) should have had the courage of its convictions and refused to perform any work beyond that which was required by the sub-contract, and if this resulted in its being put off the job should have sued the appellant for damages. It must, however, be remembered that the sub-contract was so difficult to construe that there has been a difference of judicial opinion as to its true meaning. The appellant (who held what turns out to be a mistaken view as to the meaning of the sub-contract) threatened the respondent with what might well amount to financial ruin unless it did the additional work which the sub-contrct did not obligate it to do. To say that because in such circumstances the respondent was not prepared to stop work and so risk the ruinous loss which would have fallen on it if its view of the meaning of the contract turned out to be erroneous the appellant may retain the benefit of all the additional work done by the respondent without paying for it would be to countenance an unjust enrichment of a shocking character, which, in my opinion, can and should be prevented by imposing upon the appellant the obligation to pay to which I have referred above.

JUDSON J: Had it been necessary to choose between these two views of the legal relations between the parties, I would have preferred the view of the learned trial Judge that the Eakins company was performing no more than its contractual duty. But quite apart from this, it is to me an impossible inference in this case that the parties agreed to substitute a new contract for the original one. From the very beginning, the Eakins company knew of this added term. It began to protest late in the day that the term imposed added obligations. The engineer, who had clearly defined duties under the main contract, denied any such interpretation. Nothing could be clearer. One party says that it is being told to do more than the contract calls for. The engineer insists that the work is according to contract and no more, and that what is asserted to be extra work is not extra work and will not be paid for. The main contractor tells the sub-contractor that it will have to follow the orders of the engineer and makes no promise of additional remuneration. In these circumstances the sub-contractor continued with the work. It must be working under the contract. How can this contract be abrogated and another substituted in its place? Such a procedure must depend upon consent, express or implied and such consent is entirely lacking in this case. Whatever Eakins recovers in this case is under the terms of the original sub-contract and the provisions of the main contract relating to extras. The engineer expressly refused to order as an extra what has been referred to throughout this case as 'overdriving'. The work was not done as an extra and there can be no recovery for it on that basis. When this position became clear, and it became clear before any work was done, the remedy of the Eakins company was to refuse further performance except on its own interpretation of the contract and, if this performance was rejected, to elect to treat the contract as repudiated and to sue for damages. In the absence of a clause in the contract enabling it to leave the matter in abeyance for later determination, it cannot go on with performance of the contract according to the other party's interpretation and then impose a liability on a different contract. Having elected to perform in these circumstances, its recovery for this performance must be in accordance with the terms of the contract.

Another case in which the parties differed as to whether the work was an extra was C. Bryant & Son Ltd v. Birmingham Hospital Saturday Fund (1937), on page 118. The decisive factor there was that the bills of quantities incorporated the Standard Method of Measurement which required excavation in rock to be given separately.

Even where it is clear that the work is an extra, it may be necessary to construe the contract to discover the appropriate basis of payment.

Sir Lindsay Parkinson & Co. Ltd v. Commissioners of Works

COURT OF APPEAL   [1950] 1 All ER 208

ASQUITH LJ: The question in this appeal is whether the original contract between the parties, as varied by the deed of August 18, 1937, is to be read as producing this extraordinary result, namely, first, that in no circumstances may the contractor's profit exceed £300,000, but, secondly, that nevertheless, after he has earned that profit, the Commissioners are entitled to require him to perform further additions and variations till the end of time, without paying him an extra penny in respect of them. For brevity, I will call such variations and additions 'extras'. I propose to consider, in its bearing on this problem, first the construction of the original contract and, secondly, the effect of the superimposition on it of the deed of August, 1937.

Under general condition 33 of the original contract, read literally, the commissioners were entitled to require the contractor to carry out 'extras' without limit of quantum or time. They could require him to erect a series of buildings on the site, pull them down, and put them up again, with or without alterations, as often, and for as long, as they chose. When the original contract alone was concerned, this was not such an unreasonable stipulation as on its face it appears, since two factors would in practice militate against the abuse by the commissioners of their strict legal rights. First, under that contract they would have to pay for all extras under either sub-clause (a) or sub-clause (b) of condition 35 and, secondly, pay for them out of public funds. The contractor might feel some legitimate confidence that, whatever their strict rights, the commissioners would not in practice choose, or perhaps in a democratic country be allowed, to waste the taxpayers' money wantonly. A further consideration suggests some doubt whether such a course of conduct would even be within their strict legal rights. There is a note to item 57 of the first bill of quantities. (The bills of quantities are among the documents in which the original contract was embodied.) The note in question, which my Lord has read, but which I will repeat, reads as follows: 'It is probable that further work to the value of approximately £500,000 will be ordered on a measured basis under the terms of the contract.' So far as the original contract is concerned, it is not suggested that this note was *per se* a term of that contract, in the strict sense that the commissioners were undertaking to order not less or not more than about £500,000 by value of 'extras.' The very words 'it is probable' rule out any such construction, for they necessarily imply that, in less probable events, the commissioners may order less (or more) than this quantum of extras, a possibility which would not exist if they were not entitled to do so. It was, however, suggested that the presence of this note is indirectly relevant as supporting an implied limitation of the literal scope of condition 33, a limitation to extras on some such scale as £500,000, and at all events precludes so literal a reading of condition 33 as would entitle the commissioners to require extras to be carried out to the tune of, say, £100,000,000 or indeed £2,000,000. It is not necessary to decide this point. If it were, I should for myself feel serious doubt whether any such limitation could validly be implied.

When, however, the original contract is read subject to the overriding effect (where the two are inconsistent) of the deed of August, 1937, the case for reading condition 33 as subject to an implied restriction on its literal tenor is much stronger. Two new considerations here intervene, the first derived from the terms of the deed itself; the second derived from the circumstances which surrounded and led up to its execution. To deal with these in turn, first, if the original contract plus the deed are read without any implied limitation on their literal meaning, the result, as indicated above, is that after £300,000 profit has been earned by the contractor, he can be compelled to labour like the Danaids without reward or limit, on any further 'extras' which the commissioners may elect to exact from him, 'till the last syllable of recorded time.'

The restraining practical factor that such 'extras' will have to be paid for, and paid for out of the taxpayers' funds, no longer operates, because of the proviso to clause 4 of the deed imposing a maximum profit. Only the most compelling language would induce a court to construe the combined instruments as placing one party so completely at the mercy of the other. Where the language of the contract is capable of a literal and wide, but also of a less literal and a more restricted, meaning, all relevant circumstances can be taken into account in deciding whether the literal or a more limited meaning should be ascribed to it.

The reasoning in Parkinson was applied in:

Cana Construction Co. Ltd v. The Queen

SUPREME COURT OF CANADA   (1973), 21 BLR 12

The appellants were main contractors for the construction of a new postal terminal at Edmonton, Alberta. The contract provided for the main contractor to enter into a sub-contract for the installation of the mail handling equipment with a sub-contractor to be selected by the Crown. The contractual documents required the tenderers not to include supply and installation of the mail handling equipment but to include in the tender figure for overheads, supervision and profit on this installation. The invitation to tender indicated that the estimated cost of the installation was $1,150,000. In fact the contractor was instructed to enter into a sub-contract with the selected installer at a price of $2,078,543. The contractor requested a variation order for $2,171,398 being the sub-contract tender plus 10% of the difference between the sub-contract tender and $1,150,000 to cover overheads and profits. The Crown refused to pay any increased overheads and profits.

Held: In the circumstances the contractor was entitled to a payment for overheads and profits calculated on the real sub-contract price and not on his tender figure, which was based on the estimate.

SPENCE J: Applying this same method of interpretation [i.e., that of

Asquith LJ in *Parkinson*] to the present case, I have come to the conclusion that when the parties entered into the original contract then it was on the basis that the appellants would be required to enter into a sub-contract for the installation of the mechanical handling equipment at a cost of about $1,150,000 and that therefore their bid should include overhead, supervision and profit for that amount. As it turned out, the appellant was required to enter into a sub-contract for an amount of over $2,000,000 and therefore the overhead, supervision and profit on that larger amount was not within the contemplation of either party. None of the authorities cited by the respondent during the argument have convinced me that I should not approach the interpretation of this business contract in a fashion which I regard as business-like reaching the result which I have reached.

In some cases, the architect's failure to issue variation orders may constitute breach of contract by the employer.

Holland & Hannen and Cubitts (Northern) Ltd v. Welsh Health Technical Services Organisation

QUEEN'S BENCH DIVISION (1981), 18 BLR 80

The plaintiffs were the main contractors for the erection of a hospital at Rhyl under JCT 63 (1969 revision). Crittalls, who were nominated sub-contractors for the supply and installation of the window assemblies, began work in September 1973. By January 1974 it was clear that the window assemblies were failing to keep rainwater out. In due course the judge held that the major reason for the leaks were design defects although there were also defects in workmanship.

Crittalls proposed remedial works which were discussed at site meetings. (Crittalls accepted that they would be liable either under the sub-contract, or under the direct warranty which they had given to the employer). The architects took the position that the defects were due to bad workmanship and that Cubitts should put forward Crittalls' remedial proposals as their own. (Evidence at the trial showed that at another Welsh hospital being built at the same time with the same architect and same window sub-contractor, a different main contractor had acceded to this course and a variation order had been issued.) Cubitts took the position that the defects were due to faulty design and that it was impossible to complete the contract without a variation of the design.

Held (*inter alia*): By their failure to issue a variation order, the architects had made it impossible for Cubitts and Crittalls to complete the contract works.

Note: In the judgment, the architects are referred to as 'PTP'.

JUDGE JOHN NEWEY QC: In my view, PTP, by their failure to issue a variation instruction, made it impossible for Crittalls to complete their sub-contract, and made it impossible for Cubitts to complete the main contract so far as the windows are concerned.

From then onwards, PTP kept up a public face of disapproval of Crittalls proceeding with installation of windows to a revised design, shown by the seeming clause 6(4) notices while from 6 May 1974 they permitted Cubitts to proceed with finishes, which were wholly dependent on the efficacy of the remedial measures, and repeatedly offered to issue a variation instruction if only Cubitts would recommend the changes in design. What can only be described as the hypocrisy of PTP's position is demonstrated by what happened at Gurnos, where Douglas, more accommodating than Cubitts, gave their support to Crittalls' identical proposals and a variation instruction followed.

It is quite clear from PTP's internal memoranda and from Mr Keegan's remarks at the inspection on 27 September 1978, that PTP were satisfied that Crittalls had by the 1974 and later changes in the windows, overcome both defects in design and workmanship, although they quite properly wished to preserve WHTSO's right to complain if any further defects in workmanship manifested themselves. The promise made by Mr Pickup at the meeting of 8 September 1977 that, subject to water penetration being eliminated, he would instruct PTP to release money to Crittalls indicates that he, representing the mind of WHTSO themselves, was of the same view.

I think that PTP's failure to issue a variation instruction in 1974 rendered Cubitts' and Crittalls' obligation to supply windows pursuant to the entire contracts into which they had entered impossible of performance.

Crittalls went on to provide windows 'at their own risk', meaning that, if the windows should still leak, no payment would be made to them, but in the expectation that, if the windows did not leak, they would be paid. In my view, PTP and WHTSO, by their words and conduct, agreed to that arrangement. I think that the result was a new contract between Crittalls and WHTSO, to which Cubitts were not parties, that, in consideration of Crittalls providing window assemblies for the hospital, WHTSO would pay to them a *quantum meruit*.

PTP's failure to issue a variation instruction when defects in design had become apparent, when they had come to believe that Crittalls' remedies were overcoming the defects and when they had no alternative proposals of their own, may possibly be excused on the grounds that they were labouring under a mistake of law as to Cubitts' responsibilities. However, I find it impossible to believe that architects in charge of a great building project, which has been brought to a stop by an unexpected difficulty, are entitled to adopt a passive attitude, as PTP did in this case. PTP's failures were ones of omission rather than of commission, but I think that they none the less amounted to breach of contract.

The same conclusion as I have reached in regard to the issue of a variation instruction applies, I think, to the grant to Cubitts of an extension of time.

Chapter 12

Prolongation and Disruption Claims

Standard form building contracts contain express provisions enabling the contractor to claim for loss and expense which he incurs as a result of certain specified events. In every case the exact wording of the relevant clause is important. The forms currently in use generally allow additional or alternative claims for breach of contract at common law.

Under JCT terms the phrase 'direct loss and/or expense' or 'direct loss and/or damage' means that the sums recoverable are equivalent to damages at common law.

Wraight Ltd v. P.H. & T. (Holdings) Ltd

QUEEN'S BENCH DIVISION   (1968), 13 BLR 26

The contractors agreed, on JCT 63 terms, to build properties at Sheerness in Kent. Shortly after work commenced, unsuitable soil conditions were encountered, and the architect properly directed a suspension of the works. The suspension having continued for more than the period stated in the appendix to the contract, the contractors determined their employment under clause 26(1)(c)(iv) of the contract (now JCT 80, clause 28.1.3.). There was no dispute that the employer was bound to pay the contractors an amount as direct expenditure on starting the work, but the employer disputed the contractors' claim for the profit which they would have earned had the contract been completed.

Held: The words 'direct loss/and/or damage' in clause 26 must be given the same meaning as they would have in the case of a breach of contract. The loss of profit was the direct and natural consequence of the determination of the contract, and the contractors were entitled to succeed.

MEGAW, J: For the [contractors], it is put forward that the phrase 'direct loss and/or damage' means what is its usual, ordinary and proper meaning in the law: one has to ask whether any particular matter or items of loss or damage claimed has been caused by the particular matter. Here, the particular matter is not breach of contract but determination of the contract. If it has been caused by it, then one has to go on to see whether there has been some intervention of some other cause which prevents the loss or damage from being properly described as being the direct consequence of the determination of the contract.

Causation, as is so well known, is a difficult concept with all kinds of metaphors having been used to explain it. It is, I think, for general purposes best regarded in this way: that there may be a very large number of causes, in one sense of the word, to produce any particular result, but the law has to look at the question of causation from the point of view of common sense and it has to eliminate causes which are of little real significance, though they might in one sense be called 'causes', and to find what is the real cause. The same thing applies with regard to intervening causes.

It is said here on behalf of the [contractors] that quite plainly the loss of profit to the [contractors] was, in any ordinary sense of the word, a direct consequence of the determination of the contract. If the contract had not been determined, they would have gone on and done the work, continued their legal obligations thereunder and would have had their legal right to obtain their full contractual remuneration at the end of it, and therefore they have, by this determination, lost the right to receive that sum of money which they would have received if the contract had not been determined: what can be more direct than that?

But, of course, they will also be excused by the determination from their other legal obligations which would otherwise have fallen upon them, and they cannot claim as being a loss or a partial loss or damage, or a part of damage incurred by them directly resulting from the determination, any expenses which they have been saved from incurring by reason of the limitation of their legal obligations as a result of the determination.

In my judgment, there are no grounds for giving to the words 'direct loss and/or damage caused to the contractor by the determination' any other meaning than that which they have, for example, in a case of breach of contract or other question of the relationship of a fault to damage in a legal context. Therefore it follows . . . that the [contractors] are, as a matter of law entitled to recover that which they would have obtained if this contract had been fulfilled in terms of the picture visualised in advance but which they have not obtained. . . .

'Direct damage' is damage which flows naturally from the breach or event relied on without other intervening cause.

Saint Line Ltd v. Richardsons, Westgarth & Co. Ltd

KING'S BENCH DIVISION   [1940] 2 KB 99

In a contract for the supply of engines for a ship one clause provided 'nor shall [the manufacturers'] liability ever and in any case . . . extend to any indirect or consequential damages or claims whatsoever'. The manufacturers were in breach of contract and the owners brought an action claiming damages for: (a) loss of profit for the time during which they were deprived of the use of the ship; (b) expenses of wages, stores, etc.; (c) fees paid to experts for superintendence.

Held: All these heads of claim were recoverable as direct damage and were not excluded by the clause.

ATKINSON J: What does one mean by 'direct damage'? Direct damage is that which flows naturally from the breach without other intervening cause and independently of special circumstances, while indirect damage does not so flow. The breach certainly has brought it about, but only because of some supervening event or some special circumstances. The word 'consequential' is not very illuminating, as all damage is in a sense consequential, but there is a definition in the Oxford English Dictionary to which both sides have appealed: 'Of the nature, or a consequence, merely; not direct or immediate; eventual'. It cites the definition of 'consequential damages' from Wharton as: 'losses or injuries which follow an act, but are not direct or immediate upon it'. But, apart from that, I have the guidance of the Court of Appeal as to what is meant by 'consequential'. In *Millar's Machinery Co Ltd* v. *David Way & Son* (1934), where the Court was construing a clause of the same class as this, I find this, that Maugham LJ said: 'On the question of damages the word "consequential" has come to mean "not direct".' Roche LJ agreed 'that the damages recovered by the defendants on the counterclaim were not merely "consequential" but resulted directly and naturally from the plaintiffs' breach of contract.' It is quite clear that the Court there took it for granted that the word 'consequential' meant 'merely consequential' and referred to something which was not the direct and natural result of the breach.

In my judgment, the words 'indirect or consequential' do not exclude liability for damages which are the direct and natural result of the breaches complained of. . . . What the clause does do is to protect the respondents from claims for special damages which would be recoverable only on proof of special circumstances and for damages contributed to by some supervening cause. I am satisfied that it does not protect them from the claims which are made in this case.

Croudace Construction Ltd v. Cawoods Concrete Products Ltd

COURT OF APPEAL   (1978), 8 BLR 20

Croudace were main contractors for building a school, and contracted with Cawoods for the supply and delivery of masonry blocks. Cawoods sought to limit their liability by providing in the supply contract that:

'We are not under any circumstances to be liable for any consequential loss or damage caused or arising by reason of late supply or any fault, failure or defect in any materials or goods supplied by us or by reason of the same not being of the quality or specification ordered or by reason of any other matter whatsoever'.

Croudace claimed damages for breach of contract, alleging late delivery and defects. The claim included items for loss of productivity, additional costs of delay in executing the main contract works, and an indemnity against a claim by another sub-contractor for delay to his programme.

Held: All the losses directly and naturally resulted in the ordinary course of events from Cawoods' alleged breach and were not excluded as 'consequential loss or damage'.

MEGAW LJ: Counsel for [the defendants] submitted that the word 'consequential' could have three possible meanings. The first meaning that it could have is, as he put it, 'all resulting damage'. He would have been prepared to argue in this case that that was the meaning here of 'consequential', so that the exclusion of all consequential damage for delay would have the effect of excluding all damage resulting from delay, were it not that he was precluded from putting forward that argument because of [another clause in the contract . . .]

The second possible meaning which [he] said could be given to 'consequential' in the context of this contract was one which he defined in this way:

' "Consequential" is all damage other than the normal or ordinary damage which is recoverable in the case of delay of delivery in a contract for the sale of goods.'

That was the meaning which [he] suggested was the appropriate meaning here. His submission would involve, if that meaning were accepted, that the damages to which I have referred as being heads of damage in the particulars of the statement of claim would here not be recoverable, because, he contended, such things as extra costs on site as a result of men and materials being idle as a result of delay, inflated price costs and the claim by the sub-contractor would be within this definition of 'consequential': they would be damages other than the normal or ordinary damages which resulted from delay in a sale of goods case. His submission as to what were the normal or ordinary damages which result from delay in a sale of goods case was, in effect, the difference in market value of the relevant goods – the late-delivered goods – as between the date when the goods should have been delivered under the contract and the date when they were in fact delivered. Those, [he] submitted, were the normal, or ordinary, damages which result from delay in delivery in a sale of goods case: any damages which fall outside that are not 'normal' damages; and, not being normal damages, they come within the meaning of 'consequential' damages in this clause, and hence the defendants are excused from liability in respect of them.

The third possible meaning of 'consequential' damage as put forward by [counsel] was, as he put it, some meaning referring to 'second-stage' consequences – consequences which did not follow as a first stage, or immediately, from the breach, but which were, nevertheless, damages which, in the absence of this clause, would have been recoverable on ordinary principles of the law as to damages.

[He] accepted that one way of expressing that third possible meaning was the meaning given to 'consequential' in *Millar's Machinery Co Ltd* v. *David Way and Son* (1934). In the report it is recorded that:

> 'Roche LJ agreed that the damages recovered by the defendants on the counterclaim were not merely "consequential", but resulted directly and naturally from the plaintiffs' breach of contract.'

So in essence the third meaning which [he] submitted might possibly be given to the word 'consequential' in the context of the clause with which we are concerned is that damages were not consequential if they resulted directly and naturally from the breach of contract. But [he] submitted that, in the context of this clause, the second suggested meaning ought to be preferred.

[Counsel] has put forward (if he will allow me to say so) a clear, closely-reasoned and interesting argument in support of his submission as to the meaning of the clause in this contract. He will, I am sure, understand that it is in no way disrespect to the clarity and the interesting nature of his argument if I deal with the matter quite shortly. I do that for this reason, that to my mind the decision of this court in *Millar's Machinery Co Ltd* v. *David Way & Son* is a decision, the *ratio decidendi* of which is directly applicable to the present case and which is binding on this court.

In these circumstances the references which [he] gave us to passages in textbooks, and in other cases not directly concerned with the issue with which we are concerned, are not really of assistance. It is clear that the word 'consequential' can be used in various senses. It may be difficult to be sure in some contexts precisely what it does mean. But I think that the meaning given to the word in *Millar's* case is applicable to the present case. It is binding on us in this case. Even if strictly it were not binding, we ought to follow it. That case was decided in the year 1934. It has stood, therefore, now for more than 43 years. So far as I know it has never been adversely commented upon.

'Direct loss and/or expense' includes a claim for managerial and supervisory expenses directly attributable to matters which have materially affected progress of the works.

Tate & Lyle Food & Distribution Ltd v. Greater London Council

QUEEN'S BENCH DIVISION [1982] 1 WLR 149

Although this case eventually went to the House of Lords and is reported in [1983] 2 WLR 649, the issues discussed in the judgment set out below did not arise in the House of Lords. The judgment casts doubt on the legitimacy of charging a 'head office percentage' to cover managerial time spent consequent on disruption and delay and in particular on the use of formulae to calculate such a percentage in these circumstances.

GLC constructed two new piers in the river Thames, causing heavy deposits of silt which interfered with the use of the plaintiffs' jetties. As a result, the plaintiffs incurred heavy dredging costs. It was found that the engineers who designed the piers should have foreseen this and adopted a different design.

Held: The GLC were liable. In principle the plaintiffs could properly recover damages for managerial and supervisory expenses directly attributable to GLC's wrongful act, but the claim failed because the plaintiffs could not prove their loss by proper records or otherwise. It was not permissible to allow a percentage of the other items of the claim.

FORBES J: [The] plaintiffs claim that 'they have expended managerial and supervisory resources in attending to the problems created by the infringement of their rights.' Originally, they asserted that such expenditure could not be quantified, but ... they now claim such expenditure at 2·5 per cent. of the total loss and damage. [Counsel] for the plaintiffs, accepts that there is no direct evidence, other than passing references, of the fact that such managerial and supervisory resources were expended or of the extent to which managerial time was deployed on attending to the problems involved. He takes his stand on established [Admiralty] practice. . . . Apparently, it is usual in Admiralty collision cases for the judge to decide the issue of liability and for all questions of damage to be subsequently referred to the arbitrament of the registrar sitting with 'merchants.' One of the items usually allowed, without proof, is known as 'agency.' I quote from Marsden, *British Shipping Law*, 11th edition (1961), volume 4, paragraph 520:

'In cases of damage to vessels, the most usual causes of loss to the claimant (apart from detention which will be considered later) are out of pocket expenses. These may be recovered as damages from the wrongdoer. They may be classified as follows: . . . (6) Agency, that is to say certain miscellaneous expenses and work based on the disturbance and extra work which a collision causes in the shipowner's office and those of his agents in ports, e.g. making fresh arrangements for the use of the vessel, paying accounts, etc. Agency is allowed as a single sum without strict proof, and is assessed as far as possible on the probable expense incurred and work done which would not have been necessary if there had been no collision. Underwriters' agency is not, however, recoverable in so far as differing from owner's agency. Superintendence may also be recovered.'

[Counsel] tells me that the usual figure is taken as 1 per cent of the collision damage because any exact computation of the expense due to the 'disturbance and extra work which a collision causes in the shipowner's office' is almost impossible. He also referred me to a passage in *The Liesbosch* (1933), where Lord Wright said:

' . . . compensation for disturbance and loss . . . including in that loss such items as overhead charges, expenses of staff and equipment, and so forth thrown away . . .'

I do not think, however, that *The Liesbosch* helps [him]. I think Lord Wright in the passage quoted was dealing with the expenses of retaining staff and hired equipment for the period for which they could not usefully be employed because there was no dredger available to carry on with the dredging contract, and not with the cost of staff time involved in remedying the injury.

[The defendant's counsel] suggest that an item of damages under this head and calculated in this way is unknown in a Queen's Bench action; in addition, [they maintain] that managerial expenses could only be recovered as loss of profit, and that they might be recovered if quantified by acceptable evidence, but that there is no such evidence here.

The problem, it seems to me, resolves itself into two constituents: (a) Is there any warrant for suggesting that managerial time, which otherwise might have been engaged on the trading activities of the company, had to be deployed on the initiation and supervision of remedial work (excluding anything which might properly be regarded as preparation for litigation)? And (b) if so, could this reasonably have been the subject of evidence, or is it so difficult to quantify that the application of some suitable rule of thumb is justified?

I think there is evidence that managerial time was in fact expended on dealing with remedial measures. There was a whole series of meetings, in which the plaintiffs' top managerial people took a leading role, the object of which was to find out what could be done to remedy the situation and to persuade the defendants to do something about it. In addition, while there is no evidence about the extent of the disruption caused, it is clear that there must have been a great deal of managerial time involved in dealing with the dredging required and in rearranging berthing schedules and so on ... to enable the delivery of material to the refinery to proceed without interruption. I have no doubt that the expenditure of managerial time in remedying an actionable wrong done to a trading concern can properly form the subject matter of a head of special damage. In a case such as this it would be wholly unrealistic to assume that no such additional managerial time was in fact expended. I would also accept that it must be extremely difficult to quantify. But modern office arrangements permit of the recording of the time spent by managerial staff on particular projects. I do not believe that it would have been impossible for the plaintiffs in this case to have kept some record to show the extent to which their trading routine was disturbed by the necessity for continual dredging sessions. In the absence of any evidence about the extent to which this occurred the only suggestion [counsel] can make is that I should follow Admiralty practice and award a percentage on the total damages. But what percentage? [Counsel] claims 2·5 per cent but tells me that the usual Admiralty figure is 1 per cent, and this appears to cover the work of port agents as well as that in the shipowner's office. While I am satisfied that this head of damage can properly be claimed, I am not prepared to advance into an area of pure speculation when it comes to quantum. I feel bound to hold that the plaintiffs have failed to prove that any sum is due under this head.

Loss of profit is a permissible head of claim under JCT 80 clause 26

[JCT 63, clauses 11(6) and 24(1)] but it seems that the contractor must establish proof of profitability elsewhere, i.e. that he could have employed his resources elsewhere had he been free to do so.

Peak Construction (Liverpool) Ltd v. McKinney Foundations Ltd

COURT OF APPEAL   (1970) 1 BLR 111

The facts of this case are set out on page 246.

SALMON LJ: Loss of profit can only be awarded for such a period of delay as the official referee concludes was caused by the defendants.

The way in which the claim for loss of profit was dealt with below had caused me some anxiety. The basis upon which the claim was put in the pleadings was that for 58 weeks no work had been done on this site. Accordingly, a large part of the plaintiffs' head office staff, and what was described as their site organisation, was either idle or employed on non-productive work during this period, and the plaintiffs accordingly suffered considerable loss of gross profit. It was said that what I can shortly describe as the land bank – which was a quantity of land owned by the plaintiffs which they could develop if there were no other private contracts on hand – could not avail them, because they did not know during any part of the 58 weeks when they would be able to re-commence on the East Lancashire Road site, and accordingly they had to remain ready to do so and could not undertake any further work.

[At first instance] the matter was put rather differently. The case was put on the basis that in the time during 1966 and 1967 when they were engaged in completing the construction of the East Lancashire Road project they were unable to take on any other work, which they would have been free to do had the East Lancashire Road been completed on time, and they lost the profit which they would have made on this other work. When the case was argued in this court is seemed to me that the plaintiffs were a little uncertain about which basis they were opting for. In the end, however, I think they came down in favour of the second basis: that is, the one that was argued before the deputy official referee. I think when there is a new trial every issue under this head should be open for the official referee to consider. It might be of some help to him not to be left only with the evidence of the plaintiffs' auditor on this point: possibly some evidence as to what the site organisation consisted of, what part of the head office is being referred to, and what they were doing at the material times, could be of help. Moreover, it is possible, I suppose, that an official referee might think it useful to have an analysis of the yearly turnover from, say, 1962 right up to say, 1969, so that if the case is put before him on the basis that work was lost during 1966 and 1967 by reason of the plaintiffs being engaged upon completion of this block, and therefore, not being free to take on any other work, he would be helped in forming an assessment of any loss of profit sustained by the plaintiffs. However, that is a matter for the parties.

'Direct loss and/or expense' under JCT terms is the loss and/or expense which directly and naturally results in the ordinary course of events from the matters specified. It covers financing charges where appropriate, such being a constituent part of the loss and/or expense.

F.G. Minter Ltd v. Welsh Health Technical Services Organisation

COURT OF APPEAL   (1981), 13 BLR 1

Minter was main contractor on a hospital project. The contract was in JCT 63 form. Many variations were made as the contract progressed, and these affected regular progress of the works as a whole and a nominated sub-contractor's work was materially affected by lack of necessary instructions. Claims were made and paid under clauses 11(6) and 24(1)(a) (now JCT 80, clause 26.1.) The wording of clauses 11(6) and 24(1) had been amended by the parties so as to substitute 'not more than 21 days' for the standard 'within a reasonable time'. Minter claimed that finance charges which they had incurred as a result of being kept out of their money were 'direct loss and/or expense'. WHTSO argued that these charges were not 'direct' and were 'a naked claim for interest'.

Held: The loss was 'direct'. Minter was entitled under the contract terms to recover financing charges.

Note:

See: *Rees & Kirby Ltd* v. *Swansea Corporation* (1985), Appendix p. 361.

JCT 80,, clause 26, has been re-worded so that only one written application is required to be made by the contractor in respect of loss and/or expense caused by the happening of one of the prescribed events.

STEPHENSON LJ: I approach the question of construction which we have to answer on a certain amount of common ground. It is agreed that clauses 11(6) and 24(1) must be read together and the words 'direct loss and/or expense' in them – and indeed in clause 34(2) – ought to mean the same thing. It is also agreed that if the sub-clauses can be construed without reference to other provisions of the contract there are no grounds for giving to those words any other meaning than that which they have in a case of breach of contract in a legal context. That was the opinion of Megaw LJ of the words 'direct loss and/or damage' in clause 26(2)(b)(vi) (which I have read) in *Wraight Ltd* v. *P.H. & T. (Holdings) Ltd* (1968). He there considered that loss or damage caused by the determination of the contract was direct if it would be recoverable as damages flowing naturally in the usual course of things from a determination in breach of contract, and he held that loss of 10 per cent for establishment charges and profit which the contractor would have earned if the contract had not been determined and of 12½ per cent for a proportion of the overhead costs of his business attributable to the contract were recoverable as direct loss or damage.

It is agreed that accordingly the court should apply to the interpretation of what loss or expense is direct the distinction between direct and indirect or consequential which was discussed by Atkinson J in *Saint Line Ltd* v. *Richardsons, Westgarth & Co.* (1940), in construing an exclusion of liability clause and should at least recognize that loss of profit and expenses thrown away on wages and stores may be recoverable as direct loss or expense, as he there held. So this court held that the cost of men and materials being kept on a site without work was recoverable as damages and not excluded as 'consequential' in *Croudace Construction Ltd* v. *Cawoods Concrete Products Ltd* (1978). It had held the same in *B. Sunley & Co.* v. *Cunard White Star Ltd* (1940), making there also a small award of interest on money invested and wasted.

It is further agreed that in the building and construction industry the 'cash flow' is vital to the contractor and delay in paying him for the work he does naturally results in the ordinary course of things in his being short of working capital, having to borrow capital to pay wages and hire charges and locking up in plant, labour and materials capital which he would have invested elsewhere. The loss of the interest which he has to pay on the capital he is forced to borrow and on the capital which he is not free to invest would be recoverable for the employer's breach of contract within the first rule in *Hadley* v. *Baxendale* (1854) without resorting to the second, and would accordingly be a direct loss, if an authorized variation of the works, or the regular progress of the works having been materially affected by an event specified in clause 24(1), has involved the contractor in that loss.

On reaching this point the claimants might be thought to be nearly home. Why, [ask the claimants] should they not be entitled to be repaid what they have lost as a natural as well as contemplated result of what has happened to their contract through happenings which are in no way their responsibility, if not the respondents' faults or the fault of their architect acting for them? If the respondents choose to vary the contract, or if their architect fails to give them the necessary instructions and they have to keep men and machinery idle and have to pay for them and their subcontractors, why should they be allowed the cost of the extra work but be deprived of what is part of the same loss? Why should the contract sum be adjusted to include the one but not the other? [They submit] that the whole risk of the claimants' loss, or the risk of the whole of their loss, naturally resulting from the happenings necessary to activate clauses 11(6) and 24(1), was contractually assumed by the respondents. Yet if the judge is right, the claimants are worse off as a result of actions for which the respondents are responsible and that unjust result is reached by limiting the natural meaning of direct loss and/or expense to exclude losses which were just as clearly within that meaning as the sums which the architect has allowed. There is, [they submit], nothing in the machinery provided by the two clauses or its operation in this case to break the chain of causation or to halt the operation of the root cause of these losses, even if there is any concurrent or supervening cause, or to turn an obviously direct business loss into a loss that is indirect or consequential.

In support of [this submission] we were reminded of what eminent judges had said about causation in *HMS London* (1914), *The Edison* (1932), and *Smith Hogg & Co.* v. *Black Sea & Baltic General Insurance Co.* (1970). [The claimants] conceded that the language of clauses 11(6) and 24(1), particularly clause 11(6), created a difficulty in the way of interpreting direct loss and/or expense to include the losses [they claim], in particular the use of the past tense in 'has involved' and 'having been incurred' in clause 11(6) and 'has been involved' in clause 24(1). But [they] asked us to treat the applications required by both clauses as what [they] called 'foot in the door' applications to give warning to the architect of losses both past and future.

It is the language of these clauses in which the machinery for paying the contractor's loss and expense is described, on which the respondents principally rely to defeat the claim.

[His lordship then considered and rejected the respondents' argument that interest was not recoverable because of the decision in *London Chatham & Dover Railway Co.* v. *S.E. Railway Co.* (1893), holding that JCT terms contained an implied promise to pay interest. He continued:]

    In the context of this building contract and the accepted 'cash flow' procedure and practice I have no doubt that the two kinds of interest claimed here are direct loss and/or expense unless there is something in the contractual machinery for paying direct loss and/or expense which excludes this loss and expense of interest by the claimants. It was not obviously indirect, like the loss of a generous productivity bonus paid by the contractor to his men. It was ... interest reasonably paid on capital required to finance variations or work disrupted by lack of necessary instructions and as such direct loss or expense in which the claimants had been involved.

    To see how the contractual machinery appears to work it may help to set out the temporal sequence in chronological order. . . . In a clause 11(6) situation first comes (1) a variation, then (2) a direct loss or expense, then (3) a written application ... [within a reasonable time after it has been incurred], then (4) the architect's opinion that it would or would not be reimbursed by payments in respect of a valuation made in accordance with clause 11(4): then (5) if his opinion is that it would not, his ascertainment of the amount of the loss or expense, then (6) its inclusion in the next monthly interim certificate; finally (7) its payment within 14 days. The only difference in a clause 24(1) situation is that it begins with (1) the architect's failure to instruct, followed by (2) it becoming apparent that the regular progress of the works has been materially affected by it. Then follow the same five steps or stages.

    If the clauses mean what they appear to say, the loss or expense must come before the application, not just *some* loss or expense but *the* loss or expense to be included in the certificate. For clause 11(6) requires the application to be made 'within [a reasonable time] of the loss or expense having been incurred'. which must mean the loss in which the variation 'has involved the contractor' and it is 'the amount of such loss' which the architect has to ascertain and certify. The matter is not quite so clear in clause 24(1) because there is in that clause no intermediate reference to the loss having been incurred between the loss in which 'the contractor has been involved' by the progress of the works being affected and 'the amount of such loss' which the architect has to ascertain and certify. But both [parties] agree that the clauses must be read together and given the same interpretation, so the choice would seem to be between (1) reading 'the loss or expense' in clause 11(6) as 'some loss or expense' or adding 'or being incurred' after 'having been incurred' and (2) reading into clause 24(1) after 'the work has been affected as aforesaid' some such words as 'so as to involve the contractor in such loss or expense', or, 'so that the loss or expense has been incurred.'

I can find no justification for (1) which involves rectifying clause 11(6), not construing it. I prefer (2) which does no violence to the language of clause 24(1) and merely reads into it what is already included in the natural meaning of its express language. I therefore proceed to apply that construction of the two clauses to the three periods (a), (b) and (c) in the question stated in the case. Period (a) is the interval between (2) the incurring of the loss or expense and (3) the making of an application. Period (b) is the interval between (3) the application and (5) the ascertainment of the loss or expense. Period (c) is the interval between (5) ascertainment and (6) certification.

(a) I cannot see why interest lost or expended before the application, a loss or expense necessarily not more than 3 weeks old, is not direct loss or expense. The judge held that it was not direct within clause 11(6) because it was 'directly due to the fact that they (the Claimants) have chosen to wait before making application', and they were not direct within clause 24(1) because they are 'losses resulting from the fact that the effect of the interruption was not immediately ascertainable'. I find the reasoning difficult to follow and impossible to accept. Of the clause 11(6) loss I think the judge is saying that it would be direct were it not for the intervening (and avoidable?) delay of up to 21 days which makes it indirect. I regret that I cannot agree. It would not only be reasonable for the claimants to wait 21 days; the contract contemplates that they may do so. If the loss is direct when the 21 days start to run, it is, in my opinion, still direct throughout those 21 days. Of clause 24(1) he seems to say that no loss can result from the interruption of the progress of the works by a specified event unless the loss has been incurred at the time at which it becomes apparent that the works have been affected, so that interest charges between that time and the application result directly from the 21 days (or less) delay and not from the interruption. Again with respect I cannot agree. If the charges are incurred because of the interruption during the short period allowed by the contract for making the application, why are they not a direct result of the interruption?

[As to] (b) and (c) I agree with the judge's construction of clauses 11(6) and 24(1) and with his conclusion that the architect can only ascertain and certify the amount of interest charges lost or expended at the date of the application. It is these charges which are the subject of the application; it is these charges which he has power to investigate, ascertain and certify. I respectfully agree with the judge that the architect would be exceeding his powers were he to take into account further financial charges or other losses accruing during these two periods, however long, and such further charges and losses would be recoverable only, if at all, under a subsequent application or subsequent applications – although he might obtain the respondents' approval to waiving the required applications or extending the time for making them. But again I feel bound to differ from the judge's conclusion that they are therefore not direct losses or expenses and are therefore not recoverable at all. I appreciate that a series of successive applications may increase the burden of the architect's labours, and that burden may be further increased by delay on the part of contractors like the claimants in providing information which he reasonably requires either for forming his opinion and for ascertaining amounts under those clauses. But contractors have a duty under clause 30(1)(b) of the main contract (not altered by an additional duty imposed by an amendment to clause 30) to send him 'all documents necessary for the purposes of the computations required' by the conditions of the contract; and I should not have thought, in spite of the judge's contrary opinion, that the investigation and computation of interest charges at three-weekly intervals would make his burden so intolerable as to be outside the contemplation of the parties to the contract, particularly if the charges are likely to be recurring charges assessed and paid in the same amounts at the same intervals of time. No doubt the substitution for 'a reasonable time' of a period which is less than one month will not reduce the architect's labours in doing his duty to his employers and the contractors under these clauses during these two different periods of 21 days and one month. It is, I suspect, to meet the difficulties created by that amendment and by the need for successive applications that the 1980 edition of the Standard Contract has been redrafted (clause 26.1) to require applications to be made 'as soon as it has become, or should reasonably have become apparent to him (the contractor) that the regular progress of the works or any part thereof has been or *is likely* to be affected' by specified events (including instructions requiring variation) and to state 'that he has incurred or *is likely to incur* direct loss and/or expense'; and also to require the architect to ascertain 'the amount of such loss and/or expense which has been *or is being* incurred by the contractor'.

The architect under all forms of this contract has to investigate and compute values and expenditure from time to time and to adjust the contract price by adding certified amounts as a consequence of action taken or not taken by himself and/or his employer. It is only if the duties which these two clauses on their true construction put upon him are so unreasonable, if they cover investigating and ascertaining and certifying interest charges of this kind, as to have gone beyond the contemplation of the parties to this contract that a court would be driven to hold that they are no part of the claimants' direct loss or expense.

It seems to have been the opinion of the judge that they are so unreasonable. His conclusion was: '. . . the facts that the contract provides for direct losses and expenses and that the losses and expenses here sought to be recovered are losses and expenses which arise only because the machinery for ascertainment and certification provided for by the contract involve delay, in my judgment indicates that they are indirect. The position in the end is that the parties have provided a machinery which will enable payment of the primary loss to be recovered with reasonable speed. They have, notwithstanding that both must have been well aware that the machinery would involve some delay, not provided for interest or other payment in respect of the delay. They could easily have so provided and they have not done so notwithstanding that they have amended the standard form in a number of respects.' I say 'seems to have been' because what he said . . . was that charges or losses incurred during period (b) and made the subject of a further application 'would in my judgment be demonstrably not direct loss and expense resulting from the variation. They would be indirect losses so resulting assuming the period was reasonable and if the period was unreasonable the chain of causation would be completely broken. This might give rise to a claim against the architect . . .' – indeed the respondents – 'but the charges are not in my judgment recoverable under the contract. If they were, it would follow that no claim under clause 11(6) or 24(1) could ever be completely dealt with. For as each claim was investigated, ascertained, certified and paid, a further claim would necessarily arise.'

Again I regret to say that I cannot follow the learned judge or his reasoning, if he is, as I understand that passage, saying that a direct loss can become indirect by lapse of time in applying for it to be paid or in its becoming payable, or perhaps – compared with what he said of period (a) – that a loss cannot be direct if it is not applied for when a direct loss of the same kind was applied for. What this seems to me to overlook or underestimate is that the losses and expenses claimed here are one continuous loss and expense, incurred as it were by instalments but continuing to flow from the variations or specified events in the ordinary course of things without interruption from any delay by contractor or architect. And direct loss and expense can continue and recur for an indefinite period, as many a party to an action for damages for personal injuries knows to his cost. For a continuing and recurring loss of instalments of interest charges like those claimed the machinery provided by clauses 11(6) and 24(1) is very ill-fitting. It is the continuation of the loss and expense which prevents finality – not for ever, but until completion of the contract works as varied. But I do not consider the machinery fits these charges so ill that I am driven to conclude that they must be treated for the purpose of these clauses as indirect loss and expense when they appear to be direct.

Any interest allowed as part of 'direct loss and/or expense' should be assessed at a rate equivalent to the cost of borrowing, disregarding any special position of the contractor.

Tate & Lyle Food & Distribution Ltd v. Greater London Council

QUEEN'S BENCH DIVISION [1982] 1 WLR 149

For the facts and holding see page 220.

FORBES J: Despite the way in which Lord Herschell LC in *London, Chatham and Dover Railway Co.* v. *South Eastern Railway Co.* (1893) stated the principle governing the award of interest on damages, I do not think the modern law is that interest is awarded against the defendant as a punitive measure for having kept the plaintiff out of his money: I think the principle now recognised is that it is all part of the attempt to achieve *restitutio in integrum.* One looks, therefore, not at the profit which the defendant wrongfully made out of the money he withheld – this would indeed involve a scrutiny of the defendant's financial position – but at the cost to the plaintiff of being deprived of the money which he should have had. I feel satisfied that in commercial cases the interest is intended to reflect the rate at which the plaintiff would have had to borrow money to supply the place of that which was withheld. I am also satisfied that one should not look at any special position in which the plaintiff may have been; one should disregard, for instance, the fact that a particular plaintiff, because of his personal situation, could only borrow money at a very high rate or, on the other hand, was able to borrow at specially favourable rates. The correct thing to do is to take the rate at which plaintiffs in general could borrow money. This does not, however, to my mind, mean that you exclude entirely all attributes of the plaintiff other than that he is a plaintiff. There is evidence here that large public companies of the size and prestige of these plaintiffs could expect to borrow at 1 per cent over the minimum lending rate, while for smaller and less prestigious concerns the rate might be as high as 3 per cent over the minimum lending rate. I think it would always be right to look at the rate at which plaintiffs with the general attributes of the actual plaintiff in the case (though not, of course, with any special or peculiar attribute) could borrow money as a guide to the appropriate interest rate. If commercial rates are appropriate I would take 1 per cent over the minimum lending rate as the proper figure for interest in this case.

In wholly exceptional circumstances it may be possible to adopt a 'broad-brush' approach to the assessment of individual heads of claim where proof of each and every individual item is very difficult.

Penvidic Contracting Co. Ltd v. International Nickel Co. of Canada Ltd

SUPREME COURT OF CANADA   (1975), 53 DLR (3d) 748

Contractors agreed to lay ballast and track for a railroad. The employer was in breach of its obligation to facilitate the works. The contractors had agreed to do the work for a certain sum for each ton of ballast. They claimed as damages the difference between that sum and the larger sum which they would have demanded had they foreseen the problems caused by the employer's breaches of the contract. There was evidence that the figure claimed was a reasonable estimate.

Held: Damages should be assessed on the basis claimed. Where proof of the actual additional costs caused by the breach was difficult, it was permissible to assess damages broadly.

SPENCE J: Viscount Haldane LC in *British Westinghouse Electric & Manufacturing Co. Ltd* v. *Underground Electric Railways Co. of London Ltd* (1912) said:

'The quantum of damage is a question of fact, and the only guidance the law can give is to lay down general principles which afford at times but scanty assistance in dealing with particular cases. The judges who give guidance to juries in these cases have necessarily to look at their special character, and to mould, for the purposes of different kinds of claim, the expression of the general principles which apply to them, and this is apt to give rise to an appearance of ambiguity.

Subject to these observations I think that there are certain broad principles which are quite well settled. The first is that, as far as possible, he who has proved a breach of a bargain to supply what he contracted to get is to be placed, as far as money can do it, in as good a situation as if the contract had been performed.'

The difficulty in fixing an amount of damages was dealt with in the well-known English case of *Chaplin* v. *Hicks* (1911), which had been adopted in the Appellate Division of the Supreme Court of Ontario in *Wood* v. *Grand Valley Railway Co.* (1913), where Meredith CJO, said:

There are, no doubt, cases in which it is impossible to say that there is any loss assessable as damages resulting from the breach of a contract, but the Courts have gone a long way in holding that difficulty in ascertaining the amount of the loss is no reason for not giving substantial damages, and perhaps the furthest they have gone in that direction is in *Chaplin* v. *Hicks* (1911). In that case the plaintiff, owing, as was found by the jury, to a breach by the defendant of his contract, had lost the chance of being selected by him out of fifty young ladies as one of twelve to whom, if selected, he had promised to give engagements as actresses for a stated period and at stated wages, and the action was brought to recover damages for the breach of the contract, and the damages were assessed by the jury at £100. The defendant contended that the damages were too remote and that they were unassessable. The first contention was rejected by the Court as not arguable, and with regard to the second it was held that 'where it is clear that there has been actual loss resulting from the breach of contract, which it is difficult to estimate in money, it is for the jury to do their best to estimate; it is not necessary that there should be an absolute measure of damages in each case:' *per* Fletcher Moulton LJ.

When *Wood* v. *Grand Valley Railway Co.* reached the Supreme Court of Canada, judgment was given by Davies J and . . . the learned Justice said:

> It was clearly impossible under the fact of that case to estimate with anything approaching to mathematical accuracy the damages sustained by the plaintiffs, but it seems to me to be clearly laid down there by the learned Judges that such an impossibility cannot 'relieve the wrongdoer of the necessity of paying damages for his breach of contract' and that on the other hand the tribunal to estimate them whether jury or Judge must under such circumstances do 'the best it can' and its conclusion will not be set aside even if *'the amount of the verdict is a matter of guess work.'*

[Emphasis added.]

I can see no objection whatsoever to . . . using the method suggested by the plaintiff of assessing the damages in the form of additional compensation per ton rather than attempting to reach it by ascertaining items of expense from records which, by the very nature of the contract, had to be fragmentary and probably mere estimations.

Where a contractor's claim depends on 'an extremely complex interaction in the consequences of various denials, suspensions and variations, it may be difficult or even impossible to make an accurate apportionment of the total extra cost between the several causative events' it is permissible to assess the claim on a global basis. This applies to claims for both time and money.

J. Crosby & Sons Ltd v. Portland Urban District Council

QUEEN'S BENCH DIVISION   (1967), 5 BLR 121

The dispute arose out of a contract in ICE (4th edition) form, and one aspect of the case was whether the arbitrator was entitled to make a lump sum award for overall delay and disorganisation.

Held: This was permissible, provided there was no duplication and no profit element, where this was inadmissible under the relevant contract clauses.

DONALDSON J: The claimants . . . say that where you have a series of events which can be categorised as denial of possession of part of the site, suspension of work, and variations, the result is, or may be, that the contractor incurs the extra costs by way of overhead expenses and loss of productivity: these extra costs are all recoverable [under the contract]. . . . Since, however, the extent of the extra cost incurred depends upon an extremely complex interaction between the consequences of the various denials, suspensions and variations, it may well be difficult or even impossible to make an accurate apportionment of the total extra cost between the several causative events. An artificial apportionment could of course have been made; but why, they ask, should the arbitrator make such an apportionment which has no basis in reality?

I can see no answer to this question. Extra costs are a factor common to all these clauses, and so long as the arbitrator . . . ensures that there is no duplication, I can see no reason why he should not recognise the realities of the situation and make individual awards in respect of those parts of individual items of the claim which can be dealt with in isolation and a supplementary award in respect of the remainder of these claims as a composite whole.

Merely because an item is omitted from the bills or specification does not necessarily mean that the contractor has a claim for extras. The employer is not bound to pay for 'things that everybody must have understood are to be done but which happen to be omitted from the quantities': Channell J in Patman & Fotheringham Ltd v. *Pilditch (1904).*

Williams v. Fitzmaurice

COURT OF QUEEN'S BENCH   (1858), 3 H & N 844

The plaintiff contracted to build a house 'to be completed and dry and fit for Major Fitzmaurice's occupation by 1 August, 1858'. In the specification, prepared on the defendant's behalf, no mention was made of floorboards. It stated that 'the whole of the materials mentioned or otherwise in the foregoing particulars, necessary for the completion of the work, must be provided by the contractor'. The plaintiff signed the specification and agreed 'to do all the works of every kind mentioned and contained in the foregoing particulars, according, in every respect, to the drawings furnished or to be furnished, for the sum of £1,100'. The plaintiff brought floorboards on site, but refused to lay them without extra payment.

Held: The plaintiff was not entitled to recover for the flooring as an extra, because it was included in the contract, though not mentioned in the specification.

Default by a nominated sub-contractor or nominated supplier does not give rise to any claim by the contractor against the employer.

Kirk & Kirk v. Croydon Corporation

QUEEN'S BENCH DIVISION   [1956] JPL 585

The plaintiffs contracted to build a school for the defendants, the contract being in RIBA form (1939 edition, revised 1948). The contract included a clause providing for nominated suppliers. After making enquiries of a brick manufacturer as to delivery dates, the architect instructed the plaintiffs to place an order with the manufacturer. The manufacturer delayed in supplying the bricks, thereby causing delays and involving the plaintiffs in extra costs.

Held: The defendants were not liable to the plaintiffs.

Where the employer has himself undertaken to provide goods or carry out work or procure someone else to do so, a term may be implied into the contract that such goods will be supplied at the proper time and the contractor will have an action for damages for breach of that term: Thomas Bates & Son Ltd v. Thurrock Borough Council (1975) p. 146.

The employer is not liable to the contractor for acts of his architect which are not within the scope of the architect's authority.

Stockport Metropolitan Borough Council v. O'Reilly

QUEEN'S BENCH DIVISION   [1978] 1 Lloyd's Rep 595

O'Reilly contracted to build 105 houses, garages and ancillary works for the council. The contract was in JCT 63 form. Disputes between the parties were referred to arbitration. The main points at issue were (a) whether the contract had been properly determined or repudiated and (b) the extent to which the works had been varied. The arbitrator failed in his interim award to find the extent to which the works had been varied. He confused variations under the contract with orders given by the architect which were not authorised by the contract.

Held: The interim award of the arbitrator should be set aside.

JUDGE EDGAR FAY QC: An architect's *ultra vires* acts do not saddle the employer with liability. The architect is not the employer's agent in that respect. He has no authority to vary the contract. Confronted with such acts, the parties may either acquiesce in which case the contract may be *pro tanto* varied and the acts cannot be complained of or a party may protest and ignore them. But he cannot saddle the employer with responsibility for them.

The employer may be liable to the contractor for the architect's failure properly to operate the terms of the contract.

Rees & Kirby Ltd v. Swansea City Council

QUEEN'S BENCH DIVISION   (1985), 25 BLR 129

Although this case was reversed on appeal, (see Appendix p. 361), the following extract appears to be good law.

KILNER BROWN J: The more interesting point is whether the architect was agent for the defendants making them vicariously responsible. There are two authorities which are in point, namely *Burden* v. *Swansea Corporation* (1957) and *Sutcliffe* v. *Thackrah* (1974). As I understand the law, an architect is usually and for the most part a specialist exercising his special skills independently of his employer. If he is in breach of his professional duties he may be sued personally. There may, however, be instances where the exercise of his professional duties is sufficiently linked to the conduct and attitude of the employer that he becomes the agent of the employers so as to make them liable for his default. In the instant case the employers through the behaviour of the Council and the advice and intervention of the Town Clerk were to all intents and purposes dictating and controlling the architect's exercise of what should have been his purely professional duty. In my judgment this was the clearest possible instance of responsibility for the breach attaching to the employers.

Note: See also *Croudace Ltd* v. *London Borough of Lambeth* (1985) 6 ConLR 70, Appendix, p. 358.

Chapter 13

Liquidated Damages and Extensions of Time

(i) LIQUIDATED DAMAGES

Liquidated damages are a monetary amount fixed and agreed by the parties in advance, as the damages payable in the event of a specified breach of contract. In building contracts, liquidated damages are generally payable only for the contractor's failure to complete on time. A provision for liquidated damages is enforceable if the amount fixed is a genuine pre-estimate of the loss likely to be caused by the breach. In contrast, a 'penalty clause' is invalid. If the agreed sum is extravagant in relation to the greatest possible loss it will be held to be a penalty. Liquidated damages are recoverable without proof of loss.

Dunlop Pneumatic Tyre Co. Ltd v. New Garage Motor Co. Ltd

HOUSE OF LORDS   [1915] AC 79

D contracted to sell tyres and other accessories to New Garage on terms designed to ensure that the tyres, etc. were not sold below the manufacturer's list price. New Garage agreed to pay D 'the sum of £5 for each and every tyre . . . sold or offered in breach of this agreement, as and by way of liquidated damages and not as a penalty'. New Garage sold tyres in breach of the agreement.

Held: In the circumstances of the case, the sum of £5 was liquidated damages and not a penalty.

LORD DUNEDIN: I shall content myself with stating succinctly the various propositions which I think are deducible from the decisions which rank as authoritative: (i) Though the parties to a contract who use the words penalty or liquidated damages may prima facie be supposed to mean what they say, yet the expression used is not conclusive. The court must find out whether the payment stipulated is in truth a penalty or liquidated damages. This doctrine may be said to be found passim in nearly every case. (ii) The essence of a penalty is a payment of money stipulated as in terrorem of the offending party; the essence of liquidated damages is a genuine covenanted pre-estimate of damage: *Clydebank Engineering Company* v. *Yzquierdo y Castaneda* (1905). (iii) The question whether a sum stipulated is penalty or liquidated damages is a question of construction to be decided upon the terms and inherent circumstances of each particular contract, judged of as at the time of the making of the contract, not as at the time of the breach: *Public Works Comr* v. *Hills* (1906) and *Webster* v. *Bosanquet* (1912). (iv) To assist this task of construction various tests have been suggested, which, if applicable to the case under consideration, may prove helpful or even conclusive. Such are: (a) It will be held to be a penalty if the sum stipulated for is extravagant and unconscionable in amount in comparison with the greatest loss which could conceivably be proved to have followed from the breach – illustration given by Lord Halsbury LC, in the *Clydebank Case* (1905). (b) It will be held to be a penalty if the breach consists only in not paying a sum of money, and the sum stipulated is a sum greater than the sum which ought to have been paid: *Kemble* v. *Farren* (1829). This, though one of the most ancient instances, is truly a corollary to the last test. Whether it had its historical origin in the doctrine of the common law that, when A. promised to pay B. a sum of money on a certain day and did not do so, B. could only recover the sum with, in certain cases, interest, but could never recover further damages for non-timeous payment, or whether it was a survival of the time when equity reformed unconscionable bargains merely because they were unconscionable – a subject which much exercised Jessel MR, in *Wallis* v. *Smith* (1882) – is probably more interesting than material. (c) There is a presumption (but no more) that it is a penalty when

'a single lump sum is made payable by way of compensation, on the occurrence of one or more or all of several events, some of which may occasion serious and others but trifling damages':

per Lord Watson in *Lord Elphinstone* v. *Monkland Iron and Coal Co.* (1886). On the other hand: (d) It is no obstacle to the sum stipulated being a genuine pre-estimate of damage that the consequences of the breach are such as to make precise pre-estimation almost an impossibility. On the contrary, that is just the situation when it is probable that pre-estimated damage was the true bargain between the parties: *Clydebank Case* (1905) *per* Lord Halsbury; *Webster* v. *Bosanquet* (1912) *per* Lord Mersey.

If there is a liquidated damages clause, it does not matter whether the actual loss is greater or less. The liquidated damages are recoverable whether or not the employer can prove that he has in fact suffered loss, or even if, in the event, there is no loss at all.

Clydebank Engineering & Shipbuilding Co. v. Castaneda and Others

HOUSE OF LORDS   [1905] AC 6

The Spanish government ordered four torpedo boats from Clydebank. The contract provided a 'penalty' for late delivery at the rate of £500 per week. The boats were delivered late, and the Spanish government claimed £67,500 at £500 for each week of late delivery. One of the arguments put forward by Clydebank was that after the delivery date the greater part of the Spanish fleet had been sunk by the American fleet. In effect, they argued that they had done the Spanish government a favour by being late in delivery.

Held: The sum of £500 a week was recoverable as liquidated damages. Clydebank's contentions were rejected.

LORD HALSBURY LC: Then comes the question whether, under the agreement, the damages are recoverable as an agreed sum, or whether it is simply a penalty to be held over in terrorem, or whether it is a penalty so extravagant that no court ought to enforce it. It is impossible to lay down any abstract rule as to what might or might not be extravagant without reference to the principal facts and circumstances of the particular case. A great deal must depend on the nature of the transaction. On the other hand, it is an established principle in both countries to agree that the damages should be so much in the event of breach of agreement. The very reason why the parties agreed to such a stipulation was that, sometimes, the nature of the damage was such that proof would be extremely difficult, complex and expensive. If I wanted an example of what might be done in this way, I need only refer to the argument of counsel as to the measure of damage sustained by Spain through the withholding of these vessels. Suppose there had been no agreement in the contract as to damages, and the Spanish government had to prove damages in the ordinary way, imagine the kind of cross-examination of every person connected with the Spanish administration. It is very obvious that what was intended by inserting these damages in the contract was to avoid a minute, difficult and complex system of examination which would be necessary if they had attempted to prove damage in the ordinary way.

It was also suggested that there could be no measure of damage in the case of a warship which had no commercial value at all. It is a strange and somewhat bold assertion to say that, in the case of a commercial ship, the damages could be easily ascertained, but that the same principle could not be applied to a warship as it earned nothing. The deprivation of a nation of its warships might mean very serious damage, although it might not be very easy to ascertain the amount. But is that a reason for saying they were to have no damages at all? It seems to me hopeless to advance such a contention. It is only necessary to state the assertion to show how absurd it is.

But there was a more startling suggestion still. It was argued for the appellants that, if they had acted up to their contract, and these vessels had been sent out at the specified date, they would have shared the same fate as the other Spanish warships at the hands of the American Navy. Therefore, the respondents, instead of suffering damage, had, by the action of the appellants, been saved the loss of these ships, which otherwise would have been at the bottom of the Atlantic. After considerable experience, I do not think that I have ever heard such an argument before, nor do I think that I am likely to hear it again. Nothing could be more absurd, and to give effect to it would be a striking example of defective jurisprudence. If your Lordships look at the nature of the transaction, it is hopeless to contend that the penalty was intended merely to be in terrorem. Both parties recognised that the question was one in which time was the main element of the contract. I have come to the conclusion that the judgment of the court below was perfectly right. There is no ground for the contention that the sum in the contract was not the damages agreed on between the parties for very good and excellent reasons at the time at which the contract was entered into.

There must be a definite date from which liquidated damages can run.

Miller v. London County Council

KING'S BENCH DIVISION (1934), 151 LT 425

The plaintiff contracted with the defendants for alteration works to Deptford Pumping Station. The contract provided that the whole of the work should be completed within 7 months of the engineer's order to commence, which was given on 16 April 1931. The work was not completed until 25 July 1932. There was a liquidated damages clause and also an extension of time clause which provided that 'it shall be lawful for the engineer, if he think fit, to grant from time to time, and at any time or times, by writing under his hand such extension of time for completion of the work and that either prospectively or retrospectively, and to assign such other time or times for completion as to him may seem reasonable'. On 17 November 1932 after completion of the work, the engineer issued a certificate extending time to 7 February 1932. He subsequently certified the sum of £2,625 as payable by the plaintiff to the defendant for the delay from 7 February to 25 July 1932.

Held: The words 'either prospectively or retrospectively' did not give the engineer power to fix a new date for completion after the completion of the works. They only empowered him, where a delay occurred during the progress of the work, to wait until the cause of delay had ceased to operate and then, within a reasonable time thereafter 'retrospectively' fix a new completion date. Since the power to extend time had not been properly exercised, the defendants lost the benefit of the clause. There was thus no date from which the liquidated damages could run and none could be recovered.

DU PARCQ J: The question of construction with which I am faced is not a simple one. 'Prospectively' means, 'as one looking forward,' 'retrospectively' means, 'as one looking backward.' It is plain that the engineer is entitled to defer the grant of an extension of time to a stage when he is looking backward at something. The question is – at what? Counsel for the owners says 'At the completion of the work, or that portion of it which is in question.' Counsel for the builder says that it is wrong, or at least unnecessary, to give that meaning to the word, and suggests as a possible and more reasonable construction, that the intention of the clause is to empower the engineer, instead of granting an extension of time when he orders extra work, or when, to take another case, a strike begins, to wait if he chooses until the cause of the delay has ceased to be operative, and then, looking back upon the acts or events which have caused the delay, to assign a new time for completion. He contends, however, that clause 31 does nothing to deprive the builder of the right to have the date to which he must work fixed in advance, and that it was not the intention of the parties that the builder should work in the dark and be told some months after he had completed his task what time the engineer allowed him within which to finish it.

I have come to the conclusion that the words of clause 31 do not give to the engineer the power for which the owners contend. The words 'to assign such other time or times for completion as to him may seem reasonable' are not, in my opinion, apt to refer to the fixing of a new date for completion ex post facto. I should rather paraphrase them in such some words as 'to fix a new date by which the builder ought to complete the work.' If I am right, this phrase, coming after the word 'retrospectively', throws some light upon its meaning. Next, it is important to observe that without words in the contract to make the matter clear it might be a matter of dispute whether the engineer was or was not bound to grant each extension of time at the time of the delay. The following passage occurs in *Hudson on Building Contracts* (6th Edition) at page 360:

'The extension of time need not necessarily, it would seem, be granted before the contract date for completion. If, for example, a strike were to last over the contract date, the extent of the delay might not be known, nor the necessary extension of time be granted, until after the contract time had expired. But each 'extension of time should, it would seem (depending always, however, upon the terms of the particular contract), be granted at the time of the delay, or, where the cause of delay is continuing and its extent uncertain, within a reasonable time after the delay has come to an end.'

My attention was called to a case decided in 1900 in the Court of Appeal of New Zealand – *Anderson* v. *Tuapeka County Council* (1900). In that case the court was considering a building contract which, after stipulating for the payment of liquidated damages if the works were not completed on a day fixed, went on to provide that in the event of any alterations, deviations, additions or extra work being required, the engineer should allow such an extension of time as he should think adequate in consequence thereof, and that any sum to become payable by way of damages for non-completion should be computed from the expiration of such extended time. The court held that an alleged extension was ineffective because it was not allowed when the extra works occasioning it were ordered. Sir Robert Stout CJ, said that the extended time should have been fixed when giving the orders for the extras, and the other judges took the same view. In my judgment the word 'retrospectively' may well have been intended to make it clear that an extension granted (in the words of Hudson) 'within a reasonable time after the delay has come to an end' is a valid extension. It may also be read as empowering the engineer to grant an extension after the contract date for completion has gone by, but I do not read it as meaning that the engineer may fix a new date for completion and grant extensions of time at some date subsequent not only to the contract date, but also to the substituted date. In my opinion, clause 34, which clearly contemplates that the engineer will be in a position to give a certificate of completion as soon as or very shortly after the work is complete, supports the view that it was the intention of the parties that all extensions of time should be granted before the substituted date for completion arrived. The language of clause 37 is more consistent with the view which I have adopted than with that put forward on behalf of the owners. I should perhaps add here that I also agree with counsel for the builder's contention that clause 11 of the contract does not apply to delays caused by the ordering of extra work or by lack of expedition on the part of other contractors. I deal briefly with this point, as I do not understand from his argument that counsel for the owners places any reliance upon clause 11, although it is pleaded in a rejoinder.

If I am wrong in thinking that the clause cannot properly be interpreted in the sense contended for by the owners, it is, in my judgment, at least ambiguous, and reasonably capable of the construction which the builder asks me to put upon it. I agree with counsel for the builder's contention that upon this view it must be construed according to that one of two possible meanings which is more favourable to the builder, on the ground that it was the owners who prepared and put forward the contract. It follows from the view which I have expressed that the power to extend the time was not in this case exercised within the time limited by the contract, and that the owners are not in a position to claim liquidated damages.

Where the employer is wholly or partly responsible for the contractor's failure to complete on time, he cannot recover liquidated damages, unless the contract provides otherwise.

Peak Construction (Liverpool) Ltd v. McKinney Foundations Ltd

COURT OF APPEAL   (1970), 1 BLR 111

Peak contracted with Liverpool Corporation to construct multi-storey flats. The contract was in Liverpool Corporation's own form, described by Salmon LJ as containing 'the most one-sided, obscurely and ineptly-drafted clauses in the United Kingdom', with a completion period of 24 months. The corporation's Director of Housing was appointed as architect-in-charge. Clause 22 provided that 'time shall be considered as of the essence of the contract' and provided for liquidated damages at the rate of '25s per dwelling for each and every week or part of a week' of delay. Clause 23 empowered the architect to grant extensions of time for, *inter alia*, extras and additions, force majeure and 'other unavoidable circumstances'. The defendants were nominated sub-contractors for the piling. They completed their work and left site. Subsequently, the piles were found to be seriously defective.

At a site meeting, it was agreed that work should cease pending investigation and at a later meeting it was further agreed that the problem should be submitted to one of three named consulting engineers. Everyone, including the architect, agreed to accept his opinion. A month later, the corporation's housing committee resiled on this decision and merely agreed to consider the consultant's report 'without prejudice'. The corporation did not appoint the engineer for several months and a further period elapsed whilst he completed his report. There was a further delay before the corporation authorised the report and the defendants were not able to recommence until 58 weeks after work had ceased.

Held: (1) Since part of the total delay was attributable to the corporation's fault, there should be a new trial to assess how much of that period was attributable to the defendants' breach.

(2) So far as the employers were concerned, they were not entitled to recover any liquidated damages against the plaintiffs under the terms of the main contract. The corporation could not recover liquidated damages for delay for which it was partly to blame.

SALMON LJ: A clause giving the employer liquidated damages at so much a week or month which elapses between the date fixed for completion and the actual date of completion is usually coupled, as in the present case, with an extension of time clause. The liquidated damages clause contemplates a failure to complete on time due to the fault of the contractor. It is inserted by the employer for his own protection; for it enables him to recover a fixed sum as compensation for delay instead of facing the difficulty and expense of proving the actual damage which the delay may have caused him. If the failure to complete on time is due to the fault of both the employer and the contractor, in my view, the clause does not bite. I cannot see how, in the ordinary course, the employer can insist on compliance with a condition if it is partly his own fault that it cannot be fulfilled: *Wells* v. *Army & Navy Co-operative Society Ltd* (1902); *Amalgamated Building Contractors* v. *Waltham Urban District Council* (1952) and *Holme* v. *Guppy* (1838). I consider that unless the contract expresses a contrary intention, the employer, in the circumstances postulated, is left to his ordinary remedy; that is to say, to recover such damages as he can prove flow from the contractors' breach. No doubt if the extension of time clause provided for a postponement of the completion date on account of delay caused by some breach or fault on the part of the employer, the position would be different. This would mean that the parties had intended that the employer could recover liquidated damages notwithstanding that he was partly to blame for the failure to achieve the completion date. In such a case the architect would extend the date for completion, and the contractor would then be liable to pay liquidated damages for delay as from the extended completion date.

The liquidated damages and extension of time clauses in printed forms of contract must be construed strictly *contra proferentem*. If the employer wishes to recover liquidated damages for failure by the contractors to complete on time in spite of the fact that some of the delay is due to the employer's own fault or breach of contract, then the extension of time clause should provide, expressly or by necessary inference, for an extension on account of such a fault or breach on the part of the employer. I am unable to spell any such provision out of . . . the contract [clause] in the present case. In any event, it is clear that, even if [the] clause had provided for an extension of time on account of the delay caused by the contractor, the failure in this case of the architect to extend the time would be fatal to the claim for liquidated damages. There had clearly been some delay on the part of the corporation. Accordingly, as the architect has not made 'by writing under his hand such an extension of time', there is no date under the contract from which the defendants' liability to pay liquidated damages for delay could be measured. And therefore none can be recovered: see *Miller* v. *London County Council* (1934).

EDMUND DAVIES LJ: Proceeding upon the basis that the employers contributed to the delay, it seems clear that the architect could not, pursuant to the extension of time clause, have on this account extended the time for completion, for the contractor would not have been 'unduly delayed . . . by reason of any enlargements or other additions to the works, or in consequence of any local combination of workmen or general strikes or lock-outs force majeure or other unavoidable circumstances . . .'. Delay due to the employer cannot be said to have been an unavoidable circumstance to anyone save the contractor. The stipulated time for completion having ceased to be applicable by reason of the employer's own default and the extension clause having no application to that, it seems to follow that there is in such a case no date from which liquidated damages could run and the right to recover them has gone. Accordingly, no liquidated damages were recoverable. . . .

PHILLIMORE LJ: A clause providing for liquidated damages is closely linked with a clause which provides for an extension of time. The reason for that is that when the parties agree that if there is delay the contractor is to be liable, they envisage that that delay shall be the fault of the contractor and, of course, the agreement is designed to save the employer from having to prove the actual damage which he has suffered. It follows, once the clause is understood in that way, that if part of the delay is due to the fault of the employer, then the clause becomes unworkable if only because there is no fixed date from which to calculate that for which the contractor is responsible and for which he must pay liquidated damages. However, the problem can be cured if allowance can be made for that part of the delay caused by the actions of the employer; and it is for this purpose that recourse is had to the clause dealing with extension of time. If there is a clause which provides for extension of the contractor's time in the circumstances which happen, and if the appropriate extension is certified by the architect, then the delay due to the fault of the contractor is disentangled from that due to the fault of the employer and a date is fixed from which the liquidated damages can be calculated.

An employer is not entitled to liquidated damages if by his acts or omissions he has prevented the contractor from completing the works by the completion date. This general rule may be amended by the express terms of the contract and is so amended by JCT terms.

Percy Bilton Ltd v. Greater London Council

HOUSE OF LORDS   (1982), 20 BLR 1

B contracted with the GLC to erect a housing estate. The contract was dated 25 October 1976, and was substantially in JCT 63 form. The original completion date was 24 January 1979. The nominated sub-contractor for mechanical services went into liquidation on 28 July 1978, being already 40 weeks behind programme, and B determined their employment on 31 July 1978. B had asked for a new sub-contractor to be nominated on 7 July 1978, but the sub-contractor nominated as a replacement withdrew before starting work. Ultimately a third firm was nominated on 31 October 1978, the sub-contract being eventually agreed on 22 December 1978. The ultimate sub-contractor's programme was such that the work would not be completed until 25 January 1980.

Various extensions of time were granted, and eventually the extended completion date became 1 February 1980, part of the extension of time being awarded under clause 23(f). B had not completed by that date and thereafter GLC deducted liquidated damages from money certified due to B.

B contended that the architect's clause 22 certificate of delay was invalid and argued that GLC were not entitled to deduct liquidated damages, time having become 'at large' because of GLC's failure to re-nominate so as not to delay them.

Held: Withdrawal of a nominated sub-contractor is not a fault or breach of contract on the part of the employer, nor is it covered by clause 23 of JCT 63. GLC's delay in making a timely renomination fell within JCT 63, clause 23(f) (now JCT 80, clause 25.4.6.), and B was entitled to an extension of time to cover that delay. The architect's clause 22 certificate was valid and GLC was entitled to deduct liquidated damages.

LORD FRASER OF TULLYBELTON: The true position is . . . correctly stated in the following propositions:

1. The general rule is that the main contractor is bound to complete the work by the date for completion stated in the contract. If he fails to do so, he will be liable for liquidated damages to the employer.
2. That is subject to the exception that the employer is not entitled to liquidated damages if by his acts or omissions he has prevented the main contractor from completing his work by completion date – see for example, *Holme* v. *Guppy* (1838), and *Wells* v. *Army & Navy Co-operative Society* (1902).
3. These general rules may be amended by the express terms of the contract.
4. In this case, the express terms of clause 23 of the contract do affect the general rule. For example, where completion is delayed '(a) by *force majeure*, or (b) by reason of any exceptionally inclement weather,' the architect is bound to make a fair and reasonable extension of time for completion of the work. Without that express provision, the main contractor would be left to take the risk of delay caused by *force majeure* or exceptionally inclement weather under the general rule.
5. Withdrawal of a nominated sub-contractor is not caused by the

fault of the employer, nor is it covered by any of the express provisions of clause 23. Paragraph (g) of clause 23 expressly applies to 'delay' on the part of a nominated sub-contractor but such 'delay' does not include complete withdrawal. . . .

6. Accordingly, withdrawal falls under the general rule, and the main contractor takes the risk of any delay directly caused thereby.

7. Delay by the employer in making the timeous nomination of a new sub-contractor is within the express terms of clause 23(f) and the main contractor, the appellant, was entitled to an extension of time to cover that delay. Such an extension has been given.

Where the reason for the contractor's failure to complete on time is wholly or partly the fault of the employer, time may become 'at large' and the contractor's obligation is then to complete 'within a reasonable time'.

Wells v. Army & Navy Co-operative Society Ltd

KING'S BENCH DIVISION   (1902), 86 LT 764

A building contract provided that certain matters causing delay and 'other causes beyond the contractor's control' were to be submitted to the decision of the directors of the employer. The directors were 'to adjudicate thereon and make due allowance therefor if necessary, and their decision shall be final'. There was a liquidated damages clause. There was a year's delay in completion, and the directors allowed a three-month extension for delays caused by sub-contractors. Other breaches were established, including failure by the employer to give possession of the site and failing to provide plans and drawings in due time.

Held: The words 'other causes beyond the contractor's control' did not extend to delay caused by the interference of the employers or their architect. Liquidated damages could not therefore be deducted and the contractor's obligation was to complete within a reasonable time.

Where the contract stipulates that work is to be done within a stated period, if by his conduct the employer renders it impossible or impracticable for the contractor to do his work within the stipulated time, then the employer can no longer insist upon strict adherence to the time for completion and he cannot enforce any liquidated damages.

Dodd v. Churton

COURT OF APPEAL   [1897] 1 QB 562

A contract provided for the whole of the works to be completed by 1 June 1892, with liquidated damages of £2 per week for every week that any part of the work remained unfinished after that date. There was a provision that any authority given by the architect for any alteration or addition in or to the works was not to vitiate the contract. There was no provision for extending the time for completion if additional work was ordered. Additional works were ordered involving delay in the completion of the works beyond the specified date. The works were not completed until 5 December 1892. A fortnight was a reasonable time for the doing of the additional work, and the employer, allowing a fortnight's additional time for the completion of the works, claimed £2 per week in respect of the delay of 25 weeks.

Held: By giving the order for the additional works the employer had waived the stipulation for liquidated damages in respect of non-completion of the work by 1 June.

LORD ESHER MR: Where one party to a contract is prevented from performing it by the act of the other, he is not liable in law for that default; and, accordingly, a well recognised rule has been established in cases of this kind, beginning with *Holme* v. *Guppy* (1838), to the effect that, if the building owner has ordered extra work beyond that specified by the original contract which has necessarily increased the time requisite for finishing the work, he is thereby disentitled to claim the penalties for non-completion provided for by the contract.

The architect's certificate of delay under clause 24.1. of JCT 80, is a condition precedent to the contractor's liability to pay liquidated damages. Once such certificate has been issued, the employer may deduct liquidated damages from amounts due to the contractor under interim certificates.

Algrey Contractors Ltd v. Tenth Moat Housing Society Ltd

QUEEN'S BENCH DIVISION   (1972), 1 BLR 45

A were contractors to the defendants under a contract in JCT 63 form, clause 22 of which was in similar terms to JCT 80, clause 24. The architect issued a certificate of delay under clause 22, and the defendants deducted £5,600 in respect of liquidated damages from the architect's next interim certificate. A sued to recover this amount.

Held: The defendants were entitled to make the deduction.

> FORBES J: What I have to decide is whether it is right to treat a clause such as clause 22, which permits deductions by a building owner of liquidated and ascertained damages, as one which entitles him to deduct such sums; or whether the view put forward by the Court of Appeal (in *Dawnay Ltd* v. *F.G. Minter Ltd* (1971), p. 136) means that nothing except an admitted amount can be deducted from a certificate.
>
> One has to construe the contract as a whole. If it says that the building owner is entitled to deduct sums from any monies due or to become due, it seems to me that it means what it says. As long as they are liquidated and ascertained, so that they can properly be deducted, and as long as they are also the subject of the architect's certificate, they should properly be deducted from sums due under the interim certificate.

(ii) EXTENSIONS OF TIME

Unless there is an extension of time clause, the contractor is under a strict obligation to complete on time, unless he is prevented from doing so by acts or breaches of the employer or by operation of the law.

If the contractor is so prevented from completing on time, any liquidated damages clause ceases to apply, even to subsequent delays which are the contractor's fault.

The contract time may be extended by the employer because of his own delay only if there is an express provision for extension of time and the terms of the clause must be strictly complied with.

Under JCT terms, in considering whether or not to grant an extension of time, it is the effect on completion which is critical. If the contractor is ahead of programme, the architect is entitled to take that fact into account.

London Borough of Hounslow v. Twickenham Garden Developments Ltd

CHANCERY DIVISION   (1970), 7 BLR 81

The facts of this case are set out on page 304.

MEGARRY J: Provided the contractor has given written notice of the cause of the delay, the obligation to make an extension appears to rest on the architect without the necessity of any formal request for it by the contractor. Yet he is required to do this only if in his opinion the completion of the works 'is likely to be or has been delayed beyond the date for completion', or any extended time for completion previously fixed. If a contractor is well ahead with his works and is then delayed by a strike, the architect may nevertheless reach the conclusion that completion of the works is not likely to be delayed beyond the date for completion. . . . The contractor is under a double obligation: on being given possession of the site, he must 'thereupon begin the works and regularly and diligently proceed with the same', and he must also complete the works 'on or before the date for completion', subject to any extension of time. If a strike occurs when two-thirds of the work has been completed in half the contract time, I do not think that on resuming work a few weeks later the contractor is then entitled to slow down the work so as to last out the time until the date for completion (or beyond, if an extension of time is granted) if thereby he is failing to proceed with the work 'regularly and diligently'.

Under JCT terms the architect may grant an extension of time even after completion of the works. JCT 80, Clause 25.3.3., makes express provision for the architect to review the position as regards extensions within 12 weeks of practical completion.

Amalgamated Building Contractors Ltd v. Waltham Holy Cross Urban District Council

COURT OF APPEAL   [1952] 2 All ER 452

Clause 18 of the then current RIBA Contract provided:

> 'If in the opinion of the architect the works be delayed (i) by *force majeure*, or (ii) by reason of exceptionally inclement weather . . . or (ix) by reason of labour and material not being available as required . . . then in any such case the architect shall make a fair and reasonable extension of time for completion of the works . . .'.

The date for completion was 7 February 1949, and the contractors were liable to liquidated damages in the event of delay at the rate of £50 a week. On 19 January 1949, the contractors sought a 12 months' extension of time because of labour and materials difficulties, repeating their request on 27 January. The architect merely gave a formal acknowledgment. The work was completed on 28 August 1950 and it was not until 20 December 1950 that the architect wrote extending the time for completion to 23 May 1949.

Held: The retrospective extension of time was valid.

DENNING LJ: The contractors say that the words in clause 18

'the architect shall make a fair and reasonable extension of time for completion of the works'

mean that the architect must give the contractors a date at which they can aim in the future, and that he cannot give a date which has passed. I do not agree with this contention. It is only necessary to take a few practical illustrations to see that the architect, as a matter of business, must be able to give an extension even though it is retrospective. Take a simple case where the contractors, near the end of the work, have overrun the contract time for six months without legitimate excuse. They cannot get an extension for that period. Now suppose that the works are still uncompleted and a strike occurs and lasts a month. The contractors can get an extension of time for that month. The architect can clearly issue a certificate which will operate retrospectively. He extends the time by one month from the original completion date, and the extended time will obviously be a date which is already past. Or take a cause of delay, such as we have in this case, due to labour and materials not being available. That may cause a continuous delay operating partially, but not wholly, every day, until the works are completed. The works do not stop. They go on, but they go on more slowly right to the end of the works. In such a case, seeing that the cause of delay operates until the last moment, when the works are completed, it must follow that the architect can give a certificate after they are completed. These practical illustrations show that the parties must have intended that the architect should be able to give a certificate which is retrospective, even after the works are completed. . . . *Miller* v. *London County Council* (1934) is distinguishable. I regard that case as turning on the very special wording of the clause which enabled the engineer 'to assign such other time or times for completion as to him may seem reasonable'. Those words, as du Parcq J said, were not apt to refer to the fixing of a new date for completion ex post facto. I would also observe that on principle there is a distinction between cases where the cause of delay is due to some act or default of the building owner, such as not giving possession of the site in due time, or ordering extras, or something of that kind. When such things happen the contract time may well cease to bind the contractors, because the building owner cannot insist on a condition if it is his own fault that the condition has not been fulfilled.

Failure by the employer properly to grant an extension of time may have the effect of setting time at large. Whether the completion date is set at large by delay in granting an extension of time depends on all the circumstances.

Fernbrook Trading Co. Ltd v. Taggart

SUPREME COURT OF NEW ZEALAND   [1979] 1 NZLR 556

T contracted to carry out road construction works for F. The contract period was 15 weeks, to commence 14 days after the date of the engineer's order to commence, and liquidated damages of $500 a week were payable by T for delay. The extension of time clause provided for time to be extended for, inter alia, 'delays caused by other contractors on the site, delays in installation of services or exceptional circumstances'.

The completion date was 30 December 1975. T applied for an extension of time on 18 November because, as well as being required to carry out additional work, there were delays by other contractors on the site. He received no acknowledgment until 2 March 1976 when the engineer granted an extension to 24 March. On 29 April the engineer granted a further extension to 1 June. The delays by other contractors caused a summer job to run into winter and exceptionally bad weather caused further delays; as did a drivers' strike. In addition F was in breach of contract in failing to make progress payments for a total of 23 weeks, causing T to reduce labour on the site because of financial difficulties. There was also a failure by the engineer to issue progress payment certificates for five separate months. The work was completed 46 weeks after the final extended completion date.

Held: The exercise of the engineer's discretion to extend time after the completion date was reasonable in the circumstances. Accordingly, the extensions were validly given, and the completion date was not set at large. However, the engineer had no jurisdiction to extend time on account of F's breach of contract in failing to make progress payments. This breach set the completion date at large with the effect that T was not liable to pay liquidated damages.

Note: This judgment is quoted at length as it represents an admirable summary of the law.

ROPER J: [Taggart's counsel] submitted that to be effective an extension of time must be granted within the original completion period, or within a validly granted extension of that period; or in other words, that a retrospective extension was invalid and set the completion time at large. As the author of Hudson's *Building and Engineering Contracts* (10th edition, 1970) says at page 643, 'Regrettably, the law on this subject [i.e. the time for exercise of extensions] appears to have become unnecessarily complicated'. The differing circumstances and contractual provisions under review make for difficulty in reconciling the authorities.

The decision of our Court of Appeal in *Anderson* v. *Tuapeka County Council* (1900) is the logical starting point. In that case the contract provided for extensions of time if extra works were ordered. Extras were ordered on two occasions, one being after the contract completion date, and on neither occasion was reference made to an extension of time beyond the contract date. In June, some months after the date originally fixed for completion, the engineer purported to deduct penalties indicating that the time for completion had been extended from 15 November to a date in April. Stout CJ said:

'This is not a case in which the engineer is appointed an arbitrator, as in many building cases, to decide on amounts due at the end of contract. Were it such a case, he might, at the completion of the contract, have had power to determine the amount due by either party to the other. Here he has a function to perform – namely, to grant such an extension of time as he shall think adequate if 'alterations, deviations, additions, or extra work' are ordered. It must, I think, be admitted that until 20 June, when the time was extended – if the words I have quoted can be deemed an extension – the penalties were at large. No time for finishing the contract was then in existence. The 15 November was no longer the date, and no date had been substituted for the 15 November. I do not think that once the penalties were at large they could be reimposed by the engineer. He was, in fact, creating on the 20 June a new contract not in existence, and antedating it back to the first week in May. I can find no authority for his so doing. The words "as he shall think adequate" seem to me to imply that the moment to fix the extended time is when the extra works have not been done but have to be done, and that therefore it was before the contractors began to execute the extra works that the time had to be fixed'.

Williams J said:

'If no date is specified within which the works are to be completed, how is it possible for the contractor to complete the works by a specified date? or how can he have broken a contract to complete on a specified date if he did not know beforehand what the date was on which he was under an obligation to complete? A proviso which was intended to preserve to the contractee the right to recover penalties in an event which, had it not been for the proviso, would have deprived him of that right, should be expressed in clear and unambiguous terms. If it had been intended to allow the engineer to decide ex post facto whether there was a breach of contract to complete, the contractor never having been made aware of when he was required to complete, it should have been very plainly stated. The language of the proviso seems to be altogether unadapted to carry out such an intention'.

In *Anderson* the contract provided for completion on 15 November:

'Provided nevertheless that in the event of any alterations deviations additions or extra work being required the Engineer shall allow such an extension of time as he shall think adequate in consequence thereof and any sum to become payable by way of damages for non-completion as aforesaid shall be computed from the expiration of such extended time instead of from the day hereinbefore mentioned for the completion of the works'.

Anderson's case was referred to by du Parcq J in *Miller* v. *London County Council* (1934). There the engineer granted an extension of time in November 1932, the works having been completed in July of that year, the completion date thereby being extended from 15 November 1931 to 7 February 1932. The proviso relating to extensions of time in that particular contract read:

'Provided always that if by reason of additional work or enlargements of the work or for any other just cause arising with the council or the engineer or in consequence of any strikes or combination of workmen or for want or deficiency of any orders, drawings or directions, or by reason of any difficulties, impediments, obstructions, oppositions, doubts, disputes or differences whatsoever, and howsoever occasioned the contractor shall in the opinion of the engineer have been unduly delayed or impeded in the completion of the work or any part of it, it shall be lawful for the engineer, if he shall think fit, to grant from time to time, and at any time or times by writing under his hand such extension of time for completion of the work, and that either prospectively or retrospectively, and to assign such other time or times for completion as to him may seem reasonable without thereby prejudicing or in any manner affecting the validity of the contract'.

Du Parcq J held that the words 'either prospectively or retrospectively' did not confer on the engineer a right to fix the extension of time ex post facto when the work was completed, but merely empowered him to wait till the cause of delay had ceased to operate, and then 'retrospectively' with regard to the cause of delay to assign to the contractor a new date to work to. In the result he concluded that the extension not having been granted in time no liquidated damages were payable. It appears that the cause of the delay in completion in that case was extra work ordered by the engineer.

Miller v. *London County Council* was applied and followed by Burt J in *MacMahon Construction Pty Ltd* v. *Crestwood Estates* (1971). The particular contract under review contained two clauses relating to extensions of time, the one applicable to delays from such causes as bad weather, strikes or shortage of materials, and the other where variations in the work had been ordered by the engineer. The latter read:

'Clause 40.2: *Modification of Obligations and Guarantees.* If, in the opinion of the contractor, any variation under clause 40.1 prevents him from fulfilling any of his obligations or guarantees under the contract, he shall notify the engineer in writing as soon as possible in what manner his obligations or guarantees may be affected. The engineer shall notify the contractor forthwith whether he confirms or withdraws the order for variation.

If the engineer confirms the order for variation, the contractor's obligations and guarantees shall be varied in so far as the variation under clause 40 shall in the opinion of the engineer require.'

The contract completion date was 3 December 1969. The engineer ordered a number of variations and the contractor wrote some seven letters requesting extensions for each variation, the last being dated 3 December 1969. It was not until 17 April 1970 that the engineer replied and on 23 April he purported to grant extensions. The work was completed in fact on 29 May 1970. Burt J said:

> 'An "extended time", should it exist, must be the product of the proper exercise of a power appropriate to the circumstances to be found in the contract and by a "proper exercise" I mean that if, upon the proper construction of the power to extend, it should appear that the power must be exercised within a period of time either fixed or reasonable, then a purported exercise outside that time is ineffective and there then being no date from which liquidated damages can run, the building owner loses the benefit of that provision: see [Hudson's *Building Contracts*], 6th edition page 359, and *Miller* v. *London County Council* (1934) *per* du Parcq J.'

And further:

> 'The plaintiff, on the other hand, submits that the power is to be found within clause 40 and more specifically within clause 40.2. The submission is that if the impact of a variation order upon the contractor's time schedule for the carrying out of the works is such that the contractor is unable to complete "the works" as that term is defined so as to include the variations within the stipulated time, then in the terms of clause 40.2 the "variation . . . prevents him from fulfilling . . . his obligations . . . under the contract". This submission I accept. The condition controlling the operation of the clause can, in this way, be satisfied and it was so satisfied and the engineer was notified. Confirmation of the order is, I think, a necessary implication from the facts. But no opinion was expressed by the engineer that the contractor's obligations should be varied by the time for completion being extended, and that power not having been exercised at that time cannot, in my opinion, be exercised later.
>
> In the result there is no "extended time" for the purposes of 40.1 and, therefore, no liquidated damages can be recovered.'

[His lordship then referred to *Amalgamated Building Contractors Ltd* v *Waltham Holy Cross UDC* (1952), and cited the passage from the judgment of Denning LJ quoted on p. 256, and continued:]

> In the instant case it was competent for the engineer to grant extensions for 'extra or additional work of any kind or other special circumstances' (clause 11.4 general conditions); and 'for delays caused by other contractors on the site, delays in installation of services or exceptional circumstances' (addendum to special condi-

tions). The delay with which we are concerned, (for which it is said the granted extensions were retrospective and therefore invalid) arose from delays caused by other contractors, namely the sewer contractor. We are therefore not dealing with a cause of delay attributable to Fernbrook. It appears that one of Lord Denning's reasons for distinguishing the *Miller* case was that in *Miller* the delay stemmed from an act of the owner, whereas in the *Amalgamated Building* case it did not.

In Hudson's *Building and Engineering Contracts* (10th edition, 1970) pages 644-645 the three possible interpretations of extension of time clauses are dealt with as follows:

'In the first place, the contract may contemplate that the power should be exercised at once upon the occurrence of the event causing delay. This construction may be appropriate to non-continuing causes of delays, such as the ordering of extras. Secondly, the contract may contemplate that the power should be exercised when the full effect upon the contract programme is known. This is appropriate to continuing causes of delay, such as strikes, withholding of the site, and so on, or to cases, like some extras, where precise estimation is difficult or impossible. Or, thirdly, the contract may contemplate exercise of the power at any time before issue of the final certificate. Since the case of *ABC Ltd* v. *Waltham Holy Cross UDC* (1952) a decision on the then current RIBA form of contract, which distinguished *Miller* v. *LCC* (1934) on somewhat slender grounds, it is suggested that this latter interpretation will normally prevail in the absence of clear language to the contrary, particularly as the ambit of most modern extension of time clauses usually comprehends delays due to causes of many different kinds.'

The above passage from Hudson was considered by Casey J in *New Zealand Structures & Investments Ltd* v. *McKenzie* (1979) where he said:

'In reply to the defendants' submissions that extensions could not be given after completion date, Mr Hicks referred me to page 644 of Hudson, where the cases cited are reviewed. I agree that the defendants have totally misunderstood the comments at the top of page 645, in claiming they supported the view expressed in *Miller* v. *London County Council* (1934) – namely, that the exercise of the power after completion was too late under that contract. On the contrary, Hudson's comments are clearly intended to support the last of three possible interpretations of extension clauses – that the contract may contemplate the exercise of the power at any time before issue of the final certificate (in this case the maintenance certificate of 10 June 1977). That was the effect of the judgment of the Court of Appeal in England in *Amalgamated Building Contractors Ltd* v. *Waltham Holy Cross UDC* (1952) which specifically departed from the reasoning in those earlier cases to the contrary, where they suggest that the time had to be fixed before completion date so that the contractor could know his new date in time to achieve it. (This was also the approach taken by Stout CJ in the Court of Appeal in *Anderson* v. *Tuapeka County Council* (1900).) In modern building practice such clauses are not generally used for this purpose; their principal function is to enable a date to be fixed for the calculation of damages for delay – liquidated or otherwise. As this case so clearly demonstrates, in a major contract it is virtually impossible to gauge the effect of any one cause of delay while it is still proceeding, let alone assess the consequences of concurrent or overlapping causes. Finally, any need to have a prompt decision loses some force as a factor in interpreting such a clause, when one considers the normal review and arbitration procedures, which can take a decision well beyond the final certificate date. In my view of this contract, adopting the reasoning in the *Amalgamated Building* case, and bearing in mind modern building practice, the engineer may grant the extension at any time up to the stage when he becomes *functus officio*, and he is not limited by the completion date.'

The provisions of the contract in the instant case are not appropriate to the fixing of a new date of completion ex post facto regardless of the cause of the delay. The extension clauses themselves could hardly be said to contain, in Hudson's words, 'clear language to the contrary' so that the decision in *Amalgamated Building* did not apply, but clause 11.5.1 of the general conditions obviously contemplates extensions being given, at least for some causes of delay, before the issue of the final certificate. Clause 11.5.1 provides:

'Should the rate of progress of the works or any part thereof be at any time in the opinion of the engineer too slow to ensure completion of the works by the prescribed time or extended time for completion the engineer shall so notify the contractor in writing and the contractor shall thereupon take such steps as he may think necessary and as the engineer may approve in order to expedite progress so that the works may be completed by the prescribed time or any extended time.'

It is significant I think that in the *Anderson, Miller* and

MacMahon cases the sole cause of the delay requiring an extension appeared to be the ordering of extra work, where there could be no justification for delaying a decision as to the appropriate extension, whereas in the *Amalgamated Building* case the causes of the delays did not rest with the employer and were of a nature which made the duration of the delay uncertain.

In my opinion no one rule of construction to cover all circumstances can be postulated and the best that can be said on the present state of the authorities is that whether the completion date is set at large by a delay in granting an extension must depend upon the particular circumstances pertaining.

I think it must be implicit in the normal extension clause that the contractor is to be informed of his new completion date as soon as is reasonably practicable. If the sole cause is the ordering of extra work then in the normal course the extension should be given at the time of ordering so that the contractor has a target for which to aim. Where the cause of delay lies beyond the employer, and particularly where its duration is uncertain then the extension order may be delayed, although even there it would be a reasonable inference to draw from the ordinary extension clause that the extension should be given a reasonable time after the factors which will govern the exercise of the engineer's discretion have been established. Where there are multiple causes of delay there may be no alternative but to leave the final decision until just before the issue of the final certificate.

Applying that approach to the present case I think the exercise of the engineer's discretion to extend time after the completion date was reasonable in the circumstances and I therefore agree with the arbitrator's conclusion that the extensions were validly given, and that the completion date was not set at large by that circumstance.

I therefore reject [counsel for T's] first submission.

His second submission was that the contract completion time had been set at large, and the liquidated damages clause invalidated, by Fernbrook's breach of the contract terms. The breach relied upon was Fernbrook's delay of 23 weeks (as found by the arbitrator) in meeting progress payments, and allied to that the engineer's admitted failure to issue progress payment certificates in five separate months although that failure was at a much later date in the contract. Clause 18(c) of the arbitrator's memorandum . . . refers to Taggart's difficulties because of delayed payments. He was in effect deprived of what has been referred to as the life blood of commerce, an expression which has particular significance in the contracting world. Another consequence was that the contract extended into the winter months and clause 18(b) of the arbitrator's memorandum deals with that aspect.

[Fernbrook] argued that any delays that might have been occasioned by delayed payments, or failure to issue progress certificates, could have been met by extensions of time granted by the engineer, but as Taggart had not applied for them it was not open to the arbitrator to review the matter.

On this score I am satisfied that the arbitrator fell into error on the law, and that the authorities are against Taggart's submission.

The general conditions provide for extensions of time upon the happening of certain specified events 'or other special circumstances of any kind whatsoever', and the addendum to the special conditions refers to delays caused by other contractors, or in the installation of services, 'or exceptional circumstances'.

In *Perini Pacific Ltd* v. *Greater Vancouver Sewerage and Drainage District* (1966), the British Columbia Court of Appeal held that where a building contract contains a clause providing that the owner may extend the time for completion upon application by the builder, by reason of extras or delays occasioned by strikes, lockouts, *force majeure* 'or other cause beyond the control of the contractor', the concluding words must be construed narrowly and with reference to the preceding specific causes of delay and so as not to include defaults of the owner which would unreasonably result in making him the judge of the extent of his own default.

In that case the employer, in default under the implied terms of the contract, had delivered defective machinery for incorporation in the sewage disposal plant being erected by the contractor. The necessary repairs to the machinery delayed the contractor's completion date. The court held that the contractor was thereby left at large as to the date of completion and was released of his liability to pay liquidated damages even though the contractor's own tardiness may have caused some of the delay. Davey JA said:

'The contract was an entire one providing for the construction of a number of units making up the plant, and the plaintiff had to finish all parts of the work by the due date in order to perform the contract in due time. Even if the plaintiff had not been otherwise in default, the repairs to the engines would have prevented the plaintiff from completing the power house by 10 January 1963, and so fulfilling its contract. In my opinion the defendant's default, although preventing the completion by the due date of only one unit of several covered by the entire contract, disentitles the defendant to liquidated damages for the plaintiff's failure to complete the whole plant by the due date:

Holme v. *Guppy* (1838); *Roberts* v. *Bury Improvement Com'rs* (1870); *Dodd* v. *Churton* (1897); *Russell* v. *Viscount Sa da Bandeira* (1862). Some of those cases speak of the owner's having made it impossible for the contractor to complete, or having prevented completion by the due date. Other language in those cases, sometimes by the same judge in the same case, speaks of the owner's having delayed completion as the thing that exonerates the contractor. In *Holme* v. *Guppy*, Parke B said the contractor was thereby left at large as to the date of completion.

As I understand those authorities, neither the fact that the work would not have been completed in time if there had been no delay by the defendant, nor the fact that the plaintiff's tardiness or wrongful acts caused some of the delay, prevents the plaintiff being released of its liability for the liquidated damages through the defendant's substantial defaults that delayed it in completing the contract on time.'

And Bull JA said:

'The case in England of *Wells* v. *Army & Navy Co-operative Society Ltd* (1902) has, I suggest, considerable value with respect to the matter in discussion. In this case, which is singularly similar to the one at bar, an owner claimed liquidated damages for a contractor's failure to construct certain buildings by a date which, under the contract procedures, had been extended by three months by the owner by reason of certain delays. The contractor failed to complete by the extended time claiming that it was prevented from so doing by the defaults of the owner. It was found that material delays were caused by the action or fault of the owner with respect to part of the project, but that there were other delays, including those caused by the contractor, present. Wright J, the trial Judge, said with reference to various causes of delay.

"It is very difficult to determine how far any particular defaults of this kind on the part of the defendants would entitle the plaintiffs to relief from penalties, especially when, as in this case, there were other and more important causes of delay which would not be grounds of relief in this action; but on the whole I think that the delays in giving details not merely contributed to the delay of completion, but were such as even in the absence of the other causes of delay would have prevented completion in due time, and in my view have to a great extent increased the delay of completion."

and again:

> "On the whole, I think that the conclusion must be that the defaults of the defendants were such that in their cumulative effect they were inconsistent with their claim to insist on completion within the stipulated time. The defaults were, in my opinion, sufficiently substantial to cast upon the defendants the burden of showing that the defaults did not excuse the delay. It is true that, apart from their defaults, the plaintiffs had, by the default of subcontractors, been delayed to an extent which might of itself have involved them in penalties, but in the absence of the further defaults by the defendants, it is impossible to say to what extent the liability to penalties might not have been reduced."

'On appeal, where the decision was affirmed, Vaughan Williams LJ with whom the other Lord Justices expressed agreement, dealt specifically with the matter here concerned after discussing the submission of counsel for the owners that despite the fact that the owners had delayed the work it could not be said that that *prevented* the completion in time because a contributory cause was the fault of the contractor, and said:

> "In law, I wholly deny the proposition Mr Bray put forward which was this really in effect: 'Never mind how much delay there may be caused by the conduct of the building owner, the builder will not be relieved from penalties if he too has been guilty of delay in the execution of the works.' I do not accept that proposition in law."

'Accordingly, it is my respectful view that it can be fairly said that the respondent acted in a manner in breach of its contract that *actually prevented or made it impossible for* the appellant to perform its obligation at the proper time. To hold that despite this absolute prevention, the appellant under the circumstances could not have so performed, would, I suggest with respect, have the effect of permitting the respondent to take full advantage of its own wrong, or, to put it as Mathew J did in *Wells* v. *Army & Navy Co-operative Society*, "the defaults of the defendants were such that they were inconsistent with their claim to insist upon completion within the stipulated time".'

The present case is not one where it was a question of whether the engineer should have granted an extension of time for his employer's breach of the contract as the arbitrator found (although in fairness to the arbitrator I should say that the *Perini* case was not cited to him) for in my opinion there was no jurisdiction in the engineer so to do, but rather a question of whether Fernbrook's breach relieved Taggart of liability under the liquidated damages clause. On the authority of *Perini* I conclude that there was no jurisdiction in the engineer to extend time on account of Fernbrook's breach, and that that breach set the completion date at large with effect that Taggart was not liable to pay liquidated damages in the sum of $2500, as found by the arbitrator, nor in any other sum.

(iii) INTERPRETATION OF PROVISIONS

Although in general liquidated damages and extension of time clauses in standard form contracts are to be construed strictly contra proferentem, this does not apply to negotiated contracts such as the JCT forms.

Tersons Ltd v. Stevenage Development Corporation

COURT OF APPEAL   (1963), 5 BLR 54

In a dispute arising out of the ICE conditions (2nd edition), it was argued that since the conditions had been included in the invitation to tender sent by SDC to the contractors, they should be interpreted *contra proferentem* the employer and in favour of the contractors. The court rejected this argument.

PEARSON LJ: [Counsel] has contended that the maxim *verba fortius accipiuntur contra proferentem* should be applied in this case in favour of the contractor against the corporation on the ground that the general conditions were included in the invitation to tender sent by the corporation to the contractor. In my view, the maxim has little, if any, application in this case. The general conditions are not a partisan document or an 'imposed standard contract' as that phrase is sometimes used. It was not drawn up by one party in its own interests and imposed on the other party. It is a general form, evidently in common use, and prepared and revised jointly by several representative bodies including the Federation of Civil Engineering Contractors. It would naturally be incorporated in a contract of this kind, and should have the same meaning whether the one party or the other happens to have made the first mention of it in the negotiations.

The relationship between the JCT extension of time and direct loss and/or expense provisions (JCT 80, clauses 25 and 26; JCT 63, clauses 23 and 24) is illustrated and explained by:

Henry Boot Construction Ltd v. Central Lancashire New Town Development Corporation

QUEEN'S BENCH DIVISION   (1980), 15 BLR 1

The facts are not relevant for present purposes.

JUDGE EDGAR FAY QC: Part of the contractual framework is relevant. . . . The contract provided for liquidated damages in case of the completion date not being met. In the case of such delay the contractors were to pay by way of liquidated damages the sum, among other things, of £7 per week per dwelling. That is a factor to be borne in mind in considering the background of this case. Now if there is delay in the carrying out of a construction contract there is of course loss. There is in the first instance loss to both parties. To the employer, the owner, there is loss of the return upon his investment. The day when he starts getting a return by way of rent from his property is postponed and he may well have an extended period of expenditure upon supervision and the like. Equally, there is loss upon the contractor, owing to the prolongation of the period for which he has to supply matters falling within overheads as well as other expenditure, and indeed possibly idle time as well. If the delay is the fault of the contractor, then the damages for delay, the liquidated provision which I have referred to, comes into operation. He is not entitled to any relief and he has to pay the damages at £7 per week per dwelling in this instance. But if the delay is not the fault of the contractor, or he alleges that it is not, then under the terms of the contract there are two provisions which enable him to obtain relief from the loss which he otherwise would suffer.

Those provisions are condition 23 and condition 24. . . .

The broad scheme of these provisions is plain. There are cases where the loss should be shared, and there are cases where it should be wholly borne by the employer. There are also those cases which do not fall within either of these conditions and which are the fault of the contractor, where the loss of both parties is wholly borne by the contractor. But in the cases where the fault is not that of the contractor the scheme clearly is that in certain cases the loss is to be shared: the loss lies where it falls. But in other cases the employer has to compensate the contractor in respect of the delay, and that category, where the employer has to compensate the contractor, should, one would think, clearly be composed of cases where there is fault upon the employer or fault for which the employer can be said to bear some responsibility.

The scheme can be illustrated by provisions which are not in issue in this case. Under condition 23 the first two of the paragraphs which invoke the condition are (a) *force majeure*, or (b) by reason of any exceptionally inclement weather. If the delay is caused by those matters then under condition 23 the contractor is relieved of his penalty payment by reason of not meeting the completion date, but he gets no compensation himself. The loss, in other words, lies where it falls, as one would expect from something over which neither party has any control, such as *force majeure* or exceptionally inclement weather.

Those provisions are not repeated in condition 24. There is no question in those cases of any compensation being paid by the employer to the contractor.

Note: See also *Bramall & Ogden Ltd* v. *Sheffield City Council* (1985), Appendix, p. 378.

PART III

DISCHARGE OF THE CONTRACT

Part III is concerned with cases where the contractual obligations of the parties come to an end. In some cases the contract itself is brought to an end; in other cases one of the parties is relieved from further performance.

Chapter 14

Release

The parties may agree to release each other from further performance. This presents no problem where the contract is still partly (or wholly) unperformed on both sides. More difficulty arises where one party has wholly performed and it is alleged that there is agreement that the other may not perform in full. The most obvious example in a building contract would be for the builder to have completed the work and then agreed to accept less than full payment. There is no problem if there is a genuine dispute as to whether the work is up to contract standard.

If the contractor agrees to accept part payment and to release the employer from payment of the balance, the agreement will not be binding unless at least the contractor accepts it voluntarily and (perhaps) unless the agreement is under seal or supported by fresh consideration.

D. & C. Builders Ltd v. Rees

COURT OF APPEAL   [1966] 2 QB 617

The plaintiffs were a small firm of jobbing builders who had done work for the defendant to the undisputed value of £482.13s.1d. Six months later, the money not having been paid and the plaintiffs being desperate for cash, they were offered £300 in full settlement, being told in effect that if they did not accept it, they would get nothing. The plaintiffs accepted the cheque and issued a receipt stating that it was 'in completion of the account'. They then sued for the balance.

Held: The action succeeded.

LORD DENNING MR: [The principle of equitable estoppel] has been applied to cases where a creditor agrees to accept a lesser sum in discharge of a greater. So much so that we can now say that, when a creditor and a debtor enter upon a course of negotiation, which leads the debtor to suppose that, on payment of the lesser sum, the creditor will not enforce payment of the balance, and on the faith thereof the debtor pays the lesser sum and the creditor accepts it as satisfaction: then the creditor will not be allowed to enforce payment of the balance when it would be inequitable to do so. This was well illustrated during the last war. Tenants went away to escape the bombs and left their houses unoccupied. The landlords accepted a reduced rent for the time they were empty. It was held that the landlords could not afterwards turn round and sue for the balance: see *Central London Property Trust Ltd* v. *High Trees House Ltd* (1947). This caused at the time some eyebrows to be raised in high places. But they have been lowered since. The solution was so obviously just that no one could well gainsay it.

In applying this principle, however, we must note the qualification: The creditor is only barred from his legal rights when it would be *inequitable* for him to insist upon them. Where there has been a *true accord*, under which the creditor voluntarily agrees to accept a lesser sum in satisfaction, and the debtor *acts upon* that accord by paying the lesser sum and the creditor accepts it, then it is inequitable for the creditor afterwards to insist on the balance. But he is not bound unless there has been truly an accord between them.

In the present case, on the facts as found by the judge, it seems to me that there was no true accord. The debtor's wife held the creditor to ransom. The creditor was in need of money to meet his own commitments, and she knew it. When the creditor asked for payment of the £480 due to him, she said to him in effect: 'We cannot pay you the £480. But we will pay you £300 if you will accept it in settlement. If you do not accept it on those terms, you will get nothing. £300 is better than nothing.' She had no right to say any such thing. She could properly have said: 'We cannot pay you more than £300. Please accept it on account.' But she had no right to insist on his taking it in settlement. When she said: 'We will pay you nothing unless you accept £300 in settlement,' she was putting undue pressure on the creditor. She was making a threat to break the contract (by paying nothing) and she was doing it so as to compel the creditor to do what he was unwilling to do (to accept £300 in settlement): and she succeeded. He complied with her demand. That was on recent authority a case of intimidation: see *Rookes* v. *Barnard* (1964) and *J.T. Stratford & Son Ltd* v. *Lindley* (1964). In these circumstances there was no true accord so as to found a defence of accord and satisfaction: see *Day* v. *McLea* (1889). There is also no equity in the defendant to warrant any departure from the due course of law. No person can insist on a settlement procured by intimidation.

In my opinion there is no reason in law or equity why the creditor should not enforce the full amount of the debt due to him. I would, therefore, dismiss this appeal.

WINN LJ: In my judgment it is an essential element of a valid accord and satisfaction that the agreement which constitutes the accord should itself be binding in law, and I do not think that any such agreement can be so binding unless it is either made under seal or supported by consideration. Satisfaction, viz., performance, of an agreement of accord, does not provide retroactive validity to the accord, but depends for its effect upon the legal validity of the accord as a binding contract at the time when it is made: this I think is apparent when it is remembered that, albeit rarely, existing obligations of debt may be replaced effectively by a contractually binding substitution of a new obligation.

Chapter 15

Frustration

The obligations of the parties may be brought to an end by the doctrine of frustration. Originally this doctrine applied to unexpected events which made continued performance of the contract impossible. An early example is Appleby v. Myers, *page 116, where a factory was destroyed by fire. Later cases have applied the doctrine to developments which, without making performance strictly impossible, have made it fundamentally different from that contemplated at the time the contract was made.*

Metropolitan Water Board v. Dick, Kerr & Co. Ltd

HOUSE OF LORDS [1918] AC 119

In July 1914, the respondents contracted to build a reservoir for the appellants within six years. The contract gave the engineer typical powers to award extra time. In February 1916 the Ministry of Munitions, acting under war time powers, ordered the respondents to cease work which they did. Granted the possibility of extensions of time it was not impossible to perform but the House of Lords held that a contract resumed after the war time interruption would be fundamentally different from that before the interruption.

LORD DUNEDIN: On the whole matter I think that the action of the government, which is forced on the contractor as a *vis major*, has by its consequences made the contract, if resumed, a work under different conditions from those of the work when interrupted. I have already pointed out the effect as to the plant, and, the contract being a measure and value contract, the whole range of prices might be different. It would in my judgment amount, if resumed, to a new contract; and as the respondents are only bound to carry out the old contract and cannot do so owing to supervenient legislation, they are entitled to succeed in their defence to this action.

However, it is not sufficient to show that the contract has turned out more difficult and expensive for one party to perform than he expected.

Davis Contractors Ltd v. Fareham Urban District Council

HOUSE OF LORDS [1956] AC 696

The plaintiffs contracted to build 78 council houses for the authority within a period of eight months and for a fixed price. Through no fault of either party, there was a scarcity of skilled labour and the work took 22 months to complete. The plaintiffs argued that because of the shortage of labour, the contract had been brought to an end by frustration, and that they were entitled to recover a sum well in excess of the fixed price on the basis of a *quantum meruit* (see p. 134).

Held: The claim must fail. The contract had not been frustated and the plaintiffs had not been released from its terms as regards price.

LORD RADCLIFFE: Lord Loreburn ascribes the dissolution to an implied term of the contract that was actually made. This approach is in line with the tendency of English courts to refer all the consequences of a contract to the will of those who made it. But there is something of a logical difficulty in seeing how the parties could even impliedly have provided for something which ex hypothesi they neither expected nor foresaw; and the ascription of frustration to an implied term of the contract has been criticized as obscuring the true action of the court which consists in applying an objective rule of the law of contract to the contractual obligations that the parties have imposed upon themselves. So long as each theory produces the same result as the other, as normally it does, it matters little which theory is avowed (see *British Movietonews Ltd v. London and District Cinemas Ltd* (1951), *per* Viscount Simon). But it may still be of some importance to recall that, if the matter is to be approached by way of implied term, the solution of any particular case is not to be found by inquiring what the parties themselves would have agreed on had they been, as they were not, forewarned. It is not merely that no one can answer that hypothetical question: it is also that the decision must be given 'irrespective of the individuals concerned, their temperaments and failings, their interest and circumstances'.... The legal effect of frustration 'does not depend on their intention or their opinions, or even knowledge, as to the event.' On the contrary, it seems that when the event occurs 'the meaning of the contract must be taken to be, not what the parties did intend (for they had neither thought nor intention regarding it), but that which the parties, as fair and reasonable men, would presumably have agreed upon if, having such possibility in view, they had made express provision as to their several rights and liabilities in the event of its occurrence' (*Dahl* v. *Nelson* (1881), *per* Lord Watson).

By this time it might seem that the parties themselves have become so far disembodied spirits that their actual persons should be allowed to rest in peace. In their place there rises the figure of the fair and reasonable man. And the spokesman of the fair and reasonable man, who represents after all no more than the anthropomorphic conception of justice, is and must be the court itself. So perhaps it would be simpler to say at the outset that frustration occurs whenever the law recognizes that without default of either party a contractual obligation has become incapable of being performed because the circumstances in which performance is called for would render it a thing radically different from that which was undertaken by the contract. Non haec in foedera veni. It was not this that I promised to do....

I am bound to say that, if this is the law, the appellants' case seems to me a long way from a case of frustration. Here is a building contract entered into by a housing authority and a big firm of contractors in all the uncertainties of the post-war world. Work has begun shortly before the formal contract was executed and continued, with impediments and minor stoppages but without actual interruption, until the 78 houses contracted for had all been built. After the work had been in progress for a time the appellants raised the claim, which they repeated more than once, that they ought to be paid a larger sum for their work than the contract allowed; but the respondents refused to admit the claim and, so far as appears, no conclusive action was taken by either side which would make the conduct of one or the other a determining element in the case.

Whether a particular event is a frustrating event must depend not only on the event but also on the precise terms of the contract.

Wong Lai Ying v. Chinachem Investment Co. Ltd

PRIVY COUNCIL   (1979), 13 BLR 81

The respondents, who were building two blocks of flats in Hong Kong, entered into twenty-four contracts to sell flats as yet unbuilt to the appellants. The contracts were entered into between March and November 1971. Work began in December 1971, the contractual date for completion was 17 May 1973, and time was made of the essence but with provision in certain circumstances for extension for not more than one year.

In June 1972, part of the hillside above the building site slipped, taking with it a thirteen storey block of flats, the debris of which landed on the site obliterating the building works already completed. Work stopped and, since it could not be recommenced within three months, the respondents' building permit came to an end. A new permit was not issued until November 1975.

These facts would appear to present a classic example of frustration but the appellants argued that the contract was not frustrated (and that they were therefore entitled to specific performance) because of clause 22 of the contract which provided:

'It is further agreed that notwithstanding anything herein contained . . . should any unforeseen circumstances beyond the vendor's control arise whereby the vendor becomes unable to sell the said undivided share and apartment to the purchaser as hereinbefore provided, the vendor shall be at liberty to rescind the agreement forthwith and to refund to the purchaser all instalments of purchase price paid by the purchaser hereunder without interest or compensation . . .'.

Held: Clause 22 should not be read as applying to the kind of unforeseen disaster which had happened. The contract was frustrated.

LORD SCARMAN: It was, however, urged that the language of clause 22 is wide enough to cover the event which happened. So it is. But the question is whether the general words of the clause are sufficient to support the inference that the parties must be presumed to have made provision for the event. In answering the question, the concurrent findings of fact by four judges, all of whom would be well aware of conditions in Hong Kong, must be respected by the board. The event was, admittedly, an 'unforeseen natural disaster', and [both the trial judge and the Court of Appeal] spelt out its consequences for the contract. . . . All of them were prepared to characterise it as a 'frustrating event': they differed only in their construction of clause 22.

The unforeseen character of the event is not in dispute. Clause 22 cannot, in their Lordships' opinion, be construed as making provision for the possibility of this particular unforeseen contingency. The clause, coming at the end of the contract, replete with specific provisions and time limits, was plainly intended to confer upon the vendor a remedy of rescission if a dispute arose or it became clear he could not complete in accordance with the contract, provided he acted 'forthwith' to terminate the contract. It does not follow from the provision of a summary remedy avoiding litigation in such circumstances that the parties must have agreed that their contract would continue after an unforeseen natural disaster having the consequences analysed and assessed by the judges below. The Board agrees with the view expressed by Huggins J in the Court of Appeal that:

> 'the inclusion of a clause such as clause 31 in the Bank Line charter-party [*Bank Line* v. *Capel* (1919)] or clause 22 of the agreement in this case is . . . not inconsistent with the operation of the doctrine of frustration and does not show an intention that the doctrine shall not apply'.

Strictly, of course, the issue is not whether the doctrine of frustration is excluded but whether provision was made for an event causing the circumstances of performance to be radically different from that undertaken by the contract. The word 'forthwith' is an indication, if any other than the context and circumstances of the contract is needed, that this 'frustrating event' was not contemplated; for a most significant feature of this event was that as a result of it 'the position of the parties was clouded in uncertainty' (Briggs CJ). Another indication is the bizarre consequences of holding the event covered by the clause. If the vendor should fail to act 'forthwith', and completion should become indefinitely delayed, the purchaser, exercising his right under clause 3(3), could wait as long as he pleased, collecting all the time his interest at the rate of one per cent per month. A further indication is the presence in the contract of clause 20. Requisition, if not provided for, could well be a frustrating event: indeed it is a classic instance in some circumstances of frustration. The parties by clause 20 made specific provision for it.

Note: See also *Codelfa Construction Proprietary Ltd* v. *State Rail Authority of New South Wales* (1982), 149 CLR 337, Appendix, p. 397.

Chapter 16

Illegality

Illegality does not bring the contract to an end but it may prevent one or both parties from enforcing it. A building contract might be illegal because the building owner intended to use the building for an improper purpose, for example, as a brothel. The building owner could not then enforce the contract, nor could the builder if he knew of the improper purpose.

By far the most common practical examples arose under the building licence scheme in force during the 1939-45 war and for a period thereafter under which contracts to carry out unlicensed work were illegal.

Bostel Brothers Ltd v. Hurlock

COURT OF APPEAL   [1949] 1 KB 74

The defendant had obtained a licence to carry out repairs to his house and engaged the plaintiffs to do the work. The plaintiffs innocently exceeded the amount of the licence without having applied for a supplementary licence. The plaintiffs claimed the balance in excess of the amount of the licence.

Held: The claim failed, as the prohibition on doing unlicensed work was absolute and did not depend on the builder's state of mind.

Note: This reasoning produces a result which seems unfair, the defendant having the benefit of the work without having to pay for it. The theoretical justification for the rule is that it is necessary to be unfair to individual builders in order to encourage builders as a class to comply with the rules. However where it is the duty of the employer to obtain the licence, it was held that he had given a collateral warranty that he would obtain the necessary licence or stop the work.

Strongman (1945) Ltd v. Sincock

COURT OF APPEAL   [1955] 2 QB 525

The defendant, an architect owner, engaged the plaintiffs to undertake building work and agreed to obtain the necessary licences. In fact, he allowed work considerably in excess of the licence figure to be done and then refused to pay for it.

Held: Although the builders could not maintain a direct action for the balance, they could sue the defendant upon his undertaking to obtain the necessary licences unless they were themselves morally to blame or culpably negligent. Since it was agreed that as between architect and builder the primary duty to obtain licences lay on the architect, it was held that they were not.

DENNING LJ: The third question is whether the plaintiffs were guilty of negligence. I can well see that if there was culpable negligence on the part of the persons seeking damages, he might not be entitled to recover. In *Askey* v. *Golden Wine Co. Ltd* (1948) some merchants had been guilty of culpable negligence in not taking proper steps to see whether a liquor was safe for consumption and they were not allowed to recover. I said: 'If they were allowed to be negligent and yet recover damages, it would offer an inducement to them to turn a blind eye to contamination.' So, in any of these licensing cases, if the builder were negligent and yet were allowed to recover damages, it would be an inducement to him to turn a blind eye to the statute. That is what I had in mind in *J. Dennis & Co. Ltd* v. *Munn* (1949) and the other cases to which [counsel] referred us. When a builder is doing work for a lay owner – if I may so describe him – the primary obligation is on the builder to see that there is a licence. He ought not simply to rely on the word of the lay owner. He ought to inspect the licence himself. If he does not do so, it is his own fault if he finds himself landed in an illegality. But in this case there was not a lay owner. The owner was the architect, and he himself said in evidence: 'I agree that where there is an architect it is the universal practice for the architect and not the builder to get the licence.' No fault, it seems to me, can, in these circumstances, be attributed to the builder.

It was contended before us . . . that on the facts of this case there must have been negligence and that in point of law the official referee ought to have found it. I think not. The official referee found: 'I do not consider that the plaintiffs in the present case have done an immoral act, nor were they negligent in not insisting on the production of supplementary licences. They had done a great deal of other work for the defendant without any question being raised with regard to the sufficiency of the licences.'

As I said at the beginning of this case, it comes very ill from the mouth of the defendant to raise this point as against the plaintiffs. His attitude was well shown by an observation which he made to the solicitor. . . . He said: 'If the Nicholls can be bluffed they deserve to lose their money.' In other words, he was saying: 'If they were fools enough to trust in me, they ought to lose their money.' That is a very wrong attitude for a professional man, an architect, to take up. It shows quite clearly that on his own admission he has misled them and now seeks to turn it into his own advantage. In my judgment, his objection fails. On the findings of the official referee, the plaintiffs were entirely innocent people who were led into this unfortunate illegality by the representation of the architect, amounting to a collateral contract, that he would get the licences. That contract not having been fulfilled, I see no objection in point of law to the plaintiffs recovering the damages, and I think that the appeal should be dismissed.

Chapter 17

Forfeiture Clauses, Repudiation and Determination

Building contracts usually contain clauses (sometimes called for-feiture clauses) entitling one party to bring the contract to an end in certain circumstances. Independently of such clauses contracting parties have power to determine the contract in certain circumstances under the general law of contract. In practice there is a significant area of overlap between these two possibilities.

Under the general law one party is entitled to terminate in certain circumstances if the other party has broken the contract. It is not every breach which has this result however. Basically the party seeking to determine must show that the other party has either committed a breach which goes to the root of the contract or has repudiated his obligations under it. These concepts are often difficult to apply in practice and as the following cases illustrate it sometimes happens that A commits a breach which B wrongly thinks entitles him to terminate and by taking this view commits a repudiatory breach entitling A to terminate.

Lubenham Fidelities & Investments Co. Ltd v. South Pembrokeshire District Council & Wigley Fox Partnership

QUEEN'S BENCH DIVISION   26 May 1983

There were two contracts in JCT 63 form. The original contractors went into liquidation, and the plaintiffs, who were bondsmen, elected to perform the contract themselves rather than pay on the performance bond. They engaged sub-contractors to carry out the work. Because of various building defects, Wigley Fox (the architect under the contract) wrongfully advised the council to deduct liquidated damages from interim certificates. The plaintiffs thereupon wrote terminating the contracts for alleged breach, and ordered the sub-contractors to cease work. The council thereupon served notice of determination under clause 25(1), on the ground that the plaintiffs were in default by wholly suspending the carrying out of the works without reasonable cause. Simultaneously, the plaintiffs served notices under clause 26(1) alleging that the council were in default by failing to make payment under the interim certificates; they then issued a writ.

Held: The plaintiffs had repudiated the contracts by not complying with their terms and were in breach, even though the council were wrong in deducting liquidated damages at that stage. By withdrawing the sub-contractors from site and so suspending work without reasonable cause the plaintiffs had acted contrary to clause 25(1). By serving invalid notices of termination and thereafter issuing a writ, the plaintiffs had indicated an intention not to be bound by the contracts and since the council had no alternative but to accept the repudiation, the contracts were brought to an end.

Alkok v. Grymek

SUPREME COURT OF CANADA　　　　　　　　　(1968), 67 DLR (2d) 718

The facts sufficiently emerge from the judgment.

SPENCE J: The appellant is a contractor in the City of Toronto. He entered into a contract with the respondents dated 14 May 1956, whereby he agreed to:

(a) provide all the materials and perform all the work shown on the drawings and described in the specifications entitled 'proposed residence for Mr I. Grymek' which have been signed in duplicate by both the parties and which was prepared by Edward I. Richmond, MRAIC, acting as and hereinafter entitled 'the architect' and

(b) do and fulfill everything indicated by this agreement, the specifications, and the drawings.

The contract provided that the owner would pay to the contractor $57,500 and

(b) Make payment on account thereof upon the architect's certificate (when the architect is satisfied that payments due to sub-contractors have been made), as follows:

    i)   Upon completion of the sub-floor $6,000.00;

    ii) Upon completion of the roof $12,000.00;

    iii) Upon completion of the brown coat of plaster $8,000.00;

    iv) Upon completion of the white coat of plaster, including all plumbing and electrical work $12,000.00;

    v) Upon completion of trim $8,000.00;

    vi) the sum of eleven thousand, five hundred dollars ($11,500.00) . . .

The appellant commenced work of erecting the house in accordance with the said contract and from time to time the respondents required the addition of certain extras which the appellant added and which the Court of Appeal for Ontario found were of the value of $950. The respondents paid to the appellant the whole of the first two instalments of $6,000 and $12,000, respectively, and one-half of the third instalment, i.e., $4,000, and did so without requiring that the appellant comply with the provisions of para. (b) aforesaid by satisfying the architect that the payments due to the subcontractors had been made.

Later when the appellant pressed for the payment of the balance of the third instalment, the respondents required the appellant to comply with the provisions of the contract and to so satisfy the architect. After several conferences, the appellant failed to so satisfy the architect and in addition there were complaints from the respondents and from their architect . . . that there were defects in the construction and that the construction was delayed. The solicitor for the respondents who had drafted the original contract and who had been present at the various conferences when an attempt was made to satisfy the architect that the sub-contractors had been paid, wrote to the appellant on 10 November 1956, complaining of the progress of the work and of certain defects expressing the fear that mechanics' liens would be registered against the property and concluded with this paragraph:

> Unless all of the building infractions, which are your responsibility, have been remedied, and the work carried on at a proper pace, my clients shall have no alternative than to employ their own specific trades to complete your portion of the uncompleted work, and any moneys or expenses incurred by my clients in employing tradesmen for either work done or materials supplied shall be deducted from the contract price herein.

Six days later, on 16 November 1956, the said solicitor wrote again to the appellants in which he said:

> Further to the above matter, in confirming my conversation with you yesterday, I hereby advise you on behalf of my clients Issie Grymek and Yetta Grymek that they are terminating their contract with you as of today's date.
> They intend to complete the lands and premises in the manner in which they believe the work should be done.

In accordance with that letter the appellant ceased work on the contract though the architect for the respondents has testified that the appellant was willing and anxious to continue it. The respondents proceeded to complete the building themselves through the intervention of other contractors and materialmen.

It was the view of the learned Master that the respondents were entitled to terminate the contract at the time and in the fashion aforesaid.

The Court of Appeal for Ontario, however, came to the conclusion that sufficient grounds for the termination of the contract had not been established. McGillivray JA in his reasons for judgment, quoted *Anson on Contract*, 21st edition page 424, as follows:

> 'The question to be answered in all these cases of incomplete performance is one of fact; the answer must depend on the terms of the contract and the circumstances of each case. The question assumes one of two forms – Does the failure of performance amount in effect to a renunciation on his part who makes default? Does it go so far to the root of the contract as to entitle the other to say, "I have lost all that I cared to obtain under this contract; further performance cannot make good the prior default"?'

That proposition needs no support by citation from judgments and I accept it as expressing the proper test which the Court must apply here. As pointed out by McGillivray JA, in his reasons, the contract as between the parties was for the contractor to build a house and for the defendants to pay for it. The contractor had proceeded with the building although not in accordance with the pace at which the owners believed he should be proceeding and had been guilty of what the Court of Appeal for Ontario has found were certain minor defects in construction. The owners having paid two instalments and part of the third were refusing to pay the balance of the third instalment or those which would become due thereafter, and in so doing were relying upon the provision of the contract which required the appellant as contractor to satisfy the respondents' architect that the payments to the sub-contractors had been made. They were entitled to require that the appellant continue his work upon the contract and to refuse to pay him until he did satisfy that provision. If the appellant had refused to proceed on that basis then, of course, he would have been in breach of the provision of the contract going to the root thereof and the respondent would have been entitled to terminate the contract. The appellant did not indicate by word or conduct an intention to so act or not to be bound in every way by the contract. The appellant, therefore, did not give to the respondents the opportunity to terminate the contract on the ground that the appellant had been in breach of a term going to the root of it. It was true that he was in breach of the term requiring him to satisfy the respondents' architect that the subcontracts had been paid, but, with respect, I agree with McGillivray JA's view that this was a mere ancillary term which could be enforced perfectly by the respondents simply refusing to make payments until they were satisfied, as indeed the respondents were refusing. I am, therefore, in accord with the view of McGillivray JA, that the respondents have not shown sufficient grounds to support their termination of the contract.

R.B. Burden Ltd v. Swansea Corporation

HOUSE OF LORDS   [1957] 3 All ER 243

The facts sufficiently appear from the speech of Lord Tucker.

LORD TUCKER: My Lords, the short point for decision in the appeal is whether, on the findings of the learned official referee, the respondent corporation 'interfered with or obstructed the issue of a certificate' within the meaning of clause 20 in a building contract made between the appellants as contractors and the respondents as employer, dated 20 December 1949, for the construction of a primary school at Swansea. Under the contract, the borough architect was appointed architect and surveyor for the purposes of the contract. He was subsequently replaced as surveyor by the firm of Messrs. O who were appointed in his place by the respondents as they had power to do under the contract. H was the representative of the firm who acted pursuant to this appointment.

Clause 24(a) of the contract provides that, at the period of interim certificates (i.e., monthly), interim valuations shall be made whenever the architect considers them necessary and that, subject to clause 21 (which deals with sub-contractors), the architect shall issue a certificate stating the amount due to the contractor from the employer, and the contractor shall be entitled to payment therefor within the period named in the appendix. Clause 24(b) provides that the amount so due shall, subject to clause 21(a) and to any agreement between the parties as to stage payments, be the total value of the work properly executed and of the materials and goods delivered on the site for use in the works up to, and including, a date not more than seven days before the date of the said certificate, less the amount to be retained by the employer and less any instalments previously paid. Clause 27 provides for arbitration in case of any dispute or difference between the employer or the architect on his behalf and the contractor, either during the progress or after completion or abandonment of the works, as to the construction of the contract, or arising thereunder, or in connexion therewith (including any matter or thing left to the discretion of the architect), or the withholding by the architect of any certificate to which the contractor may claim to be entitled or the measurement and valuation mentioned in clause 19 (which deals with variations) or the rights and liabilities of the parties under clause 19, clause 20 or clause 25. Clause 20(1), which is the clause directly in question, reads, so far as relevant, as follows:

'If the employer within the period which is named in the appendix to these conditions and thereafter for seven clear days after written notice from the contractor does not pay to the contractor the amount due on any certificate, or if the employer interferes with or obstructs the issue of any such certificate ... the contractor may, without prejudice to any other rights or remedies, thereupon by notice by registered post to the employer or architect determine the employment of the contractor under this contract.'

It has been found that, in the early stages of the contract, it was the practice for persons in the respondents' surveyor's department from time to time to meet a representative of the appellants to agree figures and valuations for the purpose of the issue by the architect of interim certificates. As time went on, owing to pressure of work it was found impossible for anyone to attend the site on behalf of the respondents, and the appellants' figures were checked by the office staff so far as was practicable. When the work was nearing completion, some of the respondents' officials suspected that this practice had resulted in considerable overpayments, and this was one of the reasons for the appointment, shortly before No. 21 interim certificate was due, of H of the independent firm of Messrs. O who had prepared the original bills of quantities. It is, I think, clear that H was appointed not only to act as surveyor for the limited purposes expressly laid down in the contract but also to carry out the valuations and measurements which had previously been done, with the architect's consent, by members of the respondents' permanent staff with a view to assisting the architect to obtain the necessary materials for the issue of his certificates.

The official referee has found as a fact that 'in the case of certificate 21 the architect decided that an interim valuation should be made and instructed Messrs. O to make it'. H accordingly made what he claimed to be a valuation, as a result of which the architect issued a certificate for £1,287 as against £5,785 claimed by the appellants. The official referee has found that this valuation was not properly made, and the question for decision is whether the defects and errors found constitute obstruction or interference by the employer with the issue of a certificate.

The official referee considered H an unsatisfactory and unreliable witness, but your Lordships are only concerned with his actual ultimate findings as to what was done, or left undone, by H. He found that the amount of £1,287 recommended by H was what he then thought was due to the appellants under clause 24(b), that his valuation had been made in too great a hurry without visual valuation on the site, that it omitted to include approximately £1,500 for hard core, which had been bought and used by the appellants, and that he failed to inform the architect of this omission when it was brought to his notice. H had not included certain excess payments made by the appellants during the certificate period, consequent on price variations, and had made certain miscalculations with regard to preliminaries. Finally he found that, as a result of a mistake in the respondents' surveyor's department, the certificate as issued provided for payment to sub-contractors of a sum of £650 or thereabouts out of the total certified. This sum was not, in fact, payable and the appellants never paid it but retained the full sum. It is unnecessary to refer further to this last finding, since counsel for the appellants did not contend that this mistake would, by itself, have justified determination under clause 20.

By reason of these matters, the official referee stated he had come to the conclusion, not without some hesitation, that, by producing that so-called valuation, H obstructed the issue of the 'sort of certificate' the appellants were entitled to and, consequently, they could determine their employment under clause 20.

The Court of Appeal allowed the respondents' appeal. Denning LJ, I think, assumed without deciding that H's acts and omissions amounted to obstruction or interference, but held that the respondents were not liable for the acts of the surveyor within the sphere of duties assigned to him under the contract. Morris LJ, considered the respondents not liable because H was employed as an independent consultant not subject to their control. He refers to him 'conducting himself erroneously', but expresses no view whether such erroneous conduct constituted obstruction or inter-ference with the issue of a certificate. Parker LJ, said he was prepared to accept that H was guilty of interference or obstruction, but absolved the respondents from responsibility on grounds similar to those expressed by Denning LJ.

[His lordship continued in the words set out on p. 186, and went on:]

My Lords, one of the difficulties of this case as it has come to this House is that your Lordships do not yet know whether or not there had been over-payments on previous certificates, and whether, in fact, any sum was due when certificate No. 21 was issued. An interlocutory order was made, by consent, postponing the ascertainment of these figures until after the decision on obstruction. It would be a curious result if your Lordships were to hold that the issue of 'the sort of certificate' to which the appellants were entitled has been obstructed or interefered with so as to justify their determining the contract, if it should eventually be found that they were not at the date of the certificate entitled to any payment.

These considerations tend to confirm me in the view that errors due to negligent valuation were never intended to give rise to the remedy afforded to the appellants by clause 20, but fall to be dealt with only by arbitration under clause 27. Holding as I do that the acts and omissions found against H do not amount to obstruction or interference within clause 20, it follows that the letter from the town clerk of 29 October 1951, in which he asserted that a proper valuation executed strictly in accordance with the contract had been made, cannot of itself constitute obstruction although, in fact, the valuation was not correct or properly made.

For these reasons I would dismiss the appeal.

Canterbury Pipe Lines Ltd v. Christchurch Drainage Board

COURT OF APPEAL OF NEW ZEALAND   (1979) 16 BLR 76

The facts sufficiently emerge from the judgment of Cooke and

Woodhouse JJ delivered by Cooke J. (McMullin J delivered a separate judgment differing on some points).

COOKE J: Nineteen years ago, on 27 April 1960, the Christchurch Drainage Board accepted the tender of Canterbury Pipe Lines Limited for the laying of about 63 chains of sewer pipes in the Sparks Road area and incidental work. Subject to variations or extras, the contract price was £29,228 (omitting shillings and pence). The specified date for the completion of the works was nine months from acceptance of the tender, namely 27 January 1961. The contract provided that in a number of important respects the engineer of the board should be left with the responsibility of making decisions.

Work started and carried on until September 1960. With the acquiescence of the board there was then a delay pending delivery to the contractor of a piece of heavy duty trenching equipment called a Parsons trenchliner. Once available this machine was expected to enable the contract to be completed more expeditiously. In the event its advantage in that respect proved to be accompanied by the disadvantage that more backfill had to be imported. The machine having been obtained by the contractor, work began again after an interval of about five months. Progress was then quite rapid until mid-April. Then the board's engineer took the view that an amount of some £926 allowed by the contractor in its tender for restoration of the road surface was inadequate; and conversely he maintained that other items had been treated by the contractor too generously in arriving at the total tendered price. The engineer thought that £3,200 was likely to be needed for the restoration. A difference of opinion arose between the engineer and the governing director of the contracting company as to the extent of the company's responsibility in the matter. The amount of restoration needed appears to have been much in excess of anything contemplated on either side at the time of the tender. Against that immediate background and certain difficulties that had arisen between the two men regarding other contracts being carried out by the company for the board, the engineer decided that the £3,200 should be built up in effect in the board's hands before any further monthly progress payments were made to the contractor. Normally, the next progress payment would have been made in May and would have been for about £1,600; but for the reason just mentioned the engineer did not certify for it.

Other differences arose between the parties at about the same time. They arose from claims by the contractor to be paid for the extra backfill necessitated by the machine and to be paid for some extra work at more than schedule rates. The history of the dispute and the correspondence are dealt with in more detail in the judgment of McMullin J, which we have had the advantage of reading. For our purposes it is enough to say that the contractor stopped work and that in reliance on a clause in the general conditions of the contract the engineer then took the work out of the contractor's hands. It was completed partly by two other contractors engaged by the board, partly by the board's own employees. . . .

It is unnecessary to decide whether the unjustified withholding of progress certificates by the engineer would have entitled the contractor to rescind its contract with the board. . . . [The contractor] elected to keep the contract on foot. What the contractor did was to suspend the work. The question is whether there was any right to do that.

Mersey Steel and Iron Co v. *Naylor, Benzon & Co* (1884) concerned a contract for the sale of steel to be delivered by instalments, the instalments to be separately paid for. It was argued that a contract of that kind was made on the assumption that the parcel already delivered would be paid for punctually and in time to put the manufacturer in funds to provide for the manufacture and delivery of the next parcel. The House of Lords held, however, that payment for a previous delivery was not a condition precedent to the right to claim the next delivery; and further that, although a refusal to pay could in some circumstances amount to repudiation entitling the other party to rescind, the buyers in that case had not repudiated by merely postponing payment on erroneous legal advice that the leave of court was required because the sellers were facing a winding-up petition. Ever since then it has been taken to be the law of England that except in the event of the buyer's insolvency payment for previous instalments is not normally a condition precedent to the liability to deliver, although it can be made so by express provision (*Ebbw Vale Steel Co* v. *Blaina Iron Co* (1901)), or no doubt by necessary implication in the particular contract. . . .

Obviously the argument which failed in the *Mersey Steel* case could have been advanced with no less force by a building contractor from whom progress certificates or payments had been wrongly withheld. The *Mersey Steel* decision goes far to explain the apparent absence in English building contract law of any recognition of a common law right to suspend work for wrongful withholding of a progress certificate or payment, as distinct from a right to rescind for a breach going to the root of the contract. If the contract provides for payment on interim certificates from an architect or engineer, the issue of a certificate will normally give rise to a debt payable by the employer (subject, however, to the other terms of the contract). And in some cases of wrongful failure to certify the contractor is able to sue the employer without a certificate. Apart from suing for interim payments, or requiring arbitration where that is provided for, the remedy – and apparently the only remedy – which the contractor is recognised as having at common law is rescission if a sufficiently serious breach has occurred. If he chooses not to rescind, his own obligations continue. He is bound to go on with the work. All the available English and Commonwealth textbooks on building contracts state the law consistently with this view. . . .

There is an established principle, illustrated in the building contract field by *Roberts* v. *Bury Commissioners* (1870), which is put in various ways: that no person can take advantage of the non-fulfilment of a condition the performance of which has been hindered by himself; that a party is exonerated from the performance of a contract when the performance is rendered impossible by the wrongful act of the other contracting party; or, more emotively, that a party cannot take advantage of his own wrong. Failures by an employer to make the site available or to supply plans are illustrations. All the cases cited in argument in this connection, and all others that we have found, were cases of physical impossibility or hindrance. No case was cited where the failure was solely in the payment of money. . . .

We think this court should hesitate before seeking to introduce into the building contract law of New Zealand a new general rule as to the right to suspend work. There can be no greater need now for such a rule than in the past. Building contracts have been traditionally a fertile source of disputes. Yet, as has been seen, the rule has not evolved in England or New Zealand or (apparently) Australia. *Kamlee Construction Ld* v. *Town of Oakville* (1960) suggests that it may not have evolved in Canada either. (But support for it is evidently afforded by . . . *Hoskins* v. *Barber* (1879).) On the other hand such a rule does appear to be recognised in the United States as regards building contracts and some other contracts.

One would have to bear in mind also that building contracts are commonly made in standard forms. For many years the practice has been to provide expressly for a right of suspension if that is intended. For instance the standard forms issued by the Joint Contracts Tribunal in England (the JCT forms) in the revisions reproduced in the fourth edition of Keating's *Building Contracts* (1978) give the contractor a qualified right to determine the building contract, on certain notice, for non-payment of the amount due on any certificate . . . but give a sub-contractor against the contractor a right of suspension of sub-contract works after non-payment and notice . . .

The significance of these precedents in the present context is twofold. In the first place the parties know where they stand if the rule is that there is no right to suspend work unless it is expressly given in the contract. Customarily the period of delay and the requisite notice are then specified. Some uncertainty would be inevitable if on this pattern there were now superimposed a common law right of suspension. But, bearing in mind that a tendering contractor has little bargaining-power as to the clauses in the contract, that objection need not be treated as decisive. In the second place, however, there is a much more formidable point. Where, as is common, a certificate by an architect or engineer is a condition precedent to a progress payment, we have come across no instance of a New Zealand or English form of contract clearly giving the contractor a right of suspension for non-payment *when no certificate had been issued.* No doubt the reason is that such a right may be seen as subverting the scheme of the contract in placing the architect or engineer in charge.

This case does not call for a decision on whether New Zealand law recognises a general right in the contractor to suspend work (as distinct from rescinding) by reason of substantial default by the employer in paying a certified progress payment or a progress payment that has fallen due under a contract not requiring certificates. For the reasons already given, it seems to us as well to leave that question open. Whatever the answer to it, we are against recognising such a right when the architect or engineer has declined to issue the certificate stipulated for by the contract as a condition precedent to the employer's duty to pay. It would disrupt the scheme of these contracts. It could encourage contractors to take the law into their own hands. They might stop work which in the public interest needs to be done promptly. In such cases, if the contractor cannot or does not wish to rescind and cannot prove impossibility or its equivalent, he will be left with whatever remedies regarding the recovery of progress payments may be available to him under the contract. As already indicated we do not exclude the view that a case of impossibility or its equivalent *might* be made out by proving that for want of money the contractor could not carry on or could not reasonably be expected to do so. The present case was fought in a way not truly raising this issue. . . .

In our opinion it should be held in the light of these authorities that in certifying or acting under clause 13 here the engineer, though not bound to act judicially in the ordinary sense, was bound to act fairly and impartially. Duties expressed in terms of fairness are being recognised in other fields of law also, such as immigration. Fairness is a broad and even elastic concept, but it is not altogether the worse for that. In relation to persons bound to act judicially fairness requires compliance with the rules of natural justice. In other cases this is not necessarily so. But we do not think that it can be confined to procedure. Its use in the authorities in combination with 'impartiality' suggests that it is not meant to be a narrow concept. . . .

The result, as we see it, is that the engineer was acting less than fairly and the notice was accordingly invalid. On this view it follows that, while the contractor was in the wrong in suspending the work, the employer was also in the wrong in purporting to act under clause 13 by taking the works out of the contractor's hands and reletting to other contractors. This unusual impasse is something which the parties could not reasonably have foreseen when they made their contract. To hold the parties to further performance after reletting would alter the fundamental nature of the contract. Indeed it would not have been physically possible. They are both to blame. The question is what should be done to bring the dispute to a just end.

The board had the benefit of more work from the contractor than has been paid for. The parties are agreed that at contract rates the amount unpaid is $3,336. In addition the board is holding retention monies amounting to $4,897 and the deposit of $1,172 which was required with the tender. It is reasonable that the contractor should recover these sums. [Counsel] contends, however, that the case should be remitted to the Supreme Court to allow the contractor to pursue its counterclaim for loss of profit or *quantum meruit*. He cites *Lodder* v. *Slowey* (1904). But there the contractor was not himself at fault.

Where one party is entitled to determine the contract at common law, he will also be entitled to damages to compensate him for his loss. Where a party is terminating under an express provision of the contract, he will only be entitled to such further remedy as the contract gives him.

Thomas Feather & Co. (Bradford) Ltd v. Keighley Corporation

QUEEN'S BENCH DIVISION   (1953), 52 LGR 30

Builders contracted to erect houses for the corporation. The contract contained the usual prohibition against sub-contracting without the employers' consent but also expressly provided for a right to determine in the event of sub-contracting without consent. The contractors did sub-contract without consent. The corporation determined the contract and employed other builders to finish the work at an extra cost of £21,000. An arbitrator awarded this sum to the corporation and the builders applied to the court for a review of this decision.

Held: As the breach was of a kind which would not have entitled the corporation to determine but for the express provision of the contract, the corporation were not entitled to the extra cost of completing the houses.

LORD GODDARD CJ: Much as I respect the opinion of the arbitrator in this case I have come to the conclusion that the view taken by him is wrong. I think the meaning of this clause is clear enough. Here is a particular term of a contract which, ordinarily, would not be regarded as one going to the root of the contract or being of the essence of the contract, so that a breach of it would not have given the corporation power to determine the contract or to treat that particular breach as a repudiation by the builder. It seems to me that it would be impossible to say that, in a contract of this magnitude and description, the mere fact that the contractor might put out some small portion of the work, or even a large portion of the work, to a sub-contractor would be a repudiation of the contract. He is not by so doing saying that he does not intend to perform the contract. But what he is doing is breaking a collateral term which says that he is not to sub-contract, and the fact of his sub-contracting might not cause a pennyworth of damage to the corporation. But so anxious are the corporation that there should be no sub-contracting that they have put in this special provision, that this particular requirement, namely, that there is to be no sub-contracting, is to be regarded as of the essence of the contract, which it would not otherwise have been, in my opinion, but for that provision. Then it is provided that if there is a breach of it the corporation can determine the contract, or they can claim £100. In my opinion, it is perfectly clear what it means. It means that the corporation can determine the contract, and can determine the contract without the contractor being able to claim any damages against them for so doing. But I cannot see that they need that to preserve what may be called the common law liability. If the contractor repudiates a contract, and shows by his conduct that he does not mean to be bound by it, then, of course, the building owner can recover any extra cost to which he may be put by way of damages. That was the position in the case to which my attention has been called, *Marshall* v. *Mackintosh* (1898), a case in which the contractor did nothing.

I think that the clear meaning of this contract is that if the builder breaks one of these conditions, or certainly the condition against sub-contracting – whether it applies to the earlier part of the clause, the 'Fair Wage' clause, or not, I need not decide, but I am inclined to think it does, though I am only concerned with this matter of sub-contracting – the corporation have certain rights, and that is that they can put an end to the contract once and for all. I would have expected to find, if it was intended that, in those circumstances, the contractor would be liable for damages, that there would have been an express provision put in to that effect. I think that this provision simply gives the corporation a right to terminate the contract, which they would not otherwise have had, and that it gives them nothing more.

J.M. Hill and Sons Ltd v. London Borough of Camden

COURT OF APPEAL   (1980), 18 BLR 31

The facts and issues sufficiently appear from the judgment.

LAWTON LJ: This appeal arises out of a claim made by J.M. Hill & Sons Ltd, who are building contractors, against the Mayor and Burgesses of the London Borough of Camden, who for the purposes of this case will be described as 'the employers'. The appeal in form is from an order made by Willis J on 12 December 1979, whereby he dismissed the defendants' appeal from the order of Master Lubbock dated 17 July 1979, whereby Master Lubbock gave leave to the plaintiffs to enter judgment for the sum of £84,518 with interest thereon at the rate of 12 per cent per annum from 30 March 1979 to the date of judgment

The building contract out of which the claim arose was made in 1976. It related to the development of a site in the London Borough of Camden called the New Calthorpe Development. The contract in form provided for specified works to be done and incorporated therein were standard conditions. It is described as ' "Standard Form" of Building Contract for use With Quantities for Local Authorities' Building Works'.

The plaintiffs started working under that building contract and from time to time they were given interim payments on architect's certificates pursuant to condition 30 of the conditions. As almost always happens with building contracts, from time to time work was done which the plaintiffs claimed was extra work and for which they said they were entitled to be paid sums over and above the contract price, which was just over £1,900,000. By January 1979 the plaintiffs claimed that something like £100,000 was due to them for extra work, which had not been certified as due by the architects appointed under the contract. As a result, in January 1979 the directors of the plaintiff company called upon the appropriate officials of the London Borough of Camden and had some discussion with them about the payment of monies which they claimed to be due.

It is understandable why the plaintiffs at that time had anxiety about getting money which they said was due to them, because, as is well known, the cost of borrowing money at the present time, and at all material times under this contract, had been very high indeed. Cash flow problems (as they are now called) are very important in the performance of all building contracts of the size of this one.

As a result of the representations made by the plaintiffs' directors to the London Borough of Camden, some informal arrangement seems to have been made, the effect of which was that the employers said that they would do their best to expedite payments due under architect's certificates; and they were as good as their word for the months of January and February. In both of those months the sums certified by the architect were paid promptly.

On 15 March 1979 an architect's certificate was issued for the amount claimed in this case. It was not paid as promptly as the certificates for January and February 1979 had been paid. It seems to me clear that the plaintiffs were annoyed at finding that the employers were not going to pay as they had done in the previous two months. As a result, on 23 March 1979 they decided to cut down the size of the labour force on the site and they gave notice to the employers' clerk of work that that is what they were going to do; and on that Friday they started removing from the site a number of pieces of equipment, such as dumpers and concrete mixers. They also removed the portable office of the quantity surveyor who was on the site. The clerk of works reported to the employers what was happening and they claimed that they were apprehensive as to what the intentions of the plaintiffs were, the plaintiffs having said that they could not afford to go on working on the site in the way they had previously done.

It is necessary now to look to see what happened on Monday, 26 March 1979. The plaintiffs did not abandon the site at all; they maintained on it their supervisory staff and they did nothing to encourage the nominated sub-contractors to leave. They also maintained the arrangements which they had previously made for the provision of canteen facilities and proper insurance cover for those working on the site.

The employers took the view that they were entitled, in the circumstances, to regard the plaintiffs' behaviour in cutting down the labour force on the site as a repudiatory breach of the contract. As a result, by letter dated 28 March 1979, they purported to exercise rights which they had under the contract, alleging (and at this stage I will not go into details about the specific terms of the condition) that the plaintiffs had failed to proceed 'regularly and diligently' with the works. The plaintiffs were not willing to accept this notice; and they had good reason for refusing to accept it, because it had not been issued by the architect, as the condition under which it purported to be issued specified. But what the plaintiffs did was themselves to issue a notice, the details of which I shall refer to later, but the effect of which was to say 'You have got seven days in which to pay the amount specified in the certificate and if you do not pay we shall exercise our rights under the contract to determine the employment'.

The date upon which they were entitled to determine the employment was 11 April 1979 and the letter which purported to exercise their rights of determination was received by the defendants, without any doubt at all, on 11 April. On that day the plaintiffs' solicitor telephoned the appropriate official of the Camden Council. He asked whether the employers were going to pay the amount under the certificate. He was told on the telephone that they were not. In consequence, he gave advice to his clients and on the same day the plaintiffs issued a writ claiming the amount specified in the certificate.

It is important, therefore, for this court to bear in mind that what the plaintiffs were seeking to get in this case was the value of the work properly executed and of the materials and goods delivered to or adjacent to the works for use thereon. In other words, they were asking to be paid on account for work done and material supplied. The issue in this case is whether an event happened which deprived them of that clear right to be paid for the work done and materials supplied.

The employers thought that they had a right to exercise their powers under condition 25. Sub-paragraph (1) of that condition reads as follows:

'Without prejudice to any other rights or remedies which the employer may possess, if the contractor shall make default in any one or more of the following respects, that is to say:

(a) If he without reasonable cause wholly suspends the carrying out of the works before completion thereof, or

(b) If he fails to proceed regularly and diligently with the works'

he shall be entitled to determine the contract, but that must be done under the certificate of the architect.

One of the curious features is that the architect, although he has sworn an affidavit relating to certain conversations which took place on or about 23 March 1979, has not at any time suggested, either that the plaintiffs failed regularly and diligently to proceed with the work or that the defendants have suffered any damage as a result of what happened on or about 23 March. The employers have accepted in the course of this litigation so far that the notice which they purported to give under condition 25 was ineffective for any purpose under the contract.

The condition under which the plaintiffs purported to act is condition 26. The relevant parts read as follows:

'(1) Without prejudice to any other rights and remedies which the contractor may possess, if

(a) The employer does not pay to the contractor the amount due on any certificate . . . within 14 days from the issue of that certificate and continues such default for seven days after receipt by registered post or recorded delivery of a notice from the contractor stating that notice of determination under this condition will be served if payment is not made, within seven days from receipt thereof' (then there are other situations which are not relevant to this case), 'then the contractor may thereupon by notice by registered post or recorded delivery to the employer or architect forthwith determine the employment of the contractor under this contract; provided that such notice shall not be given unreasonably or vexatiously.'

The following problems arise on the facts of this case looked at against the background of the relevant provisions of the contract. First, did the plaintiffs, by their conduct on or about 23 March 1979, unlawfully repudiate the contract? In my judgment, it is impossible to say that they did anything of the kind. The one thing they did not purport to do was to leave the site, and indeed the employers have never suggested that they did. Indeed, their subsequent conduct indicates, in my judgment, that they were treating the contract as still subsisting. All that can be said against them is that by removing men and plant from the site in the way they did they were not 'regularly and diligently' proceeding with the works. But by 11 April 1979, which is the material date for the purposes of this case, if there was any damage sustained by the employers, it must have been minimal. It has never been quantified and it follows, therefore, that it can bear no relevance whatsoever to the amount due to the plaintiffs under the certificate. I can see no reason why the defendants should say that the plaintiffs have unlawfully repudiated the contract by what they did on 23 March. The most they can say is that they may be able to prove some small amount of damage as a result of the activities of the plaintiffs at the material date. . . .

So the position in law is as follows: both sides were claiming at or about the middle of April that the other had repudiated the contract. For the reasons I have already stated in my judgment, the defendants were wrong in thinking that the events of 23 March 1979 amount to an unlawful repudiation of the contract by the plaintiffs. . . .

In my judgment, the important word in [clause 26(1)] is 'served'. 'Service' means that the notice has got to be given to the other party. It cannot be 'served' on the other party until the other party has received it; and it is after the other party has received the notice that determination of the employment comes about. On the facts of this case, the service on the employers was on 11 April 1979. That was seven days after receipt by the employers of the warning notice as to what would happen if payment was not made within due time. It follows, in my judgment, that, on the proper construction of condition 26(1), there is no room for argument as to whether the plaintiffs had properly determined the contract on 11 April. They had and the view of those who conducted the employers' case before Master Lubbock and Willis J was correct to the extent that, on this point of construction, they made no submissions.

The final matter which has to be dealt with is the rather odd proviso to condition 26(1), namely, the words 'provided that such notice shall not be given unreasonably or vexatiously'. I am far from clear as to what kind of conduct on the part of a contractor would make what he did unreasonable or vexatious. [Camden's counsel] stated what his clients allege was unreasonable about the plaintiffs' conduct. He put his submission in this way. He said that they were wrong to take some of their men off the site on or about 23 March 1979. Because they took their men improperly off the site, then the employers, in anticipation that they might have a claim against the plaintiffs for doing what they had done, decided to withhold payment of the money due under the certificate. In those circumstances, for the plaintiffs to purport to determine the employment pursuant to condition 26(1) was unreasonable.

In my judgment, it was not unreasonable, for two reasons. First, as I have already stated, at the date when all this was happening any damage which the employers had sustained would have been minimal anyway and, secondly, they had no business at that stage to withhold money for work which had already been done and materials which had already been supplied. As I have said, the very essence of the provisions of the contract about payment on the architect's certificate was to maintain the cash flow of the contractors and when the cash flow is cut off without good reason – in my judgment, it was cut off without good reason in this case – it does not lie in the mouth of the employers to say that the plaintiffs were acting unreasonably or vexatiously.

There is often serious dispute as to whether one party has in fact done or failed to do that which would entitle the other party to determine the contract. This problem has caused particular difficulty where the employer wishes to determine the contract and turn the contractor off the site. In the first of the two cases below, Megarry J held that where the employer sought the aid of the Court by way of interim injunction ordering the contractor to leave the site he must show a clear entitlement to determine under the contract. That view has been widely criticised and was not followed in the second case.

London Borough of Hounslow v. Twickenham Garden Developments Ltd

CHANCERY DIVISION   (1970), 7 BLR 81

The dispute arose under a contract in JCT 63 terms. The architect purported to give notice to the contractors under clause 25(1) stating that in his opinion they had failed to proceed regularly and diligently with the works and that unless there was an appreciable improvement within 14 days the council would be entitled to determine their employment. By a subsequent letter, the council purported to determine the contractors' employment under the contract. The contractors alleged that the notice of determination was wrongful and amounted to a repudiation of the contract. They refused to accept the alleged repudiation and elected to proceed with the works.

The council claimed an injunction to restrain the contractors from remaining on the site.

MEGARRY J: [In] this case the contract is one for the execution of specified works on the site during a specified period which is still running. The contract confers on each party specified rights on specified events to determine the employment of the contractor under the contract. In those circumstances, I think that there must be at least an implied negative obligation of the borough not to revoke any licence (otherwise than in accordance with the contract) while the period is still running. . . .

I hold that without establishing the efficacy of the notices given under condition 25, the borough is not entitled to the injunctions sought. . . .

In the event, my conclusion is that although the borough has established some sort of a case for having validly determined the contract, that case falls considerably short of any standard upon which, in my judgment, it would be safe to grant this injunction on motion. What is involved is the application of an uncertain concept to disputed facts. As is so often the case on motion, a court is faced with a choice of evils. I fully accept the importance to the borough, on social grounds as well as others, of securing the due completion of the contract, and the unsatisfactory nature of damages as an alternative. But the contract was made, and the contractors are not to be stripped of their rights under it, however desirable that may be for the borough. A contract remains a contract, even if (or perhaps especially if) it turns out badly. The borough may indeed be able at the trial to make out a formidable case in support of its contentions, just as the contractor may be able to make out a formidable case in reply; and with the full procedure of the trial, and in particular with the advantage of seeing and hearing the witnesses, and the testing process of cross-examination (not least on any expressions of expert opinion), the court may well be able to reach a firm conclusion on one side or the other. But I lack these advantages, and with so much turning on disputed questions of fact, and inferences from the facts, I cannot say that the borough has made out its case for the injunction that it seeks.

Mayfield Holdings Ltd v. Moana Reef Ltd

SUPREME COURT OF NEW ZEALAND   [1973] 1 NZLR 309

Disputes arose between the parties to a building contract. The employer refused to make further payments until alleged defects were remedied. The contractor retaliated by slowing down work, and the employer called in another contractor who was denied access by the first contractor. The owner sought an injunction against the first contractor.

MAHON J: The question at issue in the *Twickenham Garden* case was whether an interlocutory injunction in favour of the owner should be granted pending trial of the action. The plaintiff council, as owners of the building land engaged the defendant contractor to construct roads, car parks and other services for a proposed housing estate, the work to occupy a period of four years. The contract was in the standard Royal Institute of British Architects form and it contained a condition to the effect that in the event of specified default of the contractor, and in particular in the event of the contractor failing to proceed diligently with the works, the architects of the plaintiff council could give notice specifying the default and that unless the default was remedied the plaintiff could give notice determining the contract. At one stage work was halted in consequence of a long strike which itself was not a default under the contract but work on the site was continued thereafter at a pace which the plaintiff's architect thought was too slow. He gave notice of default in terms of the contract. A month later the council gave notice that the default had not been remedied and that the employment of the contractor was therefore terminated. The council then issued a writ claiming damages for trespass in failing to vacate the site, and they also asked for a perpetual injunction. They then moved for an interlocutory injunction to restrain the contractor before trial from entering, remaining or otherwise trespassing on the site or interfering with the plaintiff council's possession of the site. There were two main issues:

(1) Disregarding the question of validity of the notice were the council, as owners, entitled to determine the licence of the contractor to enter and remain on the site?

(2) Were the notices valid and effective so as to determine the contractor's employment?

Megarry J found in favour of the contractor. There appear to have been two grounds for his decision:

(1) The contract was for the execution of specified works over a period which was still continuing when the notices were given and the learned Judge held that there must have been an implied negative obligation on the part of the Borough Council not to revoke the contractor's licence (otherwise than in accordance with the contract) while the contract period was still running, and that a Court of Equity would not assist the council to break its contract.

(2) Although the Borough Council had established 'some sort of a case' for having validly determined the contract, that case fell considerably short of any standard upon which it would be safe to grant an interlocutory injunction.

In respect of the first ground of decision, Megarry J reviewed the relevant authorities in relation to licences to enter upon land and held that this was a contractual licence containing the implied negative covenant above referred to with the result that equity would restrain the Borough Council from revoking the licence in breach of the contract of which it formed part. It followed from this that the contractor could successfully resist the issue of a mandatory injunction ordering the contractor from the site unless and until the Borough Council established a material breach of contract carrying with it the right of revocation of the licence. As to the second ground, because there were substantial questions at issue as to whether any breach of contract had been committed, it therefore followed that the Borough Council was unable in the interlocutory proceedings to prove the material breach of contract which alone would have justified forfeiture of the contract and concurrent revocation of the contractor's licence. Many points of argument were presented to Megarry J but so far as the revocability of the licence was concerned there were two principal submissions made on behalf of the contractor, firstly that the licence was coupled with an interest, and secondly, in the alternative, that the contract contained an implied negative covenant on the part of the corporation not to revoke the licence in breach of the contract. Megarry J accepted the second of these submissions and did not come to a conclusion on the first. I must examine both these arguments as they are each available to the contractor in the case before me. Thus the crux of the decision on this part of the case was the finding of an implied negative covenant in the terms expressed. To this finding there are two possible objections – (a) that the existence of such an implied term might not be thought necessary to give the contract business efficacy, (b) that even if it did exist it could not be specifically enforced, with the result that the defence to the injunction claimed, being in effect a claim for specific performance, could not be maintained. I will deal with these objections in the order named.

The proposed implication of a term in a contract, in particular a commercial contract, calls for careful consideration. A Court is not entitled to reconstruct an agreement on equitable principles, or on a view of what the parties should have contemplated. The implication of the unexpressed term must be necessary to give to the transaction such business efficacy as the parties must have intended. 'The implication must arise inevitably to give effect to the intention of the parties' – *Luxor (Eastbourne) Ltd* v. *Cooper* (1941) *per* Lord Wright. With these observations in mind, what would be the presumed common intention of the parties to an ordinary building contract in the event of the development between them of a dispute amounting to an impasse relating to construction of the works? The consequences of the impasse have to be considered. The works have been part completed, and the owner claims to be entitled to terminate the contract for reasons of faulty workmanship, deviation from plans, slow progress, or the like. The owner stops progress payments and instructs his architect not to certify for further work. In theory the contractor can complete the next section of work and bring an action, with its consequential delays, for the progress payment due in respect of that work, raising default by the owner as an answer to the defence that there is no architect's certificate. But in practice, the delay involved would be insurmountable. The contractor would have to lay off his employees, or transfer them to another job. In the absence of progress payments he probably could not pay his sub-contractors. The architect's authority to co-operate in the construction would be withdrawn. Under these circumstances, it seems impossible to say that the contractor can obtain any benefit by remaining on the job. He has a claim for unpaid work to date, and a claim for loss of profit on the whole contract, and instead of working on under difficulties and suing for unpaid progress payments he surely would take the alternative course of leaving the job and suing to enforce his claims, thus recovering by action, if not in default under the contract, the moneys to which he would have been entitled if the works had been completed by him pursuant to the contract. The contractor's only interest in the transaction is to complete the specified works and at a profit to himself. The owner's interest is to secure completion in accordance with his plans and specifications. Each object will still be attained even if the owner, in breach of his contract, procures the dismissal of the contractor from the site and completes the specified works by contract with another builder. Judged from the standpoint of business efficacy, I can for myself see no difficulty in implying a term between builder and owner that in the event of a complete breakdown in their contractual relationship, the builder will surrender the site and the works, leave the owner to complete as and when he will, and enforce by action his remedy against the owner for breach of the building contract. I quote in this context the observations of Knight Bruce LJ in *Garrett* v. *Banstead and Epsom Downs Railway Co* (1864), referred to in Hudson's *Building and Engineering Contracts* (10th edition) page 712:

> 'To suppose in a case like this, where, if the company are wrong, ample compensation in damages may be obtained by the contractor, that the company are to have a person forced on them to perform these works whom they reasonably or unreasonably object to (whereas there would be no reciprocity if the wrong were on the other side), for the purpose of compelling the performance of the works, is more than I am able to do.'

The opinion expressed by Knight Bruce LJ can be seen to be inspired by the lack of mutuality of remedy but it might also be justified upon a further ground. The practical exigencies of performance of a building contract must be kept in mind. The purported forfeiture by the owner may be based on alleged departure from the contract specifications. Must the owner be compelled to stand by while his action for breach awaits trial, and watch the building being completed in a manner which may ultimately be decided to have been in breach of the contract? Certainly if the departure from the contract terms were clear and unanswerable I should imagine that nothing in the *Twickenham Garden* case would prevent the issue of an interlocutory injunction dismissing the contractor from the site, subject to correct notices of forfeiture being given, but whenever the supposed breach was a matter of genuine controversy, as it so often is, the *Twickenham Garden* decision would maintain the contractor not only in possession of the site but in continuation of the works on a disputed basis. It is difficult to accept that either contractor or owner would ever have agreed, at the formulation of their contract, to any express term carrying with it such drastic consequences. For the reasons which I have expressed I cannot agree, with respect, with the view taken by Megarry J in the *Twickenham Garden* case. I am fortified to some extent in the view which I have taken by the editorial criticism of the *Twickenham Garden* decision in *87 Law Quarterly Review* 309 where it is said, at the end of the note on the case:

> 'The contractor's sole object is to make a profit; if he gets it, and suffers no loss, why should not the owner be entitled to change his mind and exclude him from his property?'

I am clear that this is the sense of the unexpressed term which the law must imply where an impasse between owner and building contractor is envisaged. A continuation of the impasse could only be to the detriment of both parties. Surely the contractor would say: 'If a final breakdown in our relationship occurs, or if you decide to cancel the contract without due cause, then I agree that you go ahead and finish the job yourself, and I will sue you for my lost profit.' I am therefore not prepared to hold in the case before me that there was a covenant implied in the contract binding the owner not to revoke the contractor's licence in breach of contract.

Chapter 18

Limitation

'Limitation' is the name given by lawyers to the rules prescribing within what periods actions must be started. Failure to start the action in time does not theoretically extinguish the obligation but the court will not give its assistance to the prosecution of the claim. (So if the defendant pays up after the limitation period, the payment is valid and cannot be recovered).

The governing statute is currently the Limitation Act 1980 the most important sections of which in relation to building contracts are:

THE LIMITATION ACT 1980

S.2 Time limit for actions founded on tort

An action founded on tort shall not be brought after the expiration of six years from the date on which the cause of action accrued.

S.5 Time limit for actions founded on simple contract

An action founded on simple contract shall not be brought after the expiration of six years from the date on which the cause of action accrued.

S.8 Time limit for actions on a specialty

(1) An action upon a specialty shall not be brought after the expiration of twelve years from the date on which the cause of action accrued.

(2) Subsection (1) above shall not affect any action for which a shorter period of limitation is prescribed by any other provision of this Act.

S.32 Postponement of limitation period in case of fraud, concealment or mistake

(1) Subject to subsection (3) below, where in the case of any action for which a period of limitation is prescribed by this Act, either –

    (a) the action is based upon the fraud of the defendant; or

    (b) any fact relevant to the plaintiff's right of action has been deliberately concealed from him by the defendant; or

    (c) the action is for relief from the consequences of a mistake;

the period of limitation shall not begin to run until the plaintiff has discovered the fraud, concealment or mistake (as the case may be) or could with reasonable diligence have discovered it.

References in this subsection to the defendant include references to the defendant's agent and to any person through whom the defendant claims and his agent.

(2) For the purposes of subsection (1) above, deliberate commission of a breach of duty in circumstances in which it is unlikely to be discovered for some time amounts to deliberate concealment of the facts involved in that breach of duty.

(3) Nothing in this section shall enable any action –

    (a) to recover, or recover the value of, any property; or

    (b) to enforce any charge against, or set aside any transaction affecting, any property;

to be brought against the purchaser of the property or any person claiming through him in any case where the property has been purchased for valuable consideration by an innocent third party since the fraud or concealment or (as the case may be) the transaction in which the mistake was made took place.

(4) A purchaser is an innocent third party for the purposes of this section –

    (a) in the case of fraud or concealment of any fact relevant to the plaintiff's right of action, if he was not a party to the fraud or (as the case may be) to the concealment of that fact and did not at the time of the purchase know or have reason to believe that the fraud or concealment had taken place; and

(b) in the case of mistake, if he did not at the time of the purchase know or have reason to believe that the mistake had been made.

It is not always easy to establish 'the date on which the cause of action accrued'. In contract *cases it will normally be the date on which the contract was broken. The next two cases concern claims in* tort *and vividly illustrate the difficulties which arise where the defect is covered up.*

Anns v. London Borough of Merton

HOUSE OF LORDS (1978), 5 BLR 1

It was alleged that in 1962 a builder had erected a block of maisonettes with defective foundations and that the defendant council had, through its inspector, either failed to inspect the foundations or had inspected them carelessly. The plaintiffs had taken long leases of some of the maisonettes. In February 1970 structural movement began, resulting in cracks. In February 1972 the plaintiffs issued writs against the builder and against the council. This action was brought to decide whether the proceedings against the council were barred under s.2(1)(a) of Limitation Act 1939. [There is no relevant difference between the Limitation Act 1939 and Limitation Act 1980]. Obviously if the plaintiffs' cause of action against the council accrued in 1962, it was too late to start an action in 1972. In fact although the appeal was nominally on the limitation issue, the speeches in the House of Lords were mainly concerned with the more fundamental question of whether an action would lie against the council at all on such facts.

Held: Such an action would lie: it would not be statute barred.

LORD WILBERFORCE: When does the cause of action arise? We can leave aside cases of personal injury or damage to other property as presenting no difficulty. It is only the damage for the house which required consideration. In my respectful opinion the Court of

Appeal was right when, in *Sparham-Souter* v. *Town and Country Developments (Essex) Ltd* (1976), it abjured the view that the cause of action arose immediately on delivery, i.e. conveyance of the defective house. It can only arise when the state of the building is such that there is present or imminent danger to the health or safety of persons occupying it. We are not concerned at this stage with any issue relating to remedial action nor are we called on to decide on what the measure of the damages should be; such questions, possibly very difficult in some cases, will be for the court to decide. It is sufficient to say that a cause of action arises at the point I have indicated.

The Limitation Act 1939. If the fact that defects to the maisonettes first appeared in 1970, then, since the writs were issued in 1972, the consequence must be that none of the present actions are barred by the Act.

Pirelli General Cable Works Ltd v. Oscar Faber and Partners

HOUSE OF LORDS   (1982), 21 BLR 99

The defendants, a firm of consulting engineers advised the plaintiffs on the design and erection of a boiler flue chimney. The chimney was negligently designed and in due course cracks occurred in the internal lining of the chimney. The chimney was built in June and July 1969. The plaintiffs issued their writ in October 1978. The trial judge found that the cracks occurred not later than April 1970, that the plaintiffs did not discover the cracks till November 1977 and that it was not proved that the plaintiffs could with reasonable care have discovered the cracks before October 1972. On these facts the action was time-barred if time started to run when the cracks occurred, but not if it only started to run when the cracks could with reasonable care have been discovered.

Held: That the limitation period began when the cracks occurred and that the action was therefore statute-barred.

LORD FRASER OF TULLYBELTON: Part of the plaintiffs' argument in favour of the date of discoverability as the date when the right of action accrued was that that date could be ascertained objectively. In my opinion that is by no means necessarily correct. In the present case, for instance, the judge held that the plaintiffs as owners of a new chimney, built in 1969, had no duty to inspect the top of it for cracks in spring 1970. But if they had happened to sell their works at that time, it is quite possible that the purchaser might have had such a duty to inspect and, if so, that would have been the date of discoverability. That appears to me to show that the date of discoverability may depend on events which have nothing to do with the nature or extent of the damage.

Counsel for the plaintiffs argued that in *Anns* v. *Merton London Borough* (1978) Lord Wilberforce, and the other members of this House who agreed with his speech, had approved of the observations in *Sparham-Souter* (1976) to the effect that the discoverability date was the date when the cause of action accrued. But I do not so read Lord Wilberforce's speech. He simply narrated the conflict between the cases of *Dutton* (1973) and *Sparham-Souter* (1976) without indicating any preference. He posed the question 'When does the cause of action arise?' and he answered it as follows:

'In my respectful opinion the Court of Appeal was right when, in *Sparham-Souter* v. *Town and Country Developments (Essex) Ltd*, (1976) it abjured the view that the cause of action rose immediately on delivery, i.e., conveyance of the defective house. It can only arise when the state of the building is such that there is present or imminent danger to the health or safety of persons occupying it.'

The only express approval in that passage is to the Court of Appeal's decision that the cause of action did *not* arise immediately the defective house was conveyed. His Lordship did not say, nor in my opinion did he imply, that the date of discoverability was the date when the cause of action accrued. The date which he regarded as material (when there is 'present or imminent danger . . . health or safety') was, of course, related to the particular duty resting on the defendants as the local authority, which was different from the duty resting on the builders or architects, but I see nothing to indicate that Lord Wilberforce regarded the date of discoverability of the damage as having any relevance. He was not considering the question of discoverability, no doubt because the main issue in the appeal by the time it reached the House was whether any duty at all was incumbent on the local authority. Three other noble and learned Lords expressed agreement with Lord Wilberforce. Only Lord Salmon delivered a separate reasoned speech and he clearly considered that the cause of action could arise before damage was discovered or discoverable, although he recognised that proof might be difficult. He said:

'Whether it is possible to prove that damage to the building had occurred *four years before it manifested itself* is another matter, but it can only be decided by evidence.' (My emphasis.)

Neither Lord Salmon nor the other Lords seem to have considered that they were dissenting from the majority view on that matter. In these circumstances I do not think that the majority in *Anns* are to be taken as having approved the discoverability test applied in *Sparham-Souter.*

In this case their lordships recommended statutory amendment of the law. This suggestion was adopted by the Latent Damage Act 1986, the full text of which is given in the Appendix, p. 402. Note that this Act does not reverse the *Pirelli* case but in certain circumstances introduces a secondary time limit based on discoverability.

PART IV

SUB-CONTRACTS AND
SUB-CONTRACTING

Chapter 19

Assignment and Sub-letting

Building contracts commonly contain clauses restraining assignment and sub-letting. Even without such clauses a party would not be entitled to assign his obligations without the other party's consent but he would be entitled to assign his rights. At one time contractors commonly obtained finance by assigning the right to receive payment. If the contract forbids assignment, such an assignment is ineffective to transfer rights to the assignee against the employer.

Helstan Securities Ltd v. Hertfordshire County Council

QUEEN'S BENCH DIVISION   [1978] 3 All ER 262

Messrs Renhold contracted with the council for road works on ICE, 4th edition. Condition 3 provided: 'The contractor shall not assign the contract or any part thereof or any benefit or interest therein or thereunder without the written consent of the employer.' Similar wording is to be found in most building contracts. Renhold purported to assign their rights to payment under the contract to the plaintiffs.

Held: The assignment was ineffective.

CROOM JOHNSON J: I find no ambiguity such as would lead me to consider the background against which the contract was made as an aid to interpretation. If I did, the background would not help the plaintiffs. The clause is obviously there to let the employer retain control of who does the work. Condition 4, which deals with subletting, has the same object. But closely associated with the right to control who does the work, is the right at the end of the day to balance claims for money due on the one hand against counterclaims, for example, for bad workmanship on the other. The plaintiffs say that such a counterclaim may be made against the assignees instead of against the assignors. But the debtors may only use it as a shield by way of set-off and cannot enforce it against the assignees if it is greater than the amount of the debt: *Young* v. *Kitchin* (1878). And why should they have to make it against people whom they may not want to make it against, in circumstances not of their choosing, when they have contracted that they shall not?

Note: JCT 80, clause 19.1., is a current example of a clause against assignment. In contrast, the proviso to clause 9.1. of the ACA Form of Building Agreement, (2nd ed. 1984) makes it clear that the parties 'may assign by way of charge any monies due or to become due under' the contract without consent.

Chapter 20

Domestic Sub-contractors

A major practical problem in relation to sub-contracts is the way in which the main contract and sub-contract fit together. Provisions of the main contract cannot be read into the sub-contract unless they have been expressly incorporated.

Smith and Montgomery v. Johnson Bros Co. Ltd

ONTARIO HIGH COURT (1954), 1 DLR 392

The defendants were main contractors to construct a tunnel sewer for the City of Hamilton through the Hamilton mountain. The plaintiffs, who were miners, undertook a sub-contract for tunnelling 'according to the dimensions and specifications as set forth in the contract between the City of Hamilton and the [defendants]'.

Held: That these words were not apt to incorporate the terms and conditions of the main contract into the sub-contract.

SCHROEDER J: It is, of course, well established that plans and specifications, if not contained in a contract itself but referred to therein or annexed thereto, must be construed therewith when identified. Wherever the plans and specifications are referred to for a specific purpose only, however, they become part of the contract for that purpose only and should be treated as irrelevant for all other purposes. The covenant of the plaintiffs to do the tunnelling 'according to the dimensions and specifications as set forth in the contract between the City of Hamilton and the [Johnson company]' would undoubtedly require the plaintiffs to dig out a tunnel 10 ft. 6 ins. by 12 ft. 6 ins. in dimensions, and any other provision affecting the physical characteristics of such tunnel no doubt would be applicable to and binding upon the plaintiffs, but I am not prepared to hold that the reference to the specifications as expressed in this context imports into the contract between the plaintiffs and the defendant, so as to make binding upon the plaintiff-subcontractors, who operate in a very small way, such sweeping and extraordinary powers as are exercisable by the city engineer. Only that part of the general specifications is to be read into the contract in question that relates to the specific purpose for which the same were referred to. Counsel have been unable to refer me to any authority in which all the general provisions of specifications forming part of a main contract became part of a subcontract, unless the latter by its terms made very clear the intention of the parties that all terms and provisions of the principal contract were to form part of the agreement between the principal contractor and the subcontractor. The general clauses under which the city engineer ordered a stoppage of the work and a change in the method of procedure are more in the nature of such provisions as one is likely to find in specifications furnished for the guidance of bidders for the original contract. It does not seem to me to be clear from the situation and the circumstances of the parties that it was their intention that, except in the restricted sense already mentioned, the specifications forming part of the original contract should be a part of the subcontract.

In the absence of express undertakings the main contractor does not necessarily undertake to have the site ready for a sub-contractor by a particular date (but only within a reasonable time).

Pigott Construction Co. Ltd v. W.J. Crowe Ltd

ONTARIO COURT OF APPEAL   (1961), 27 DLR (2d) 258

The appellant was appointed main contractor on an extensive government contract. Because of shortages of steel there was no date fixed for the start of plastering work.

The appellant started work on the main contract in September

1955 and the sub-contract was undertaken in the same month. The appellant then expected that it would take about two years to complete the main contract and the respondent expected that the plastering would take about 44 weeks. In September 1956 the respondent was told that the site would be ready for plastering within a few days but shortly afterwards that it would not be ready till the spring of 1957. In April 1957 the appellant informed the respondent that the works were now ready. The respondent then took the position that it would not now do the work unless it was paid extra and claimed that it was so entitled on the ground that the appellant was in breach by not having been ready for the sub-contract work earlier. The trial judge held that there was an implied term in the sub-contract that the building would be ready for plastering not later than 1 January 1957.

Held: There was no such implied term.

LAIDLAW JA: Looking at the contract between the appellant and the Crown and also the subcontract between the appellant and the respondent in a reasonable and business way, there is no necessary implication that the appellant agreed that the male infirmary or any other building to be erected by it would be ready for the commencement of the work of lathing and plastering not later than 1 January 1957. While the care and control of the work generally was the responsibility of the appellant and it was provided in the general instructions and requirements made part of the general contract and subcontract that the work was to be carried forward 'in a proper manner and as expeditiously as possible' nevertheless, it is my view that in the circumstances existing at the time the contract was entered into the appellant would not have assumed an obligation to have any of the buildings to be erected by it, ready for commencement of the work of lathing and plastering within any specified period of time. It would have been utter folly and gross recklessness for it to have done so. It knew that the progress of the work of construction to be done by it was dependent upon the supply of necessary steel and upon other uncertain factors and conditions. It could not foresee whether the delay occasioned by that shortage might continue for a period of weeks or possibly months. Again, the work of 'plumbing and drainage, heating and ventilating, electrical work and elevator and dumb waiter installations' was excluded from the work to be done by the appellant and was provided for under separate contracts. The appellant was not responsible for the commencement or progress of work to be done under those contracts and it realized that commencement of the work of lathing and plastering depended thereon. It is inconceivable that in the particular circumstances any building contractor and especially one possessing the great experience of the appellant, would undertake to have any one part of the work of construction ready for commencement thereof at any particular time or within any prescribed period. Therefore, having these contingent and uncertain matters and conditions clearly and fully in mind, it was the positive and calculated intention of the parties to omit from the terms and conditions of the subcontract an express stipulation and any provision from which it might be implied that the appellant was bound to have any buildings ready for commencement of the work of lathing and plastering at any stated time or within any prescribed period.

The learned trial Judge found contrary to the evidence of the respondent that there was no assurance by the appellant that the respondent could start work in September 1956. Likewise, in my opinion, he ought to have concluded from the evidence and the particular circumstances in this case that it was the intention of both the appellant and the respondent that the appellant was not bound to have any part of the premises ready for commencement of lathing and plastering within any particular period of time after the commencement of the work of construction. It is my respectful view that the term implied by Gale J, is directly contrary to the intention of the parties at the time they entered into the sub-contract and that an obligation was improperly created and imposed on the appellant by the Court. The authorities to which I have referred make it plain that the paramount consideration governing the decision as to whether or not a term should be implied in a contract is in the last analysis the intention of the parties at the time the contract was entered into and the Court should not add to the contract an obligation which clearly was not and could not have been intended by the parties.

Chapter 21

Nominated Sub-contractors

The use of nominated sub-contractors is extensive in modern building contracts but although convenient in some respects, it has generated more than its fair share of legal problems. While lawyers tend to start from the position that a nominated sub-contractor is just another sub-contractor for whom the main contractor is responsible, there is no doubt that, commercially, nominated sub-contractors often look for business to employers and not to contractors and that contractors feel that failures by nominated sub-contractors should be laid at the employer's door. Neither the standard forms nor the cases have fully resolved these tensions.

The contractual position between contractor and nominated sub-contractor must be found in the sub-contract.

A. Davies & Co (Shopfitters) Ltd v. William Old Ltd

QUEEN'S BENCH DIVISION   (1969), 67 LGR 395

The defendant was the main contractor for the erection of a new store on JCT 63 terms, which provided for certain work to be sub-contracted to a sub-contractor nominated by the architect. The architect obtained a tender for this work from the plaintiff and instructed the defendant to accept it. The defendants sent an order to the plaintiff on their standard printed form, which contained on its reverse printed conditions which included a 'pay when paid' clause. The plaintiffs wrote thanking the defendant for the order and carried out the work. The employer became insolvent before having paid for all the work.

Held: The contract between plaintiff and defendant was on the basis of the defendant's printed conditions, which the plaintiff had accepted. The defendant was only liable to pay for the work in so far as he had himself been paid by the employer.

The existence of a contract between the employer and a nominated sub-contractor, as is the case where the JCT Standard Form of Employer/Nominated Sub-Contractor Agreement is executed, does not affect the application of the main contract.

George E. Taylor & Co. Ltd v. G. Percy Trentham Ltd

QUEEN'S BENCH DIVISION   (1980), 16 BLR 15

Taylor were nominated sub-contractors to Trentham, and also had a contract with the employer whereby they warranted due performance of the sub-contract works so that the main contractors should not become entitled to an extension of time. The employer paid Trentham only £7,526.87 against an interim certificate for £22,101, the amount witheld being the balance payable to the sub-contractors after deduction of the main contractor's claim against them for delay.

Held: The employer was not entitled to withhold the money, the contract between employer and nominated sub-contractor being *res inter alios acta.*

JUDGE WILLIAM STABB QC: The architect has issued an interim certificate whereby he has certified that £22,101 is due and payable by the employers to the main contractors, and, by clause 30, the main contractors are entitled to be paid that amount, subject at all time to the employers' common law or equitable rights of set-off (see *Gilbert Ash (Northern) Ltd* v. *Modern Engineering (Bristol) Ltd* (1973). It is not suggested here by the employers that they have any right of set-off as against the main contractors so as to justify the retention of any part of that sum so certified to be due and payable to the main contractors. What the employers contend is that part of that certified sum is due to be paid by the main contractors to the sub-contractors under the provisions of clause 27(b) of the main contract, and that, since they (the employers) have a claim or counter-claim against the sub-contractors under the separate, so-called 'direct' contract, they should be permitted to withhold payment of that sum, certified to be payable under the RIBA contract between them and the main contractors, in order to satisfy wholly or in part damage which they are counter-claiming from the sub-contractors under the 'direct' contract between themselves and the sub-contractors.

So far as sub-contractors and main contractors are concerned, it has been said that the contract between main contractors and employers is *res inter alios acta*. In my view, but even more plainly, it would seem that, so far as employers and main contractors are concerned, the direct contract between employers and sub-contractors is also *res inter alios acta*.

Most building contracts contain some limited provision for direct payment of nominated sub-contractors by the employer. It has been further held that under JCT 63, and the 1963 NFBTE/FASS form of sub-contract, the employer holds retention money on trust for the nominated sub-contractor so that in the event of the contractor's insolvency, the nominated sub-contractor may recover these monies in full. The position is the same under JCT 80.

Re Arthur Sanders Ltd

CHANCERY DIVISION   (1981), 17 BLR 125

The relevant facts are set out in the judgment of Nourse J.

NOURSE J: This is an application by the Greater London Council in the liquidation of a company called Arthur Sanders Limited, which formerly carried on business as a building contractor. Shortly stated, the question is whether the council is entitled to claim a particular set-off under s.31 of the Bankruptcy Act 1914, which is introduced into companies' insolvent winding-up by s.317 of the Companies Act 1948. There is no difficulty over the terms or effect of s.31. The question is what, under two contracts made between the council and the company and those made between the company and six nominated sub-contractors in relation to one of the main contracts, is the extent of the set-off to be made under the section. In other words, the question is more one on the law of building contracts.

As I have said, the council had two contracts with the company, the first dated 9 July 1969 for the execution of work at Wellingborough and the second dated 2 October 1970 for the execution of work at Bletchley. The two contracts were, for present purposes, in identical terms. They incorporated the council's 'Standard Conditions' for use in contracts for works with quantities, which were based on the 1963 edition (as then revised) of the RIBA Standard Conditions applicable to a case where the employer is a local authority. I was told that there has been no material alteration to the RIBA Conditions since 1970 and, further, that there is no material distinction between a case where the employer is a local authority and one where he is a private employer. The six sub-contracts were all in the standard form of sub-contract (1963 edition as then revised) issued under the sanction of the National Federation of Building Trades Employers and the Federation of Associations of Specialists and Sub-Contractors. That is usually known as the FASS sub-contract. Again I was told that there has been no material alteration to that since 1970. It therefore appears that the decision in this case will apply to any building contract for works with quantities which incorporates the RIBA Conditions 1963 edition (as since revised) where there is a FASS sub-contract 1963 edition (as since revised). The particular provisions with which this case is concerned are clauses 25, 27 and 30 of the RIBA Conditions and clause 11 of the FASS sub-contract.

On 13 April 1971 the Company resolved to go into creditors' voluntary winding-up and [a liquidator was appointed]. The effect of that was automatically to determine the employment of the company under both its contracts with the council pursuant to clause 25(2) of the RIBA Conditions. At that date the works under the Wellingborough contract were nearly completed, but those under the Bletchley contract had a long way to go. In neither case was the employment of the company reinstated and the works were completed by other contractors. The result was that the council was left holding net retentions under the Wellingborough contract of £11,086.77, included in which was a sum of £1,374.56 in respect of work carried out by the six nominated sub-contractors, whereas under the Bletchley contract the extra cost to the council for getting the works completed was £98,155. Under the Wellingborough contract the Council owes £11,086.77, but under the Bletchley contract it is owed £98,155. Those figures are not in dispute.

The council contends that it is entitled under section 31 to set off the £11,086.77 owed by it under the Wellingborough contract against the £98,155 owed by the company under the Bletchley contract, and to prove in the liquidation for the balance of £87,068.23. The liquidator accepts that the council is entitled to set off £9,712.21 of the £11,086.77 against the £98,155, but he disputes its right to set off the balance of £1,374.56. He contends that that sum should be paid by the council to him in full, not for the benefit of the general body of creditors, but as trustee for the six nominated sub-contractors. Accordingly, says the liquidator, the council may prove in the liquidation for £88,442.79, but must pay to him the sum of £1,374.56. I should add that the liquidator has so far been able to pay dividends to unsecured creditors amounting to 15p in the £.

. . .

In order to analyse the effect of these provisions, it seems to me that I must start with clause 30(4)(a) of the RIBA Conditions. The effect of that is to provide that the employer shall hold the retentions under interim certificates on trust for the contractor, whose beneficial interest therein is expressed to be subject only to a right in the employer to which I shall return later. The first point to note about that provision is that there is no express requirement for the retentions to be funded. However, in *Rayack Construction Ltd* v. *Lampeter Meat Co Ltd* (1979), Vinelott J held that clause 30(4)(a) does impose an obligation on an employer to appropriate and set aside the retentions as a separate trust fund and he made a mandatory order to that effect on motion. That question does not arise in this case, but the employer there was not a local authority but a private company which it was recognised could have got into financial difficulties. The learned judge proceeded on the footing that if no fund were set aside and the employer went into liquidation the contractor would rank as an unsecured creditor, save only for its lien on unfixed goods and materials under the proviso to clause 26(2) of the RIBA Conditions. That may well be so, but I agree with [counsel for the liquidator] that in the case of a solvent employer equity, looking on that as done which ought to be done, will not allow him any advantage from his failure to set aside a fund. Accordingly, it seems to me that I must proceed on the footing that the council did from time to time set aside as a separate fund sums equivalent to the retention money under the interim certificates. I did not understand [counsel for the GLC] to argue to the contrary. Next, it seems to me that the sums thus notionally set aside became held by the company as a trustee. That would seem to me to be a very obvious consequence of the first step, but it was disputed by [the council] on the ground that the provisions of the main contract as a whole render the subject-matter of the trust uncertain. I reject that argument and I do not think that I need say more about it than that. . . .

I must now deal with what I have already said appears to me to be the most important question. In my view it is clear that clause 30(4)(a) creates in the first instance a trust of the retention moneys in favour of the contractor. It is then necessary to go to clause 27, which deals with nominated sub-contractors. Although the employer's agent nominates the sub-contractors, it is the contractor who contracts with them. But the proviso to sub-clause (a) erects the two safeguards in favour of the contractor to which I have already referred, the second being that no sub-contractor can be nominated who will not (save where the parties otherwise agree) enter into a sub-contract which provides for the matters set out in the nine paragraphs of sub-clause (a). The only one of those paragraphs with which this case is directly concerned is (viii), which provides that the contractor shall hold a due proportion of the retentions on trust for the sub-contractor. That gives to the sub-contractor a right corresponding to that given to the contractor by clause 30(4)(a) of the RIBA Conditions.

It follows from all this that unless his agent agrees to the contrary the employer will know that any sub-contract will necessarily include a provision of the kind required by (viii) of clause 27(a). Putting it at its lowest it seems to me to be clear that the employer requires and authorises the contractor to enter into a sub-contract in that form and that the employer has notice of its contents. The sub-contract therefore takes effect as an assignment to the sub-contractor, made with the authority and with the knowledge of the employer, of a due proportion of the contractor's beneficial interest in the retention moneys under clause 30(4)(a) of the main contract. In my judgment there can be no doubt that that is the effect of clause 11(h) of the FASS sub-contract. If authority be needed for that it can be found in the decision of Wynn-Parry J in *Re Tout & Finch Ltd* (1954). . . .

In other words, the position as between employer, contractor and sub-contractor under the present RIBA Conditions and the FASS sub-contract is that the employer holds a due proportion of the retentions on trust for the contractor as trustee for the sub-contractor.

If the foregoing analysis is correct it would seem to follow that as and when the retentions fall to be released the sub-contractor can require the contractor to recover the former's due proportion from the employer. Moreover, whether he is required to do so or not, the contractor can of his own volition take steps to recover all the retentions (including those payable to sub-contractors) from the employer. Applying that to the present case it would mean, other things being equal, that the liquidator was entitled, and for practical purposes bound, to require payment from the council of the £1,374.56 due to the sub-contractors.

'Delay on the part of nominated sub-contractors or nominated suppliers' in JCT 80, clause 25.4.7. (JCT 63, clause 23(g)) means a delay on the part of the nominated sub-contractor before the date of completion. It is apparent completion which matters and if the sub-contractor's work proves to be defective, the delay caused by his subsequently returning to site is not within the provision.

City of Westminster v. J. Jarvis & Sons Ltd and Peter Lind Ltd

HOUSE OF LORDS (1970), 7 BLR 64

Jarvis contracted with Westminster for the erection of a multi-storey car park, with flats, offices, showrooms and ancillary works. The contract was in JCT 63 form. Lind were nominated sub-contractors for the piling work. Lind carried out the work and purported to complete by the sub-contract completion date. Lind left the site. Some weeks later, it was discovered that many of the piles were defective, either as a result of bad workmanship or because of poor materials. Lind carried out remedial works. The main contract works were delayed for some 21½ weeks. The main contractor (Jarvis) claimed an extension of time under clause 23(g).

Held: Jarvis was not entitled to an extension of time since the works had apparently been completed by the due date. 'Delay' within the meaning of the clause occurs only if, by the sub-contract completion date, the sub-contractor has failed to achieve such completion of his work that he cannot hand it over to the main contractor.

LORD WILBERFORCE: [What] is to be made of the clause? One thing seems clear, that 'delay' does not, as the appellant at one time contended, mean 'sloth' or 'dilatoriness' on the part of the sub-contractor. There are at least two good reasons against this meaning; in the first place, it would put an impossible burden on the architect if he were required to form an opinion that the sub-contractor had not worked as fast or diligently as he might have done, and to measure the extent to which time could have been saved had he done so. This part of the contract would, in practice, become unworkable. And, secondly, it is contractually irrelevant whether the sub-contractor could have worked faster: what matters is whether he has done what he agreed to do in the contractual time. If he has, it does not matter, for the purposes of the contract, whether he achieved his target by leisurely methods: if he has not, it does not matter with what feverish energy he set about his work. This suggested test, then, is both unworkable and irrelevant.

That leaves two alternatives. The sub-contractor contends that he is in delay whenever in any respect he fails to fulfil in time his contractual obligation. The employer contends that there is only delay if, by the sub-contract date, the sub-contractor fails to achieve such completion of his work that he cannot hand over to the contractor. Or, putting it negatively, that the sub-contractor is not in delay so long as, by the sub-contract date, he achieves such apparent completion that the contractor is able to take over, notwithstanding that the work so apparently completed may in reality be defective. This, on the employer's argument, may involve a breach of contract, but does not involve delay.

The sub-contractor's contention has at least the merit of simplicity and comprehensiveness: its counsel admitted, indeed claimed, that it brings within the clause any breach of contract, any failure timeously to perform its total obligation, whether by repudiation, or insolvency, or departure from the site, provided always of course that such breach or failure led to delay in the progress of the main contracts works. So it was in delay here on and after 20 June 1966, because it had not constructed sound piles by that date. This contention, it was admitted, might seem a little extreme, or even in some cases absurd, but the extremity or absurdity lay in the inclusion of the clause in the contract and this once swallowed the consequences should not be strained at.

My Lords, if such an interpretation were imposed by the words used, it would have to be accepted whatever (short of completely frustrating the contract) the consequences might be. Within the limit I have mentioned the parties must abide by what they have agreed to and it is not for the courts to make a sensible bargain for them. But the words used do not suggest that this is the right meaning: for if it were, why should the word 'delay' be used? Why not frankly exonerate the contractor for any delay in completion due to any breach of contract or failure, *eo nomine*, of the sub-contractor? Add to this difficulty the unattractive consequences which follow and the objections become overwhelming: for the clause would amount to little better than a charter for bad work, wilful failure and default of every kind. The worse the sub-contractor's conduct of his sub-contract and the greater impact it has upon his timeous fulfilment and that of the contractor, the more complete is his escape from liability, the more firmly the loss so occasioned is made to rest on the helpless employer.

The employer's construction at least avoids the worst of these consequences, for it relates the impact of the clause to such default as prevents the contractor from taking over and proceeding with his work. It fits in reasonably well with those provisions in the contract which distinguish between such completion as enables the contract to proceed, and such final and verified completion as enables certificates to be given, final payment to be made and the party cleared of all obligation under the contract.

It is not without its difficulties. . . . It is only necessary to point to the fact that if the defects in the piles had been discovered before the sub-contract completion date, and work had been at once put in hand to remedy them – thereby producing a similar period of delay in the completion of the main contract, the clause would, it seems, have applied, but it does not do so if the work was 'complete' (though defective) on that same date so that the contractor could take over. One must set against this the advantage that, if the sub-contract work is apparently completed and handed over, and some defects appear very much later, but before the contract date, as they might in a large contract, this would not, on the employer's construction, be a case of delay, though it might be so on the sub-contractor's. But even so the first type of difficulty is a very grave defect and a serious reflection on the clause: indeed, I cannot believe that the professional body, realising how defective this clause is, will allow it to remain in its present form. But in my opinion, though it is never agreeable to have to choose the lesser of two incongruities, we have to do so here and I find the employer's version qualifies for this not very flattering description.

So finally, how does this construction fit the facts? It is common ground that the sub-contractor 'purported to complete' the piling works by 20 June 1966, and left the site. We know, too, that 'further construction work on the site was then carried out by the contractor'. On 13 July 1966, the architect wrote to the contractor a letter in which he said that 'further to the completion of the [piling] work' the sub-contractor might request the release of retention money against a guarantee. He did, it is true, go on to record suspicion as to the soundness of the piles and to suggest an increase in the guarantee, but the letter was consistent only with his belief that the sub-contract work was 'completed'. The sub-contractor had recorded that this was its position on 9 June 1966. On 16 August 1966, after defects had been revealed, the architect wrote again to the contractor stating that 'as the piling works . . . have not been completed to our satisfaction within the period of the sub-contract' they certified that they ought reasonably to have been completed within that time. This I do not consider alters the view of the matter which all concerned held on 20 June 1966, that the work was completed – though it might turn out to be defective: it is saying no more than that it is now seen to be defective and that the architect was not satisfied with it. The learned judge made no explicit finding but his judgment proceeded on the basis that the sub-contractor had achieved apparent completion, and handed over to the contractor on 20 June 1966. I think that this was correct in fact and in law.

JCT 63 does not deal explicitly with the question of renomination where a nominated sub-contractor repudiates the sub-contract. It has been held however that it is implicit in the contract that the employer comes under a duty to renominate. JCT 80, clause 35.2.4. imposes an express duty of re-nomination should a nominated sub-contractor fail.

North-West Metropolitan Regional Hospital Board v. T.A. Bickerton & Son Ltd

HOUSE OF LORDS   [1970] 1 All ER 1039

The facts are set out in the speech of Lord Reid.

LORD REID: My Lords, the appellants employed the respondents as contractors to erect certain buildings at their hospital at Abbotts Langley, and they nominated a company called Speediwarm Ltd as the sub-contractor who was to provide the heating system. Speediwarm then contracted with the contractor but very soon went into liquidation and the liquidator refused to carry out the contract. Then, without prejudice to their respective rights, the employers and contractor arranged that the contractor should do this work. It did so and there is no complaint about its work. But the work cost substantially more than the price at which Speediwarm had agreed to do it. The employers contend that the contractor is only entitled to be paid Speediwarm's price. The contractor contends that, when Speediwarm fell out, the employers ought to have nominated another sub-contractor and paid its price: and that, as they did not do so, they must pay the contractors on the basis of *quantum meruit*.

I must first explain the nature of the contract between the employers and the contractors. The employers put out for tender bills of quantities and drawings. With regard to the greater part of the work the bills were detailed and the contractor filled in its prices. But certain parts of the work were reserved for sub-contractors to be nominated by the employers. With regard to these parts no details were given, and sums known as prime cost sums were inserted by the employers as estimates of what those parts were likely to cost. So the tendering contractor had no concern either with the details of this work or the price to be paid for it. The work was to be part of the contract work and so the contractor's tender was made up of the sums for which it offered to do its part of the work together with the prime cost sums settled by the employers. The employers obtained tenders from specialists, selected by them for the prime cost work, and then, when they had made their contract with the contractor, they instructed the contractor to enter into a contract with the sub-contractor whom they nominated in terms which they dictated, having settled those terms with the nominated sub-contractor. Then, if the sum to be paid to the nominated sub-contractor differed from the estimated prime cost sum, as it almost certainly would, the contractor's contract price for the whole work was adjusted to take account of this difference.

The contractor's contract with the employers included bills of quantities, contract drawings and the conditions contained in the RIBA form of contract – a form published by the Royal Institute of British Architects. So the question at issue in this case must be answered by construing those conditions. Condition 1(1) is in these terms:

'The Contractor shall upon and subject to these Conditions carry out and complete the Works shown upon the Contract Drawings and described by or referred to in the Contract Bills and in these Conditions in every respect to the reasonable satisfaction of the Architect/Supervising Officer.'

The works there referred to clearly include prime cost work to be done by nominated sub-contractors so the principal contractor is in breach of its contract with the employers if these works are not duly carried out and completed. But it will be able to sue the nominated sub-contractor for breach of the sub-contract. Admittedly the principal contractor is liable if the prime cost works are defective. How much farther its obligation goes must depend on the true interpretation of later conditions.

The main condition with regard to nominated sub-contractors is condition 27. The relevant provisions are as follows:

'The following provisions of this Condition shall apply where prime cost sums are included in the Contract Bills, or arise as a result of Architect's/Supervising Officer's instructions given in regard to the expenditure of provisional sums, in respect of persons to be nominated by the Architect/Supervising Officer to supply and fix materials or goods or to execute work.

(a) Such sums shall be deemed to include 2½ per cent cash discount and shall be expended in favour of such persons as the Architect/Supervising Officer shall instruct, and all specialists or others who are nominated by the Architect/Supervising Officer are hereby declared to be sub-contractors employed by the Contractor and are referred to in these Conditions as "nominated sub-contractors". Provided that the Architect/Supervising Officer shall not nominate any person as a sub-contractor against whom the contractor shall make reasonable objection, or (save where the Architect/Supervising Officer and Contractor shall otherwise agree) who will not enter into a sub-contract which provides (*inter alia*):–

(i) That the nominated sub-contractor shall carry out and complete the sub-contract Works in every respect to the reasonable satisfaction of the Contractor and of the Architect/Supervising Officer and in conformity with all the reasonable directions and requirements of the Contractor.

(ii) That the nominated sub-contractor shall observe, perform and comply with all the provisions of this Contract on the part of the Contractor to be observed, performed and complied with (other than clause 20[A] of these Conditions, if applicable) so far as they relate and apply to the sub-contract Works or to any portion of the same.

(iii) That the nominated sub-contractor shall indemnify the Contractor against the same liabilities in respect of the sub-contract Works as those for which the Contractor is liable to indemnify the Employer under this Contract.'

Conditions 11(3) and 30 also throw some light on the matter. But I do not think that they affect the proper construction of condition 27, and I need not set them out. The crucial problem is to discover the meaning of the first part of condition 27(a). 'Such sums' must refer back to 'prime cost sums' in the preceding sentence; '. . . such persons as the Architect . . . shall instruct' must be the same as 'persons to be nominated by the Architect . . . to execute work', i.e. nominated sub-contractors. And no light is thrown on the problem by the initial words of condition 27(a) 'Such sums shall be deemed to include 2½ per cent cash discount'. I did not understand any of that to be disputed in argument. So I can rewrite the crucial words – 'prime cost sums shall be expended in favour of persons nominated by the Architect to execute work'. This is meaningless if 'prime cost sums' has its ordinary meaning of sums entered or provided in the bills of quantities for work to be executed by nominated sub-contractors. Such sums are never expended; they are only estimates of the sums which will later appear in the sub-contract between the contractor and nominated sub-contractors as the prices to be paid to the nominated sub-contractors for the prime cost work to be done by them. Once these sub-contracts are made, the prime cost sums have no further part to play except in accounting as provided in condition 30(5)(c).

In order to give this sentence an intelligible meaning, either 'prime cost sums' or 'expended' must be given an unusual meaning. It was argued for the employers that 'expended' could mean 'allocated'. But that would not help, because prime cost sums are never allocated to a nominated sub-contractor or to anybody else, and in the end counsel for the employers were unable to suggest any word or phrase the substitution of which for the word 'expended' would make sense of this sentence. On the other hand, if 'prime cost sums' can be read as meaning the sums which become payable for prime cost work then the meaning of the sentence becomes clear. And that does not appear to me to be doing too great violence to the words in the clause. I would, therefore, read this clause as directing that sums payable in respect of prime cost work 'shall' be expended in favour of nominated sub-contractors and no one else. 'Expended in favour of' is a rather odd expression but it is, I think, accounted for by the fact that payments for work done by the nominated sub-contractor are paid in the first instance by the employers to the principal contractor, but they are earmarked and are then paid by the contractor to the nominated sub-contractor.

The RIBA form of contract makes no express provision for an event which cannot be very uncommon – that for one reason or another the contract between the principal contractor and the nominated sub-contractor is terminated before the sub-contract work is completed. The problem is what is to happen in that event. There appear to me to be three possibilities, and no more were suggested in argument. The first is that the employers must then make a new nomination, or, if they do not wish the work to be continued, give an instruction to that effect by way of variation. If they do make a new nomination, then as in the case of the first nomination they must make a bargain with the new nominees and then instruct the principal contractor to make a sub-contract in terms of that bargain. Then, if the price in the second sub-contract exceeds the price in the first sub-contract which has come to an end, the employers must bear the loss. That is the contention of the contractor. The second possibility is that, when the contract with the nominated sub-contractor is terminated, it becomes the right and duty of the principal contractor to do the prime cost work itself at the price fixed in the sub-contract which has been terminated. The third possibility is that the principal contractor must be held to have undertaken that the nominated sub-contractor will complete the work so that, when, by reason of the termination of the sub-contract, it becomes impossible for that obligation to be performed, there is an irremediable breach of contract by the principal contractor. The employers argued that one or other of these interpretations is correct.

It appears to me that there are insuperable objections to both of the employers' contentions. The scheme for nominated sub-contractors is an ingenious method of achieving two objects which at first sight might seem incompatible. The employers want to choose who is to do the prime cost work and to settle the terms on which it is to be done, and at the same time to avoid the hazards and difficulties which might arise if they entered into a contract with the person they have chosen to do the work. The scheme creates a chain of responsibility. Subject to a very limited right to object, the principal contractor is bound to enter into a contract with the employers' nominee, but it has no concern with the terms of that contract, for those terms are settled by the employers and their nominee. I can find nothing anywhere to indicate that the principal contractor can ever have in any event either the right or the duty to do any of the prime cost work itself. That would, I think, be contrary to the whole purpose of the scheme, and it would be strange if the contractor could have to do work for which it never tendered and at a price which it never agreed. Moreover, if I have correctly construed condition 27, its provision that payment for prime cost work shall be made in favour of nominated sub-contractors and, therefore, cannot be made to the contractor on its own account necessarily involves the conclusion that the contractor is not to do prime cost work itself. So I reject the employers' first contention.

The employers' alternative contention involves similar difficulties. If they could establish that, under their contract, there can never be any duty on the employers to nominate a second sub-contractor if the first nominee drops out, then it could be said that, when the principal contractor undertook to complete the prime cost work – as it did in condition 1 – it undertook that it would be completed by the first nominated sub-contractor, for it was not entitled to do the work itself. But if, on a proper construction of the contract, the employers are bound to renominate if the first nominee drops out, then there is no question of guaranteeing the first nominated sub-contractor. The contractor's obligation is simply duly to hand over the completed work. If it does not do so, it is in breach of contract. If the fault is its own, it bears the loss, but if the fault is that of the nominated sub-contractor it can sue the sub-contractor for the damages it has to pay to the employers.

There is a further objection to this contention of the employers. If the contractor is bound to complete the prime cost work but cannot do it itself under the contract and it cannot be done by a nominated sub-contractor because the first has dropped out and there is no provision for a second, then there is deadlock. There is no way in which the work can be done under the contract. It is said that the contractor could then go in and do the work itself in order to mitigate damages. But then the work would not be done under the contract and the employers and their architect would lose all control over it. That cannot have been the intention, and I gravely doubt whether the contractor could have any such right. So, unless the contractor's contention is correct, there will be grave and perhaps insuperable difficulties when any nominated sub-contractor refuses to complete its contract

But I see no great difficulty in holding that the contract requires a second nomination if the original nominated sub-contractor drops out. It is said that there is no express provision for this. But then neither is there any express provision for the first nomination. Indeed, the absence of any such provision has led to a view, which has attracted considerable support, that the employers have no duty to make any nomination but only have a right or option to do so. But that cannot be right. The contract provides that the prime cost work shall be done, and it may be that the contractor cannot do or at least cannot finish its own work until it is done. The prime cost work is not even defined until the nominated sub-contractor is brought in. No one suggests that the principal contractor has any concern with prime cost work until it is required to make a contract with a nominated sub-contractor. It would be a clear breach of contract by the employers if their failure to nominate a sub-contractor impeded the contractor in the execution of its own work.

Once it is accepted that the principal contractor has no right or duty to do the work itself when the nominated sub-contractor drops out any more than it had before the sub-contractor was nominated, then equally it must be the duty of the employers to make a new nomination when a nominated sub-contractor does drop out. For otherwise the contract work cannot be completed. Moreover, condition 27 requires that payment for prime cost work shall be expended in favour of nominated sub-contractors. So if the first drops out, no payment for the prime cost work still to be done can be made under the contract unless that work is done by a second nominated sub-contractor.

Perhaps I should add that there was an argument that, if my view is right, there will be grave difficulty if a contractor wrongfully terminates the sub-contract – it may be because it thinks erroneously that the sub-contractor is in fundamental breach of the sub-contract. Then, in my view, the contractor would be in breach of his contract with the employers. A new sub-contractor would have to be nominated. But the contractor would have to pay damages for its breach of contract including any loss caused to the employers by that breach.

Although I have come to a clear conclusion that there was in this case a duty to renominate, the provisions of the RIBA form of contract are so confused and obscure that no conclusion can be reached without a long and complicated chain of reasoning. The RIBA form of conditions sponsored by the Institute is in very common use. It has been amended from time to time. For a long time it has been well known that the question at issue in the present case has given rise to doubt and controversy. It could have been set at rest by a small amendment of these conditions. But the Institute have chosen not to do that, and they have thereby caused the long and expensive litigation in the present case.

I would dismiss this appeal.

The reasoning in the *Bickerton* case rests on the holding that the contractor is neither bound nor entitled to do the nominated sub-contract work himself. On the other hand it is clear that, in principle, the contractor is liable to the employer for bad work by the nominated sub-contractor and himself has an action against the nominated sub-contractor. In *Bickerton* the nominated sub-contractor did no work at all. Is the position different where the nominated sub-contractor withdraws in mid-contract? It has been held that in this case, the employer remains bound to re-nominate and is therefore responsible for any loss due to delay in renomination, but the loss arising from the nominated sub-contractor's withdrawal falls on the contractor. See also *Fairclough Building Ltd* v. *Rhuddlan Borough Council* (1985), Appendix, p. 383.

Percy Bilton Ltd v. Greater London Council

HOUSE OF LORDS   (1982), 20 BLR 1

The facts of this case are set out on page 249, above.

The case arose from the withdrawal of a nominated sub-contractor. The parties agreed that the sub-contractor's withdrawal had caused both direct delay and also led to further delay arising from the GLC's delay in making an effective re-nomination.

Held: Although the second category was the employer's responsibility, the first was the responsibility of the contractor.

LORD FRASER OF TULLYBELTON [having stated the general position as set out on pages 250-51 continued:] These principles do not, in my opinion, operate in any way harshly or unfairly against the appellant. The so-called 'dropping out' of the nominated sub-contractor was not merely unilaterial action by him. The mechanics of the matter were that on 26 July 1978 a firm of building and quantity surveyors, acting on the instructions of the receiver and manager of Lowdells, wrote to the appellant intimating that their labour would be withdrawn by 28 July 1978. That was a notice of intention to repudiate the sub-contract. It was in effect, accepted by the appellant in a reply dated 31 July, in which it purported to 'determine' the sub-contract under clause 20(a)(i) which entitled it to determine the sub-contract if the sub-contractor 'wholly suspends the carrying out of the Sub-Contract Works' and continues in such default for ten days. But the clause was not applicable because the appellant's letter of 31 July was sent before the default had lasted for ten days. It must, in my opinion, be read simply as an acceptance of the repudiatory breach by the receiver on behalf of the sub-contractors. Accordingly, the sub-contract was terminated by the appellant's acceptance of repudiation, which it was under no obligation to give. If it had withheld acceptance of the repudiation, the sub-contract would have remained alive; it is possible that the appellant might then have been entitled to claim an extension of time under clause 23(g) on account of the sub-contractor's 'delay'. But this point was not fully argued and I express no concluded view on it. There was also another course open to the appellant. It could have exercised its right of 'reasonable objection' under clause 27(a) to prevent the nomination of any new sub-contractor who did not offer to complete his part of the work within the overall completion period for the contract as a whole. In fact, the appellant accepted the nomination, and relied on the probability that it would be allowed an extension of time. Its reliance was not misplaced, for, as already mentioned, it was allowed 14 weeks for the delay arising from the withdrawal of the sub-contractor. . . .

It is common ground between the parties that the delay which followed the dropping out of Lowdells should be divided into two parts – first, the part arising directly from the withdrawal, and secondly, that arising from the failure of the respondent to nominate a replacement with reasonable promptness. The respondent was clearly responsible for the second part, and there is no doubt that, if it had been the only delay, the appellant would have been entitled to a reasonable extension of time to allow for it in accordance with clause 23(f). The dispute centres on the consequence of the first period of the delay which, as parties are agreed, does not fall within any of the provisions of clause 23. The appellant contends that the loss directly caused by the withdrawal of the nominated sub-contractor must fall on the respondent, on the ground that it has a responsibility not only to nominate the original sub-contractor and any necessary replacement, but to maintain a sub-contractor in the field so long as work of the kind allotted to him needs to be done. This is said to flow from the decision of your Lordships' House in *Bickerton* (1970). What was actually decided in that case was that, where the original nominated sub-contractor has gone into liquidation and dropped out, the main contractor had neither the right nor the duty to do any of the sub-contractor's work himself, and that it was the duty of the employer to make a new nomination. Consequently (so it was argued for the appellant), if the nominated sub-contractor withdraws at a time when his withdrawal must inevitably cause delay, the main contractor is disabled from performing his obligations for want of a sub-contractor whom only the employer can provide, and the main contractor is thus 'impeded' from working . . . in these circumstances, it was said that the contractual time limit ceases to apply, the time for completion becomes at large, and the employer cannot rely on the provisions for liquidated damages . . . I cannot accept the argument. I respectfully agree with the passage in the judgment of the Court of

Appeal to the effect that

> 'insofar as delay was caused by the departure of Lowdells . . . it was a delay which is not within any of the provisions of clause 23 and therefore the plaintiff was not entitled to any extension in respect of it, with the result not that time became at large but that, at that stage, the date for completion remained unaffected by that period of delay. The delay that followed, a delay caused by the failure of the employer for several months to renominate, [was] quite separate and clearly within the provisions of clause 23(f).' *per* Stephenson LJ.

The appellant submitted further that clause 23(f) did not apply even to the second part of the delay, because it can only apply where the main contractor makes his application for instructions at a time when the employer, through the architect, could by timeous nomination have avoided the delay altogether. Clause 23(f) applies to delay caused by the contractor not having received instructions 'in due time', and the argument was that those words meant 'in time to avoid delay'. I do not agree. In my opinion, the words mean 'in a reasonable time' and they are therefore applicable in a case such as the present where some delay is inevitable but it has been increased by the employer's fault. Accordingly, I agree with the Court of Appeal that the second part of the delay does fall within clause 23(f).

Where the works are defective as a result of the fault of the nominated sub-contractor, the contractor will normally be liable to the employer and will have an action against the sub-contractor. However the employer may also have a direct action against the nominated sub-contractor, either in contract, if a collateral warranty has been executed (as where the JCT Agreement NSC/2 is used), or in tort if the nominated sub-contractor has been negligent.

Independent Broadcasting Authority v. EMI Electronics and BICC Construction Ltd

HOUSE OF LORDS   (1980), 14 BLR 1

The first defendants were the main contractors and the second defendants the nominated sub-contractors, for the erection of a television mast 1,250 ft. high erected on Emley Moor in Yorkshire. The mast was handed over in November 1966. It collapsed in March 1969. The plaintiffs sued both defendants and the first defendants sued the second defendants. The trial judge held that the cause of the collapse was primarily vortex shedding and secondarily asymmetric ice-loading and that the second defendants who had designed the mast were negligent in not having adequately taken these factors into account.

Held: (1) EMI had impliedly undertaken that the mast was reasonably fit for the purpose for which it was erected.
   (2) BICC were similarly in breach of their sub-contract with EMI.
   (2) BICC were liable in tort to IBA as they had been negligent in assuring them in correspondence in 1964 that the design was satisfactory.
   (4) The assurance given in this correspondence did not amount to a direct contract between BICC and IBA.

Note: These assurances were given some two years *after* the letting of the contracts. If they had been given before the nomination of BICC the argument that they were contractually binding would have been much stronger.

LORD FRASER OF TULLYBELTON: The contract was constituted by EMI's tender and IBA's acceptance. The main part of the tender consisted of a quotation for work which, as I have said, included the design of the mast. What IBA accepted was 'your quotation'. Responsibility by EMI for the design was thus in my opinion, expressly made part of their contract with IBA. Moreover EMI's references to 'our' line diagram drawing tend to show that they were putting forward the drawing as their own. The tender and acceptance superseded the original invitation to tender which I regard merely as part of the prior communings leading up to the contract. I therefore would agree with O'Connor J and with the Court of Appeal, both of whom held EMI were under a contractual obligation to IBA in respect of the design of the mast as well as of its supply and delivery.
   If the terms of the contract alone had left room for doubt

about that, I think that in a contract of this nature a condition would have been implied to the effect that EMI had accepted some responsibility for the quality of the mast, including its design, and possibly also for its fitness for the purpose for which it was intended. The extent of the responsibility was not fully explored in argument, and, having regard to the decision on negligence in the design, it does not require to be decided. It is now well recognised that in a building contract for work and materials a term is normally implied that the main contractor will accept responsibility to his employer for materials provided by nominated subcontractors. The reason for the presumption is the practical convenience of having a chain of contractual liability from the employer to the main contractor and from the main contractor to the sub-contractor – see *Young & Marten Ltd* v. *McManus Childs Ltd* (1969). Of course, as Lord Reid pointed out in that case, 'No warranty ought to be implied in a contract unless it is in all circumstances reasonable.' In most cases the implication will work reasonably because if the main contractor is liable to the employer for defective material, he will generally have a right of redress against the person from whom he bought the material. In the present case it is accepted by BIC that, if EMI are liable in damages to IBA for the design of the mast, then BIC will be liable in turn to EMI. Accordingly, the principle that was applied in *Young & Marten Ltd* in respect of materials, ought in my opinion to be applied here in respect of the complete structure, including its design. Although EMI had no specialist knowledge of mast design, and although IBA knew that and did not rely on their skill to any extent for the design, I see nothing unreasonable in holding that EMI are responsible to IBA for the design seeing that they can in turn recover from BIC who did the actual designing. On the other hand it would seem to be very improbable that IBA would have entered into a contract of this magnitude and this degree of risk without providing for some right of recourse against the principal contractor or the sub-contractors for defects of design.

LORD SCARMAN: The finding of negligence in the design of the mast makes it unnecessary for the House to determine the extent of the contractual responsibility for the design assumed by EMI to IBA (and by BIC to EMI). As my Lord Fraser of Tullybelton observes,

> 'it must at the very least have been to ensure that the design would not be made negligently.'

But I would not wish it to be thought that I accept that this is the extent of the design obligation assumed in this case. The extent of the obligation is, of course, to be determined as a matter of construction of the contract. But, in the absence of a clear, contractual indication to the contrary, I see no reason why one who in the course of his business contracts to design, supply, and erect a television aerial mast is not under an obligation to ensure that it is reasonably fit for the purpose for which he knows it is intended to be used. The Court of Appeal held that this was the contractual obligation in this case, and I agree with them. The critical question of fact is whether he for whom the mast was designed relied upon the skill of the supplier (i.e. his or his sub-contractor's skill) to design and supply a mast fit for the known purpose for which it was required.

Counsel for the appellants, however, submitted that, where a design, as in this case, requires the exercise of professional skill, the obligation is no more than to exercise the care and skill of the ordinarily competent member of the profession. Although it might be negligence to-day for a constructional engineer not to realise the danger to a cylindrical mast of the combined forces of vortex shedding (with lock-on) and asymmetric ice loading of the stays, he submitted that it would not have been negligence before the collapse of this mast: for the danger was not then appreciated by the profession. For the purpose of the argument, I will assume (contrary to my view) that there was no negligence in the design of the mast, in that the profession was at that time unaware of the danger. However, I do not accept that the design obligation of the supplier of an article is to be equated with the obligation of a professional man in the practice of his profession. In *Samuels* v. *Davis* (1943), the Court of Appeal held that, where a dentist undertakes for reward to make a denture for a patient, it is an implied term of the contract that the denture will be reasonably fit for its intended purpose. I would quote two passages from the judgment of du Parcq LJ:

> '... if someone goes to a professional man ... and says: "Will you make me something which will fit a particular part of my body?" ... and the professional gentleman says: "Yes," without qualification, he is then warranting that when he has made the article it will fit the part of the body in question.'

And he added:

> 'If a dentist takes out a tooth or a surgeon removes an appendix, he is bound to take reasonable care and to show such skill as may be expected from a qualified practitioner. The case is entirely different where a chattel is ultimately to be delivered.'

I believe the distinction drawn by du Parq LJ to be a sound one.

In the absence of any terms (express or to be implied) negativing the obligation, one who contracts to design an article for a purpose made known to him undertakes that the design is reasonably fit for the purpose. Such a design obligation is consistent with the statutory law regulating the sale of goods: see Sale of Goods Act 1893, the original s.14, and its modern substitution enacted by s.3, Supply of Goods (Implied Terms) Act 1973.

Until recently, it was generally assumed that liability in tort for negligence on the part of the nominated sub-contractor would not extend to cases where the result of the negligence was that the building was sub-standard but not dangerous to person or property. However in the case below the House of Lords held that this was not so.

Junior Books Ltd v. The Veitchi Co. Ltd

HOUSE OF LORDS   (1982), 21 BLR 66

[This case came to the House of Lords on appeal from Scotland. Under the relevant Scottish procedure the legal validity of the plaintiff's claims was being argued before the trial of the facts. At this stage therefore it was assumed that the plaintiff's allegations could all be proved.]

The respondents, building owners, engaged a building company to build a factory for them and the appellants were nominated as flooring sub-contractors. Two years after the floor had been laid it developed cracks. It was not alleged that this made the floor dangerous but it did involve the respondents in serious additional expense and the respondents brought an action in tort against the appellants. [It was not alleged that there was any direct warranty nor, apparently, did the respondents sue the main contractors.]

Held: The appellants owed the respondents a duty of care since they knew that the respondents relied on their skill and experience to lay the floor and that financial loss to the respondents would arise directly from their failure to take such care.

LORD ROSKILL: My Lords, to my mind in the instant case there is no physical damage to the flooring in the sense in which that phrase was used in *Dutton* v. *Bognor Regis UDC* (1972), *Batty* v. *Metropolitan Property Realizations Ltd* (1978) and *Bowen* v. *Paramount* (1977) and some of the other cases. As my noble and learned friend Lord Russell said during the argument, the question which your Lordships' House now has to decide is whether the relevant Scots and English law today extends the duty of care beyond a duty to prevent harm being done by faulty work to a duty to avoid such faults being present in the work itself. It was powerfully urged on behalf of the appellants that were your Lordships so to extend the law a pursuer in the position of the pursuer in *Donoghue* v. *Stevenson* (1932) could in addition to recovering for any personal injury suffered have also recovered for the diminished value of the offending bottle of ginger beer. Any remedy of that kind it was argued must lie in contract and not in delict or tort. My Lords, I seem to detect in that able argument reflections of the previous judicial approach to comparable problems before *Donoghue* v. *Stevenson* was decided. That approach usually resulted in the conclusion that in principle the proper remedy lay in contract and not outside it. But that approach and its concomitant philosophy ended in 1932 and for my part I should be reluctant to countenance its re-emergence some fifty years later in the instant case. I think today the proper control lies not in asking whether the proper remedy should lie in contract or instead in delict or tort, not in somewhat capricious judicial determination whether a particular case falls on one side of the line or the other, not in somewhat artificial distinctions between physical and economic or financial loss when the two sometimes go together and sometimes do not (it is sometimes overlooked that virtually all damage including physical damage is in one sense financial or economic for it is compensated by an award of damages) but in the first instance in establishing the relevant principles and then in deciding whether the particular case falls within or without those principles. To state this is to do no more than to restate what Lord Reid said in *Dorset Yacht Co. Ltd* v. *Home Office* (1970) and Lord Wilberforce in *Anns* v. *Merton LBC* (1978). Lord Wilberforce in the passage I have already quoted enunciated the two tests which have to be satisfied. The first is 'sufficient relationship of proximity', the second any considerations negativing, reducing or limiting the scope of the duty or the class of person to whom it is owed or the damages to which a breach of the duty may give rise. My Lords, it is I think in the application of those two principles that the ability to control the extent of liability in delict or in negligence lies. The history of the development of the law in the last fifty years shows that fears aroused by the 'floodgates' argument have been unfounded. Cook J in *Bowen* described the 'floodgates' argument as specious and the argument against allowing a cause of action such as was allowed in *Dutton*, *Anns* and *Bowen* as 'in terrorem or doctrinaire'.

Turning back to the present appeal I therefore ask first whether there was the requisite degree of proximity so as to give rise to the relevant duty of care relied on by the respondents. I regard the following facts as of crucial importance in requiring an affirmative answer to that question: (1) the appellants were nominated sub-contractors; (2) the appellants were specialists in flooring; (3) the appellants knew what products were required by the respondents and their main contractors and specialised in the production of those products; (4) the appellants alone were responsible for the composition and construction of the flooring; (5) the respondents relied on the appellants' skill and experience; (6) the appellants as nominated sub-contractors must have known that the respondents relied on their skill and experience; (7) the relationship between the parties was as close as it could be short of actual privity of contract; (8) the appellants must be taken to have known that if they did the work negligently (as it must be assumed that they did) the resulting defects would at some time require remedying by the respondents expending money on the remedial measures as a consequence of which the respondents would suffer financial or economic loss.

My Lords, reverting to Lord Devlin's speech in *Hedley Byrne & Co. Ltd* v. *Heller & Partners Ltd* (1963), it seems to me that all the conditions existed which give rise to the relevant duty of care owed by the appellants to the respondents.

I then turn to Lord Wilberforce's second proposition. On the facts I have just stated, I see nothing whatever to restrict the duty of care arising from the proximity of which I have spoken. During the argument it was asked what the position would be in a case where there was a relevant exclusion clause in the main contract. My Lords, that question does not arise for decision in the instant appeal, but in principle I would venture the view that such a clause according to the manner in which it was worded might in some circumstances limit the duty of care just as in the *Hedley Byrne* case the plaintiffs were ultimately defeated by the defendants' disclaimer of responsibility. But in the present case the only suggested reason for limiting the damage (*ex hypothesi* economic or financial only) recoverable for the breach of the duty of care just enunciated is that hitherto the law has not allowed such recovery and therefore ought not in the future to do so. My Lords, with all respect to those who find this a sufficient answer I do not. I think this is the next logical step forward in the development of this branch of the law. I see no reason why what was called during the argument 'damage to the pocket' simpliciter should be disallowed when 'damage to the pocket' coupled with physical damage has hitherto always been allowed. I do not think that this development, if development it be, will lead to untoward consequences. The concept of proximity must always involve, at least in most cases, some degree of reliance; I have already mentioned the words 'skill' and 'judgment' in the speech of Lord Morris in *Hedley Byrne*. These words seem to me to be an echo, be it conscious or unconscious, of the language of s.14(1) of the Sale of Goods Act 1893. My Lords, though the analogy is not exact, I do not find it unhelpful for I think the concept of proximity of which I have spoken and the reasoning of Lord Devlin in the *Hedley Byrne* case involve factual considerations not unlike those involved in a claim under s.14(1); and as between an ultimate purchaser and a manufacturer would not easily be found to exist in the ordinary everyday transaction of purchasing chattels when it is obvious that in truth the real reliance was on the immediate vendor and not on the manufacturer.

Note: See also *Greater Nottingham Co-operative Society Ltd* v. *Cementation Piling & Foundations Ltd* (1986), Appendix, p. 394.

Appendix

We include here a number of cases which are supplementary to the main text.

Chapter 7

THE EMPLOYER'S OBLIGATIONS

Payment and set-off

Payment of the amount of an interim certificate is not a condition precedent to arbitration under contracts in JCT 80 and similar forms. If the employer raises a bona fide *arguable contention that a payment certificate may have been overvalued he is entitled to have that issue arbitrated and the contractor is not entitled to summary judgment on the certificate.*

C. M. Pillings & Co. Ltd v. Kent Investments Ltd

COURT OF APPEAL   (1986), 4 ConLR 1

The defendants let a contract to the plaintiffs under the JCT Fixed Fee Form of Prime Cost Contract, (1967 edition, revised 1976) for the extension of a large house, the provision of an indoor swimming pool and the addition of a new master bedroom suite. The estimated prime cost written into the contract was £300,500. As work progressed, a large number of variations were ordered by the architect and delays to completion occurred. The estimated final cost was in the order of £630,000.

A certificate of practical completion was issued by the architects on 14 September 1984. On 5 October 1984 the architects issued an interim certificate for £110,000. This was subsequently revised to £101,529. The defendants disputed the sum certified and wished to challenge the accuracy of the certificate. They refused to honour the certificate, and when the plaintiffs issued a writ and sought summary judgment for £101,529, the defendants asked for an order that the action should be stayed under section 4 of the Arbitration Act 1950 and referred to arbitration.

Held:

(1) There was just sufficient evidence to entitle the employers to raise a *bona fide* arguable contention about the accuracy of the interim certificate.
(2) There is a right to raise cross-claims in any action for payment under a certificate unless the terms of the contract expressly exclude that right. On its true construction, the arbitration clause made it clear that such cross-claims could be raised by its specific reference to disputes as to 'whether or not a certificate has been improperly withheld or is not in accordance with this agreement'.

KERR LJ: The decision of the House of Lords in *Modern Engineering (Bristol) Ltd* v. *Gilbert Ash (Northern) Ltd* made it clear that, unless the contract on its true construction had the effect that the contractor was entitled to immediate payment, cross-claims which would be admissible as equitable set-offs could be raised by the employer, or head contractor as the case may be, as a reason for withholding amounts certified as being payable by him. In the ultimate analysis the House of Lords was concerned with a particular clause in the sub-contract which, on its true construction, as I think some of their Lordships put it, not only did not exclude the right to raise cross-claims against certified amounts, but made it quite clear, albeit unnecessarily as a matter of law, that such cross-claims could be raised.

In the present case the cross-claim in question – the challenge to the accuracy of the certificate – cannot be questioned as an admissible cross-claim; indeed it is expressly referred to in clause 31A. The only question is therefore whether the contract on its true construction somehow provides that such cross-claim cannot be asserted without the contractor having first been paid under the interim certificate. As I have already said, I cannot find any provision which has that effect. That appears to have been the conclusion of the majority of their Lordships in *Gilbert-Ash*, to the extent to which they referred to the head contract in that case, which appears to have been indistinguishable in principle from the contract which he have here, although it was not a prime cost plus fee contract. The numbering of the contract was clearly different, but it is not in dispute between counsel, who are very experienced in these matters, that the substance of the provisions there referred to was substantially the same as in the present case. Indeed, one can identify a number of identical provisions referred to in those speeches of their Lordships which I have also mentioned earlier in this judgment to which they draw attention in connection with the present issue. The upshot of those speeches is that, in order to exclude the right to assert cross-claims admissible as equitable set-offs, and thus to exclude rights of deduction or non-payment of certified amounts based on such cross-claims, it is necessary to find some clear and express provision in the contrast which has that effect. There was none in the head contract in *Gilbert-Ash*, and there is none here.

The learned editor of Hudson's *Building Contracts* (10th ed, supplement, paragraphs 492–494) takes the view, with which I respectfully agree, that standard contracts such as the present contain no sufficient provision which has that effect. We were also referred to the decision of Tudor Evans J at Leeds Crown Court, in *Vickers* v. *Lynn & Jones* [1981], CLY 188, where he reached precisely the same conclusion.

On the other side of the line, this court recently had before it a contract which expressly excluded those consequences. I refer to *Tubeworkers Ltd* v. *Tilbury Construction Ltd* (1985). It was concerned with the FASS Form of sub-contract. There were certain provisions there which unmistakably provided that the *Gilbert-Ash* situation, if I may so put it, should be altered by express terms. That standard form was clearly revised for that purpose. There is nothing of that kind in the present contract sufficiently strong to displace reliance upon cross-claims admissible by way of set-off, as upheld in *Gilbert-Ash*.

In general, the fact that main contractors have a bona fide *and arguable counterclaim does not entitle them to set-off its amount against payments due to nominated sub-contractors. This matter is always to be determined on the interpretation of the actual wording of the sub-contract.*

Tubeworkers Ltd v. Tilbury Construction Ltd

COURT OF APPEAL (CIVIL DIVISION)   (1986), 4 ConLR 13

The defendants were main contractors for the construction of a swimming pool. The plaintiffs were sub-contractors under the FASS form of sub-contract (1978 edition) for the structural steel work.

Under clause 11 of that form the contractor is required to pay the sub-contractor within 14 days of the receipt of a relevant architect's certificate, but under clauses 13 A and 13 B there are provisions for the contractor to set-off counterclaims in certain circumstances.

On 13 September 1983 the architect issued a certificate that £54,728 was payable by the defendants to the plaintiffs. The defendants alleged that they had a counterclaim against the plaintiffs for delay and on 9 November 1983 the architect issued a certificate under clause 8(a) of the sub-contract to that effect. The defendants assessed the value of their counterclaim at £97,069. This amount did not fall within the contractual provisions for set-off under clause 13 A(2)(c) because notice of it had not been given to the contractor 'not less than seventeen days before the money from which the amount is to be set off becomes due and payable to the sub-contractor'.

On 21 November 1983 the plaintiffs issued a writ for the certified sum and applied for summary judgment under RSC Order 14. The then Senior Official Referee awarded judgment to the plaintiffs but granted the defendants a stay of execution under RSC Order 47, r.l. The plaintiffs appealed.

Held: Although the court's jurisdiction to order a stay of execution was wide, it ought not normally to be exercised where to do so would overturn the contractual scheme for set-off. The mere fact that the defendants had a good or arguable counterclaim could not be a 'special circumstance' in the context of the FASS sub-contract.

> KERR LJ: [What] the learned judge has done, in effect, is that he has replaced the contractual machinery agreed between the parties by the process of the courts in ordering a stay under Order 47. In my view this is *prima facie* wrong. I say *prima facie* because, as already mentioned, there might be exceptional circumstances to justify it. But, generally speaking, I think that the relevant terms of the contract between the parties must be paramount and take precedence over the court's discretion under Order 47.

Of all the authorities to which we were referred, I found the most helpful one in this connection, in principle, to be the decision of this court in *Northern Regional Health Authority* v. *Derek Crouch Construction Co. Ltd*, (1984). That was not concerned with the present form of sub-contract, but it was, among other matters, concerned with the question whether the court could substitute its process by opening up an architect's certificate for the powers given by the contract to an arbitrator in that regard. It was held that there was no power to do so. For arbitrator in that context one can, I think, read adjudicator in the present context. The passages relied upon are to be found in the judgments of Dunn LJ and Browne-Wilkinson LJ in particular. Dunn LJ said: 'The parties have agreed that disputes as to anything left to the discretion of the architect should be referred to arbitration, and clause 35 gives wide powers to the arbitrator to review the exercise of the architect's discretion and to substitute his own views for those of the architect. Where parties have agreed on machinery of that kind for the resolution of disputes, it is not for the court to intervene and replace its own process for the contractual machinery agreed by the parties.' It is in particular that last sentence which I find of assistance in the present context.

Similarly, Browne-Wilkinson LJ arrived at a similar conclusion, and one finds the following sentence: 'The court's jurisdiction does not include a right to modify contractual rights.' The word jurisdiction in the present context should not be interpreted strictly, but as a pointer to the way in which the court's discretion under Order 47 should be exercised.

Against that background I return to the judgment. All that the judge took into consideration here (as shown by the passage which I have already read) was that the defendants had a good or arguable counterclaim. That in my view is incapable of constituting special circumstances in relation to the present form of sub-contract, because the very basis of clause 13A is that a counterclaim or set-off is raised against the sub-contractor. But, as has been held, the defendants cannot rely on clause 13A to protect them in that situation, and it is through no fault of the plaintiffs that they are unable to do so. It follows, so far as the judge's reason for the exercise of his discretion is concerned, i.e. the mere existence of an arguable counterclaim, that there cannot in my view be any doubt that this cannot be supported as a special circumstance. However, [counsel] then submitted that there were nevertheless special circumstances in this case, taking the evidence as a whole which was placed before the judge, because the defendants' inability to rely on clause 13A was not their fault but the fault of the architect. However, how can that possibly matter so long as it was not the plaintiffs' fault? The position is simply as follows:

(1) Under the agreed contractual machinery the plaintiffs became entitled to payment of the certified sum on 4 October 1983.

(2) The plaintiffs did nothing to interfere with the operation of the contractual machinery.

(3) This envisages that the defendants may have a good counter-claim, which is then to be dealt with under clauses 13A and B, with the immediate decision as to what is to happen pending arbitration being that of an adjudicator under clause 13B.

(4) Without any other facts than those mentioned above, the stay under Order 47 r.1 accordingly usurped the function of the adjudicator, which the defendants were unable to invoke through no fault of the plaintiffs.

(5) In these circumstances the mere existence of a counterclaim and of the facts whereby the defendants were unable to rely on the contractual machinery without fault on the part of the plaintiffs cannot in my view constitute special circumstances for a stay under Order 47, for such a stay contradicts the terms of the sub-contract and substitutes the exercise of the court's powers for those of the adjudicator on which the defendants *ex hypothesi* are not in a position to rely.

Chapter 10

APPROVAL AND CERTIFICATES

Where there is an arbitration clause in general terms in a building contract, the court has no power to open up, review and revise the architect's certificates and opinions.

Oram Builders Ltd v. M. J. Pemberton and C. Pemberton

QUEEN'S BENCH DIVISION   (1985), 2 ConLR 94
(OFFICIAL REFEREES' BUSINESS)

The plaintiffs (building contractors) entered into a contract with the defendants dated 14 July 1982. The contract was in the JCT Agreement for Minor Building Works form (1980 edition, October 1981 revision), article 4 of which is an arbitration agreement in general terms. Clause 3.6 of the contract empowers the architect/supervising officer to order variations. The supervising officer issued instructions to the plaintiffs which constituted a substantial variation of the contract. The plaintiffs claimed to have done the extra work so required and sought a certificate for its payment. This was not forthcoming and the plaintiffs claimed the extra cost of the extra work, or alternatively damages. The defendants disputed the plaintiffs' claim and counterclaimed for alleged breaches of contract. The preliminary issue, with which this judgment is concerned, was whether the court had power to review the exercise of the supervising officer's discretionary power under the contract and open up his certificates.

Held: The court has no jurisdiction to go behind a certificate of the architect/supervising officer. 'Where there is an arbitration clause in general terms referring any dispute or difference between the parties concerning the contract to the arbitrator … then even on a narrow interpretation of the reasoning of the Court of Appeal in *Northern Regional Health Authority* v. *Derek Crouch Construction Co. Ltd* [1984] QB 644; [1984] 2 All ER 175, the High Court has no jurisdiction to go behind' an architect's certificate.

HIS HONOUR JUDGE DAVID SMOUT QC: As I understand it, the Court of Appeal [in *Crouch*] was accordingly of the opinion not that the arbitration clause deprived the court of jurisdiction but rather that it gave to the arbitrator a jurisdiction to open up and review certificates which the court does not possess. On that interpretation it appears that whether or not there is any arbitration clause the court cannot go behind an architect's certificate. However, I do not have to go that far in the instant case. I am satisfied that where there is an arbitration clause in general terms referring any dispute or difference between the parties concerning the contract to the arbitrator – such as arises in the instant case – then even on a narrow interpretation of the reasoning of the Court of Appeal in *Crouch*, the High Court has no jurisdiction to go behind a certificate of an architect or supervising officer.

Chapter 12

PROLONGATION AND DISRUPTION CLAIMS

Croudace Ltd v. London Borough of Lambeth

Under a contract in JCT 63 terms, the architect's failure to ascertain, or instruct the quantity surveyor to ascertain, the amount of the direct loss and/or expense suffered or incurred by the contractor, as required by clauses 11(6) and 24(1) of the contract is a breach of contract for which the employer is liable in damages.

A contractor's claim under the contract for direct loss and/expense cannot be maintained by legal action in the absence of an architect's certificate. However, sums may be so claimed by way of an action for damages for breach of contract provided it is established that the contractor has suffered some damage. Unless the employer establishes that there are no matters in respect of which the contractor is entitled to claim under clauses 11(6) and 24(1) it necessarily follows that the contractor must have suffered some damage as a result of the failure to ascertain amounts due.

On 12 October 1979 Croudace Ltd (Croudace) contracted with the Council (Lambeth) in JCT 63 terms, to erect houses and ancillary buildings at Stockwell. The contract provided that 'the architect' and 'the quantity surveyor' meant the chief architect and chief quantity surveyor of Lambeth respectively, though these officers delegated their day to day duties under the contract to outside firms. From the date of the contract until his retirement on 31 May 1983, Lambeth's chief architect was Mr W. Jacoby.

Croudace commenced work in July 1979. A mains electrical cable was discovered on 17 December 1979, and the architect instructed Croudace to revise their order to the London Electricity Board to provide for the diversion of the cable. The diversion was not completed until 11 June 1980. Croudace gave notice of loss and/or expense and also made application for an extension of time, alleging that the delay was

Lambeth's responsibility, though this was later denied by Lambeth. Croudace also alleged that they were delayed by Lambeth's failure to conclude their negotiations with a scaffolding contractor who was in possession of part of the site.

Although the contract completion date was 19 October 1981, practical completion was not achieved until 22 July 1982, and on 2 February 1983, the chief architect awarded extensions of time under clause 23(e), totalling 31 weeks and 3 days, and also issued his certificate of delay under clause 22, ending his letter by stating that 'it will be necessary for you to submit a formal claim for any loss or expense that you may consider is due to you, to be ascertained by the quantity surveyor.

On 12 February 1983 Croudace submitted a draft claim to the outside quantity surveyors, who replied that 'all payments relating to loss and expense' claims must be approved by Lambeth, to whom they had forwarded a copy of the submission. Despite letters chasing the money, no payment was in fact made, and on 25 May 1983 Croudace submitted a final claim for loss and expense in the amount of £624,655.54. On the following day there was a meeting between the outside quantity surveyors and a representative of Croudace, and the quantity surveyors agreed that a payment of £100,000 would be fair and reasonable as an interim payment on account.

On 31 May 1983 Mr Jacoby retired and Lambeth did not appoint anyone to succeed him. Although Croudace continued to press Lambeth both orally and in writing, they received no further payment. They issued a writ on 22 February 1984 and sought summary judgment against Lambeth under order 14 for £100,000 plus interest, or alternatively for damages and interest, in both cases to be assessed. Lambeth sought to stay the action under s.4 of the Arbitration Act 1950. His Honour Judge John Newey QC found in favour of Croudace and ordered that they be at liberty to enter judgment for damages to be assessed and that Lambeth should make an interim payment to Croudace within 14 days of £100,000 pursuant to Order 29, rule 11 of the Rules of the Supreme Court. He dismissed Lambeth's application for a stay of proceedings: see (1985) 1 ConLR 12. Lambeth appealed against the trial judge's orders.

Held: (dismissing the appeal): Lambeth's acts and omissions after 2 February 1983, including but not limited to their failure to appoint a successor architect, amounted to failure by them to take such steps as were necessary to enable Croudace's claim to be ascertained, and as such amounted to a breach of contract.

BALCOMBE LJ: Although he did not formally concede the point, Mr Desmond Wright QC, for Lambeth, did not seriously contest the judge's finding that there was implied in the contract an obligation on the part of Lambeth to nominate a successor to Mr Jacoby when he retired as Chief Architect, and that their failure to do so constituted a breach of contract on their part. In my judgment Lambeth's acts and omissions after 2 February 1983, including but not limited to their failure to appoint a successor to Mr Jacoby, amounted to a failure by them to take such steps as were necessary to enable Croudace's claim for loss and expense to be ascertained, and as such amounted to a breach of contract on their part: *Cf: Smith* v. *Howden Union & Anr.* (1890), 2 Hudson's BC (4th ed.) 156.

The real issues in this case are whether Croudace can establish that it has suffered any damage resulting from that breach of contract and, if so, whether the amount of that damage should be determined by the court or by arbitration.

Unless it can be successfully maintained by Lambeth that there are *no* matters in respect of which Croudace are entitled to claim for loss and expense under clauses 11(6) and 24(1)(e), it necessarily follows that Croudace must have suffered *some* damage as a result of there being no one to ascertain the amount of their claim. Without going into the details of the many matters raised by Croudace, I am satisfied that they can establish that they have suffered some damage in this connection even though, if Lambeth are right on the Bill 1/12A point, those damages may be less than the sum of £664,994.63 which they claim for loss and expense. Further, when damages come to be assessed it will be open to Lambeth to take the point that Croudace's applications under clauses 11(6) and 24(1) were not made within a reasonable time, since the judge's statement of his view that 'the application made under both conditions was made within a reasonable time' was not a finding necessary for the orders which he made. In my judgment, therefore, the judge was entitled to make the order of 18 October 1984 for damages to be assessed, unless it was a case for a stay of proceedings pending arbitration.

Application under JCT 1963 Clauses 11 and 24 must make it clear, if that is the case, that an element of the claim is for financing charges. The claim must be presented within a reasonable time but an employer may be precluded by his conduct from objecting to late claims. Financing charges run up until the last application before the issue of a relevant certificate. Financing charges cannot be recovered for periods when non payment arose from an independent cause which was not direct loss or expense. In calculating interest which compensates a contractor for having a bigger overdraft than he would otherwise have had, it is proper to take account of the fact the banks charge compound interest with periodic rests.

Rees & Kirby Ltd v. Swansea Corporation

COURT OF APPEAL   (1985), 6 ConLR 34
(CIVIL DIVISION)

The plaintiffs were engaged to construct a housing estate for the defendants for a fixed price of £845,301 under a contract on JCT 1963 terms dated 1 February 1972. The date fixed for practical completion was 6 July 1973. There were a large number of variations ordered under Clause 11 and there were a number of delays in issuing instructions within clause 24(1). Practical completion in fact took place on 4 July 1974. In June 1979 an extension of time under the contract of 52 weeks was agreed.

During 1972 there was very sharp escalation in wage rates in the construction industry and it soon appeared that this would convert the present contract into a substantial loss maker. Business relationships between the plaintiffs and the relevant offices of the Council were at this stage excellent and it appeared that the Council would approach the plaintiff's difficulties sympathetically. Further in October 1973 the Minister for Housing issued a statement that councils would be entitled to consider making *ex gratia* payments to contractors who suffered losses on fixed price contracts. In the light of this the plaintiff's contractual claims were put on one side while the parties negotiated for an *ex gratia* settlement. However by the end of 1976 the plaintiffs had concluded that no such settlement was possible and reverted to their claim under the contract. They also indicated to the council that their losses were significantly increased by interest charges arising from their having been out of their money for so long. The plaintiff's full claim when submitted in 1978 was for £178,780.67 exclusive of interest. Interim certificates were issued by the Architect for £60,000 and £25,000 respectively on 14 February 1979 and 2 April 1979 and a 'final interim certificate' for £156,762.19 on 30 August 1979. None of these sums included any element of interest.

The plaintiffs issued their writ on 8 March 1982 claiming interest on the sum of £156,762.19 from the date of practical completion until the date of payment under the final interim certificate (12 September 1979) and compound interest on the accrued interest so claimed until the date of judgment.

Held:

(1) The contractors application for payment under clauses 11(6) and 24(1)(a) of JCT 1963 must contain a reference sufficient to make it clear that it includes some loss or espense incurred by reason of his being out of pocket.

(2) Such notice must ordinarily under the terms of these clauses be given within a reasonable time of the loss or expense having been incurred. However the defendants in the present case could not rely on the delay between practical completion in 1974 and the formal application of June 1978 since they were estopped from enforcing their strict legal rights.

(3) Under the structure of the clauses there is no cut-off point at the date of practical completion and no reason why financing charges should not continue to constitute direct loss or expense until the last written application before the relevant certificate was issued.

(4) However the delay in payment between the date of practical completion and 11 February 1977, when it became clear that the respondent's claim must be advanced under the terms of the contract, was not direct loss or expense since it arose from an independent clause.

(5) The plaintiffs were entitled to financing charges to cover the period from 11 February 1977 to 10 August 1979 when they appended their signature to the draft final account.

(6) Since the plaintiffs were operating on the basis of a substantial overdraft at their bank and since banks calculate interest with periodic rests, it was proper to take this into account.

ROBERT GOFF LJ: Thus an appeal by Swansea City Council against a decision by Kilner Brown J in which he awarded to the respondents, Rees & Kirkby Ltd, the sum of £255,195.12 as due and owing by the appellants to the respondents under a building contract in the JCT Form (1963 edition) in respect of what I will broadly call financing charges. This money was held to be due to the respondents under clause 11(c), or alternatively clause 24(1)(a), of the contract, on the principles established by this Court in *F.G. Minter Ltd* v. *Welsh Health Technical Services Organisation* (1980). An alternative claim by the respondents for damages for breach of contract was rejected by the judge.

Before I turn to the facts of the case, it is desirable that I should first set out the most relevant provisions of the contract, and refer also to *Minter*'s case. Clause 11 of the contract is concerned with variations, provisional and prime cost sums. Under clause 11(1), the architect is empowered to issue instructions requiring a variation; and clause 11(4) sets out the manner in which (*inter alia*) work executed pursuant to a variation order shall be valued. Clause 11(6) provides as follows:

'If upon written application being made to him by the Contractor, the Architect/Supervising Officer is of the opinion that a variation or the execution by the Contractor of work for which a provisional sum is included in the Contract Bills (other than work for which a tender made under clause 27(g) of these Conditions has been accepted) has involved the Contractor in direct loss and/or expense for which he would not be reimbursed by payment in respect of a valuation made in accordance with the rules contained in sub-clause (4) of this Condition and if the said application is made within a reasonable time of the loss or expense having been incurred, then the Architect/Supervising Officer shall either himself ascertain or shall instruct the Quantity Surveyor to ascertain the amount of such loss or expense. Any amount from time to time so ascertained shall be added to the Contract Sum, and if an Interim Certificate is issued after the date of ascertainment any such amount shall be added to the amount which would otherwise be stated as due in such Certificate.'

It was the submission of the respondents in the present case that the relevant financing charges incurred by them were recoverable from the appellants as direct loss and/or expense in which they were involved by reason of variation orders made under the contract, for which they were not reimbursed by the valuations made under clause 11(4); or alternatively that they had a claim under clause 24(1)(a). The material parts of clause 24 provide as follows:

'(1) If upon a written application being made to him by the Contractor the Architect/Supervising Officer is of the opinion that the Contractor has been involved in direct loss and/or expense for which he would not be reimbursed by a payment made under any other provision in this Contract by reason of the regular progress of the Works or of any part thereof having been materially affected by: (a) The Contractor not having received in due time necessary instructions, drawings, details or levels from the Architect/Supervising Officer for which he specifically applied in writing on a date which having regard to the Date for Completion stated in the appendix to these Conditions or to any extension of time then fixed under clause 23 or clause 33(1) of these Conditions was neither unreasonably distant from nor unreasonably close to the date on which it was necessary for him to receive the same; or ... and if the written application is made within a reasonable time of it becoming apparent that the progress of the Works or of any part thereof has been affected as aforesaid, then the Architect/Supervising Officer shall either himself ascertaion or shall instruct the Quantity Surveyor to ascertain the amount of such loss and/or expense. Any amount from time to time so ascertained shall be added to the Contract Sum, and if an Interim Certificate is issued after the date of ascertainment any such amount shall be added to the amount which would otherwise be stated as due in such Certificate.'

The alternative claim of the respondents was that the relevant financing charges constituted a direct loss and/or expense in which they were involved by reason of the regular progress of the works or part of them having been materially affected by their not having received the necessary instructions in due time, for which they had not been reimbursed by a payment made under any other provision of the contract.

The sums claimed by the [contractors] as due and owing under clause 11(6) or clause 24(1)(a) were, as I have said, claimed by them under the principles stated by this Court in *Minter*'s case. That case was also concerned with a claim under clauses in the same form, except that the period for the making of written applications under the clauses was not within a reasonable time of the loss or expense having been incurred but within 21 days of the loss or expense having been incurred. In *Minter*'s case, the contractors were paid amounts under the two clauses in respect of direct loss and/or expense in which they had been involved; but they claimed that the amounts so paid to them were insufficient. Their claim was that, since the amounts in question had not been certified and paid until long after the time when they were involved in the relevant loss or expense, the sum certified ought to have included amounts in respect of the loss and/or expense in which they had been involved by way of finance charges and/or being stood out of their money for those long periods. This claim raised the question whether loss or expense of this kind was recoverable under these clauses. This court, reversing the decision of Parker J. (as he then was), held that they were so recoverable. Judgments were delivered by both Stephenson LJ and Ackner LJ; Sir David Cairns agreed. In relation to the meaning of the word 'direct' in the expression 'direct loss and/or expense', the court concluded (following the decision of Megaw J. (as he then was) in *Wraight Ltd.* v. *P.H. & T. Holdings Ltd.* (1968), that loss or expense should be regarded as direct, for the purposes of both clauses, if it would be recoverable as damages flowing naturally in the usual course of things, within the meaning of those words as used in the first rule in *Hadley* v. *Baxendale* (1854). This was common ground between the parties. It was also common ground that since, in the construction industry, cash flow is vital to contractors and delay in payment results in the ordinary course of things in the contractor being short of working capital, the loss of interest which he has to pay on capital which he has borrowed and the loss of interest on capital which he is not free to invest does fall within the first rule in *Hadley* v. *Baxendale*, and so is direct loss and/or expense. Furthermore, it made no difference that the law has declined to award, by way of damages, interest on debts which have been paid late. That has always been regarded as an anomalous exception to the general principles on which damages are assessed (see *London Chatham & Dover Railway Co.* v. *South Eastern Railway Co.* (1893); and the existence of this anomalous exception to the recovery of damages should not inhibit the court from holding that such loss or expense is direct loss or expense when considering the meaning of those words used in the two clauses. As Ackner LJ put it, the contractors were seeking to claim not interest on a debt, but a debt which had, as one of its constituent parts, interest charges which had been incurred. The court, in holding that the contractors' claim was recoverable, also rejected certain arguments (which had impressed Parker J) that the two clauses contained provisions which were, upon examination, inconsistent with such a claim being recoverable under either of the clauses.

Now the claim of the [contractors] in the present case is of a similar kind to, though substantially greater in amount than, the claim successfully advanced by the contractors in *Minter*'s case. In order to explain how the claim arose in the present case, it is now necessary for me to set out the history of the matter in some detail.

The [contractors'] tender having been accepted on 30 December 1971, they took possession of the site on 5 January 1972. The contract was dated 1 February 1972: it provided for the construction of a housing estate for the [Council], at a fixed cost of £845,301. Under the contract, the nominated architect was the City Architect, who later became known as the Director of Environment. The date fixed for practical completion was 6 July 1973. In fact, practical completion was about one year late, being achieved on 4 July 1974; and it is not disputed that there were a large number of variations ordered under clause 11, and also a number of delays in issuing instructions which fell within clause 24(1). It was however not until 30 August 1979 that the architect issued a so-called "final interim certificate", allowing a sum of £156,762.19 in respect of direct loss for expense under clauses 11 and 24, including nothing for interest; interim certificates had however previously been issued on 14 February and 2 April 1979 for sums of £60,000 and £255,000 to be paid on account. So a period of over five years elapsed between practical completion on 4 July 1974 and the issue of the final interim certificate of 30 August 1979. The balance due under the certificate was paid on 12 September 1979, within the time fixed by the contract for honouring the certificate, i.e. 14 days (since it was expressed to be an interim certificate). The respondents' claim in the action consists of compound interest on the sum of £156,762.19 from the date of practical completion until the date of payment under the final interim certificate, 12 September 1979; they have also claimed compound interest on the accrued interest so claimed, from that date until the date of judgment. The compound interest so claimed has been calculated on the basis of quarterly rests. The [contractors'] writ in the action was issued on 8 March 1982; by then the accrued interest so claimed amounted to about £207,000. By the date of judgment, 15 March 1983, the total amount claimed and adjudged due to the [contractors] was £255,195.12, including a sum of £500 agreed retention.

I have already recorded that work began under the contract in January 1972. That year saw a very substantial property boom, and a very sharp escalation in wage rates in the construction industry. It soon became plain that the respondents were being very badly hit by this wage escalation, and that they were going to suffer a very substantial loss under their fixed price contract with the [Council]. Already by November 1972 they estimated their loss as likely to be £90,000; it later proved to be substantially higher. Mr Joseph, the [contractors'] managing director, enjoyed an excellent relationship of long standing with the relevant officers of the Council. On 18 November 1972 he wrote to the City Architect, explaining his serious situation and stating that [his company] had no alternative but to seek help from the Council; his approach was received with sympathy and understanding. Subsequently, on 4 July 1973 (in fact, the day before the date fixed for practical completion under the contract) the [contractors] wrote asking for an extension of time, and in their application listed numerous causes of delay, including over 150 site instructions or variations issued by that date, and delays in receiving instructions. On 2 August 1973 the architect wrote, acknowledging [that] letter, and making certain comments; he also asked for futher details. On 18 December 1973, the [contractors] sent the additional information so requested. Their letter concluded as follows:

> 'We trust that this information clearly shows where and why the delays have occurred. From these details we are able to establish that many of the delays run concurrently, and thus the overall delay on Phase I has ben reduced to 37 weeks, seven of which were due to labour and material shortages. On completion of Phase I we therefore consider that there should be an extension

of 30 weeks under clauses 23(e), (f), (h) and recovery of loss and expense under clauses 11(6) and/or 24(a), (c) and (d) plus 7 weeks under clause 23(j). Similar delays have occurred during Phases II and III. These have been minimised but we now assess that the delay amounts to 51 weeks. We therefore wish to claim an extension of 51 weeks for delays that have occurred up to 30 November 1973 41 weeks under clauses 23(e), (f) and (h) and recovery of loss and expense under clauses 11(6) and/or 24(a), (c) and (d) plus 10 weeks under clause 23(j). The anticipated completion date will now be 28 June 1974.'

Meanwhile, as a result of the nationwide problem created by delays and escalations in construction costs, the Minister for Housing issued in October 1973 a statement that local authorities would in appropriate cases be entitled to consider making *ex gratia* payments to contractors to take account of losses sustained on fixed price contracts, As a result of this announcement, the [contractors'] claim under clauses 11(6) and 24(1)(a) was almost immediately put on one side, while the parties considered the possibility of an *ex gratia* payment; and in June 1974 the council confirmed a resolution of the Housing Committee, made on the recommendation of the Town Clerk, that an *ex gratia* payment be made to [the contractors] towards their loss then estimated at £200,000, on the terms (*inter alia*) that [the contractors'] right to all claims under the contract (which would include their claims under clauses 11(6) and 24(1)(a)) be forfeited. Unfortunately, this came to nothing. Prolonged negotiations took place, but following a change in the control of the Council, and adverse publicity in the local press concerning such payments to contractors, Mr Joseph came to the conclusion that his company could not longer accept an *ex gratia* payment. This he confirmed in a letter to the Town Clerk dated 24 December 1976; in the same letter he drew attention to the fact that [his company's] losses had been aggravated by interest charges on the money outstanding for their claims under the contract. The question then arose whether the contract should be converted from a fixed price to a fluctuating contract under what is called the NEDO formula, or whether [his firm] should process their contractual claims. These two alternatives were referred to by Mr Joseph in his letter dated 24 December 1976, and again in a letter to the architect dated 21st January 1977. In the latter letter, Mr Joseph stated that the contractual claims, of which he then gave formal notice, amounted to £180,000. The letter continued:

'The points of claim will be as stated in our letter of 4th July, 1973 with obvious extensions covering the extended contract period. In this event all claims settled to attract interest charges, at 1% above bank rate, from the date of origin of claim. Dispute to be settled by immediate arbitration.'

On 11 February 1977, the architect wrote to [the contractors] as follows:

'I have now formally reported to Housing Committee the withdrawal of your request for an *ex gratia* payment on the above contract.

I have also reported your offer to re-negotiate the contract on a fluctuating basis, but, for various reasons, which in no way reflect upon your company, I was unable to recommend acceptance of your offer. It was explained to committee that it now rests with me to determine the additional amounts, if any, to which you are entitled under the terms of the original contract, and to certify accordingly. In normal circumstances I would be unable to take any action until after the next Council meeting, but the matter has already been reported, to a degree, in the press, and Council can do very little to alter the situation.

I therefore feel that an early meeting between [us] could be held to discuss, in the first instance in general terms, the procedures and methods whereby settlement can best be achieved. I shall be glad if you will contact [me] to arrange the meeting.'

The architect reported on the matter to the Housing Committee. The committee accepted his report; but on 15 March 1977 the Council amended the committee's resolution that the report be accepted by adding the words 'on the understanding that the council is in no way committed to any financial settlement in respect of the claim'. This resolution, which was no doubt intended to do no more than record the obvious fact that the council was not yet committed to pay any particular sum in respect of the [contractors'] contractual claims, was not happily phrased and so led to understandable concern on the part of Mr Joseph and to negotiations on the contractual claim being conducted with the professional officers at arms' length ...

At all events, on 8 August 1977 the [contractors] informed the Council's quantity surveyors that they were preparing their claim in full detail. By 1 November 1977, after an alternative appointment had fallen through, [they had instructed a quantity surveyor] to prepare their claim, and the Council had been informed of this. His report was not however produced until 30 June 1978, when it was forwarded to the [Council]; the Town Clerk sent it on to Mr Davies, of the architect's department, for consideration. The [contractors' claim, as calculated by the quantity surveyor], totalled £178,780.67 exclusive of any interest. It concluded with the following passage:

'Now it must be obvious that Rees and Kirby Limited have been out of funds on this Contract for a considerable period of time.

May we quote an extract from *Building Contracts* by Donald Keating just this year published by Sweet and Maxwell.
Local Authorities:
'when settling a claim for damages, or for money due by a contractor, local authorities sometimes take the point that they cannot include a sum in respect of interest because such payment is contrary to law rendering them liable to the possibility of surcharge. It is submitted that, provided the local authority acts bona fide in the interests of the rate-payers and with reasonable business acumen the inclusion of such a sum is not contrary to law, and further, if its exclusion results in litigation they would ordinarily be acting against the interest of the rate-payers. This is because, quite apart from the possibility of a claim for interest as damages, interest is ordinarily awarded under the Law Reform (Miscellaneous Provisions) Act 1934, so that the contractor would be justified in refusing the sum offered and proceeding to trial in order to obtain such interest, and would normally recover costs.'
It would appear therefore that Rees and Kirby are entitled to consideration for interest on monies arising from the breach which we have not calculated in the total above.
We have included our fee in preparation of this claim because we consider this to be an additional cost arising from the breach.
We conclude that Rees and Kirby have just contractual entitlement to settlement of the value we have assessed, plus interest and we submit that we have fairly excluded any element from this claim that may be stated as due to inefficiency on their part or National shortages of directly employable labour.'

Four months passed by; and Mr Joseph, having heard nothing, wrote to the Town Clerk on 25 October 1978 asking for [the claim] to be given urgent consideration. On 5 December he wrote again asking for urgent action: he enclosed a summary of [the company's] interest charges, amounting to £122,254.59, and stated that interest was at present accruing at the rate of £3,457 per month. On 27 February 1979, following a meeting [between the parties], the City Architect provided a detailed response to the [contractors'] claim. It was recognised that an extension of time would have to be made; this was then proposed at 57 and a half weeks, but was later fixed at 52 weeks. The council's quantity surveyors were instructed to value the claim. The architect's letter concluded as follows:

'13. There is no contractual obligation for a contractor to prepare a claim; it was Rees & Kirby's own decision to do so.
The costs of claim preparation, although they may be acceptable as 'damages' in breach (which is denied), cannot be admitted in settling under Clauses 11 or 24.

14. The matter of interest, is one for the employer. Committee has already been informed that you are claiming reimbursement, but it will be for Chief Executive and Town Clerk and City Treasurer to advise thereon.'

On 13 March 1979, Mr Joseph acknowledged receipt of the letter, reminding the architect that his company was being kept out of funds and asking for the claim to interest to be considered with urgency. On 12 April 1979 the Town Clerk acknowledged Mr Joseph's letter, but stated that he did not consider that the [contractors] were entitled to interest. He left it to Mr Joseph to consider whether he wished to have a joint meeting between the [two quantity surveyors]; Mr Joseph wrote asking for such a meeting. A meeting was finally held on 13 June 1979, when the extension of time was agreed at 52 weeks: the valuation of the claim was left to consideration by the two quantity surveyors. On 17 July, following a further repetition of the claim to interest, Mr Joseph wrote to the Town Clerk stating that he had taken advice and was satisfied that:

'... we are entitled to receive interest or damages for breach of contract or the same monies as direct loss and expense within the contract and that this right has been recognised by other public authorities.

Accordingly, following receipt of the architect's certificate covering the final account and contractual claim, we will update our figures of claims on an interest basis and submit it for payment and, unless our claim in this respect is satisfied, we will put the whole matter in the hands of the National Federation of Building Trades Employers' Legal Services Department.'

This appears to have been the first occasion on which the claim to interest was put forward on the basis of direct loss or expense under the contract. Ultimately the final account was submitted to Mr Joseph on 9 August, the claim (excluding interest) being assessed at £156,762.19. Mr Joseph returned the final account on 10 August, signed by him but qualified to reserve [this company's] 'outstanding claim for damages/interest'. At the same time, he enclosed an updated computation of his claim for interest. Thereafter [Rees & Kirby] placed their claim to interest in the hands of their legal advisers, and after the decision of the Court of Appeal in *Minter*'s case on 14 March 1980 received favourable legal advice. However, the Town Clerk wrote on 21 May 1980 rejecting a claim founded on *Minter*'s case. As I have previously recorded, the writ was issued on 8 March 1982.

It was submitted to the judge, on behalf of the appellants, that no written application within clauses 11(6) and 24(1)(a) was made within a reasonable time, as required by both clauses; and that on that ground alone the respondents' claim must fail. The judge rejected that contention. He accepted the [contractors'] submission that the claim need not be in any set form, and concluded that in all the circumstances the claim had been made within a reasonable time. He was satisfied that the claim fell within the principle in *Minter*'s case, and allowed it in full. On the alternative claim for damages for breach of contract, he held that the architect plainly failed to comply with the time fixed in the contract for measuring, valuing and issuing certificates. He rejected a submission that there had been acquiescence by the [contractors] in the delay which had occurred. He furthermore held that, as the council and the Town Clerk were to all intents and purposes dictating and controlling the exercise by the architect of his duty, the [Council] were responsible to the [contractors] for the architect's breach of the contract. However, he dismissed the claim to interest as damages on the ground that he was precluded from awarding interest as damages by the decision of the House of Lords in *London Chatham & Dover Railway Co.* v. *South-Eastern Railway Co.*

Before this court, [Counsel for Swansea] accepted that we were bound by the decision in *Minter*'s case; accordingly, his argument was directed towards challenging the conclusion of the judge that his clients were liable, under the principles stated in that case, in respect of financing charges from the date of practical completion. He submitted (1) that the respondents had failed to make the requisite written application within a reasonable time, as required both by clause 11(6) and clause 24(1)(a), and that for that reason alone their claim should fail; (2) that in any event there was a 'cut-off date' as at the date of practical completion, so that financing charges incurred after that date were not recoverable; (3) that, in the circumstances, the financing charges were incurred not by reason of the variations and delays in giving instructions, but by reason of other events; and, (4) that the claim should be calculated not on the basis of compound interest, but of simple interest.

In considering these submissions, we start from the fact that it was decided by the Court of Appeal in *Minter*'s case that financing charges are recoverable under the clauses in respect of the period between 'the loss and/or expense being incurred and the making of the written application for reimbursement of the same as required by the Contract': *per* Ackner LJ. However, the application of the clauses in the circumstances of the present case has been very much complicated by a number of matters. First, almost immediately after intimating that they were making claims under clauses 11(6) and 24(1)(a), the [contractors] entered upon prolonged negotiations with the [Council] for an *ex gratia* payment, so that the whole matter of proceeding with contractual claims was, so to speak, put on ice. Second, even when those negotiations had collapsed, the parties took a little time to pick up the threads of proceeding with claims under the contract; and, since by then the date of practical completion had long since passed, in practical terms the matter which then fell to be considered was the amount of money which was to be included in the final certificate in respect of those claims. Finally, *Minter's* case was not decided by the Court of Appeal until March 1980, which was six months after payment had been made by the [Council] under the final interim certificate; so, understandably, although the [contractors] were expressing periodic concern about the delay which was continuing and about the fact that they were suffering by being out of their money, their claim, in so far as it was formulated, was expressed primarily as a claim to interest on monies outstanding rather than as a claim to financing charges as direct loss and/or expense; though, with some prescience, it was so formulated on 17 July 1979.

Faced with these somewhat unusual circumstances, we have, I think, to regard with some caution submissions which are founded upon suggestions that the [contractors] failed strictly to comply with the contractual provisions.

With these preliminary observations in mind, I turn to [the] first submission, which was that [Rees & Kirby] failed to make the requisite written application within a reasonable time.

There was some argument before us about the form which a written application under the two clauses should take. [Counsel for the contractors] adopted as part of his argument a passage from *Keating's Building Contracts* (4th edn), at p. 56) in which the learned author has this to say about the contents of an application under clause 11(6):

'*Contents of the application.* A consideration of the effect of the Minter case raises some questions about what an application should contain in order to be valid under clause 11(6). No form is required, but it must, it is submitted, be expressed in such a way, or made in such circumstances, as to show that the architect is being asked to form the opinion referred to in the sub-clause and identify the variation or provisional sum work relied on. It is thought that great particularity is not contemplated by the sub-clause. The architect must know the variation of work relied on; he must also either be told in the application, or must be taken to know from his knowledge of the circumstances, sufficient to enable him to form the opinion that the contractor has been involved in direct loss and/or expense for which he would not be reimbursed under sub-clause (4). He does not have to have at the stage of forming his opinion sufficient details to ascertain the amount; it is sufficient if he has enough to form the view that there must be some loss. Thus, it is thought, providing the principles just suggested are met, an application is valid to found a claim at the ascertainment stage that interest should be included even though the application itself did not expressly refer to interest.'

I have much sympathy with the general approach adopted in this passage. It seems to me that, in the ongoing relationship that exists between a contractor and an architect carrying out their functions under a contract in this form, a sensible and not too technical attitude must be adopted with regard to the form of such an application. But, having regard to the form which clauses 11(6) and 24(1)(a) take in the contract presently before us, I do not feel able to say that an application need not expressly refer to what Mr Keating calls 'interest'. This is because the Court of Appeal's decision in *Minter's* case proceeded on the basis that a claim to financing charges cannot be regarded simply as parasitic to the primary expense incurred as a contractor by reason of, for example, a variation; and that financing charges are recoverable under either of the clauses between the date when the loss or expense is incurred and the date when the written application for reimbursement is made, so that, in respect of a continuing loss or expense such as financing charges, successive applications will have to be made from time to time. On that basis, it is difficult to avoid the conclusion that the application in question, if it is to be read as relating to financing charges, must make some reference to the fact that the contractor has suffered loss or expense by reason of being out of his money. I observe however in parenthesis that the clauses in question have since been revised to allow for applications in respect of loss or expense which the contractor has incurred *or is likely to incur*; and it may be (though I express no opinion upon it) that, under the clauses as so revised, there need be no express reference to interest as financing charges. But, for the purposes of the present case we must, I consider, proceed on the basis that some reference is necessary. Even so, I do not consider that more than the most general reference is required, sufficient to give notice that the contractor's application does include loss or expense incurred by him by reason of his being out of pocket in respect of the relevant variation or delayed instruction, or whatever may be the relevant event giving rise to a claim under the clause.

Even so, I feel unable to accept [the Council's] submission that the [contractors] claim must fail in the present case because they failed to give the requisite notice within a reasonable time of the loss or expense having been incurred, as required by the clause. In so far as the loss or expense in question consists of the primary loss or expense incurred by the relevant variations or delayed instructions, it appears to me that the [Council] are estopped from enforcing their strict legal right to have such a notice given within a reasonable time of such loss or expense having been incurred, on the principle stated by Lord Cairns LC in *Hughes* v. *Metropolitan Railway Co.* (1877). The passage in Lord Cairns' speech is so well known, that I trust that I will be excused from setting it out in this judgment. The present case seems to me to be a classic example of parties entering 'upon a course of negotiation which has the effect of leading one of the parties to suppose that the strict rights arising under the contract will not be enforced'. Here, as soon as the [contractors] raised the question of a claim under either clause 11(6) or clause 24(1)(a), the parties entered into the negotiations for an *ex gratia* settlement, negotiations which lasted over a period of years. Even after those negotiations had come to an end, the [Council] expressed no objection to the [contractors] instructing their own quantity surveyor to prepare their claim, nor did they give any notice to the effect that they intended to resume their right under either clause to require a written application within any particular period of time. Furthermore, once the [contractors] had submitted their claim ... on 30 June 1978, they had given what I regard to be a sufficient intimation of the fact that they were out of pocket by reason of delay in payment of their claims under the two clauses, and they continued periodically thereafter to give further intimations of that fact almost up to the date of the issue of the final interim certificate. In these circumstances, I am satisfied that the first submission of the Appellants is not well-founded.

Nor, in my judgment, is there any substance in the second of the [local authority's] submissions, which is that there is a 'cut-off point' at the time of practical completion. As I read the clauses, given that (on the clauses in the form which they take in the contract now before us) successive applications are made at reasonable intervals, I can see no reason why the financing charges should not continue to constitute direct loss or expense in which the contractor is involved by reason of, for example, a variation, until the date of the last application made before the issue of the certificate issued in respect of the primary loss or expense incurred by reason of the relevant variation. At the date of the issue of the certificate, the right to receive payment in respect of the primary loss or expense merges in the right to receive payment under the certificate within the time specified in the contract, so that from the date of the certificate the contractor is out of his money by reason either (1) that the contract permits time to elapse between the issue of the certificate and its payment, or (2) that the certificate has not been honoured on the due date. But I can for my part see no good reason for holding that the contractor should cease to be involved in loss or expense in the form of financing charges simply because the date of practical completion has passed.

There is however in my judgment much more force in the third submission of the [Council], that part of the delay in payment in respect of the primary loss or expense incurred by the [contractors] in the present case is attributable to an independent cause, so that financing charges incurred by them during the period when such independent cause operated should not be recoverable by them under either clause 11(6) or clause 24(1)(a). As I see it, the whole of the period of delay which occurred by reason of (1) the negotiations for an *ex gratia* settlement irrespective of the parties' strict contractual rights, and (2) the proposal by the [contractors] that the contract might be converted from a fixed price contract to a fluctuating contract under the NEDO formula, must be regarded as delay attributable to those matters, so that financing charges incurred by [them] during such period cannot be direct loss or expense in which [they] were involved by reason of the variations or delayed instructions. The [contractors] have made no claim in respect of financing charges incurred by them before practical completion: and the period of delay resulting from the above two matters must have run on past practical completion until the architect's letter dated 11 February 1977, after which it became clear that [their] claim must be advanced under the terms of the contract. However, from the date of that letter, I can discern no cause of the [contractors] incurring the financing charges which form the subject matter of their claim, other than the fact that they incurred the primary loss or expense incurred by them by reason of the variations or delayed instructions under one or other of the two clauses. It is true that [their] detailed claim was not submitted until 30 June 1978; but I do not consider that the time which passed in the meanwhile should be regarded as the product of any other cause so independent of the primary loss or expense as to relieve the [Council] from liability in respect of financing charges incurred by the [contractors] during that period. Once [the contractors' quantity surveyor's] report had been delivered to the [Council], the financing charges must have continued to constitute direct loss or expense up to the date of the issue of the final interim certificate on 30 August 1979, subject to the requirement, established in *Minter*'s case, that such loss or expense is only recoverable in respect of a period ending with a written application made under the relevant clause. In my judgment, written applications containing a sufficient reference to such loss or expense were made not only by the delivery of [the quantity surveyor's] report, with its reference to the fact that the [contractors] had been out of funds, and that they had 'a just contractual entitlement to settlement of the value we have assessed, plus interest'; but also by the letters claiming interest, sent at periodical intervals, on 5 December 1978, 13 March 1979 and 17 July 1979, and by their signature on the draft final account on 10 August 1979 reserving their right to interest. It follows, in my judgment, that [the contractors] have established a right to recover financing charges under the clauses in respect of the period from 11 February 1977 to 10 August 1979, on the principle in *Minter*'s case.

There remains to be considered the question whether [they] are entitled to recover their financing charges only on the basis of simple interest, or whether they are entitled to assess their claim on the basis of compound interest, calculated at quarterly rests, as they have done. Now here, it seems to me, we must adopt a realistic approach. We must bear in mind, moreover, that what we are here considering is a debt due under a contract; this is not a claim to interest as such, as for example a claim to interest under the Law Reform Act, but a claim in respect of loss or expense in which a contractor has been involved by reason of certain specific events. [Rees & Kirby], like (I imagine) most building contractors, operated over the relevant period on the basis of a substantial overdraft at their bank, and their claim in respect of financing charges consists of a claim in respect of interest paid by them to the bank on the relevant amount during that period. It is notorious that banks do themselves, when calculating interest on overdrafts, operate on the basis of periodic rests: on the basis of the principle stated by the Court of Appeal in *Minter*'s case, which we here have to apply, I for my part can see no reason why that fact should not be taken into account when calculating the [contractors'] claim for loss or expense in the present case. It follows that, in order to calculate [their] contractual claim it will be necessary to calculate it with reference to the total sum of £156,762.19 in relation to the period I have indicated, viz., 11 February 1977 to 10 August 1979, taking into account (1) the two payments made on account during that period; (2) the rates of interest charged by the bank to [them] at various times over that period; and (3) the periodic rests on the basis of which the bank from time to time added outstanding interest to the capital sum outstanding for the purposes of calculating interest thereafter ...

Finally, I must briefly mention an alternative basis on which the [contractors] sought to recover their financing charges as damages for breach of contract. The breach of contract alleged by them was that there was a failure by the architect to measure and to make valuations and to issue certificates in respect of the relevant variations and delayed instructions at the times specified in the contract (other than in the final interim certificate itself). In my judgment, however, this claim must fail. The time which passed between the [contractors'] letter of 18 December 1973, in which they first advanced the claim under clauses 11(6) and 24(1)(a), and the issue of the final certificate on 30 August 1979, can be sub-divided into the following periods:

(1) 18 December 1973–24 December 1976:
    negotiations over a possible *ex gratia* payment.
(2) 24 December 1976–11 February 1977: considering the [contractors'] suggestion that the contract should be converted into a fluctuating contract on the NEDO formula.
(3) 11 February 1977–8 August 1977:
    a period of quiescence.
(4) 8 August 1977–1 November 1977:
    time occupied by the [contractors] in instructing a quantity surveyor.
(5) 1 November 1977–30 June 1978:
    preparation of the [contractors' claim].
(6) 30 June 1978–30 August 1979:
    consideration by the [Council of the claim], and the issue by them of the final certificate.

None of the periods of delay can possibly be attributed to the [Council], except that it can be argued that they took too long to consider the [contractors'] claim during the period 30 June 1978 to 30 August 1979. But, bearing in mind the time allowed in the contract, under clause 30(5) and (6) and the Appendix to the contract, for the preparation and issue of a final certificate after practical completion, I do not consider that it would be right, in all the circumstances of this case, to hold that there was any breach of contract by the [Council] in this respect. It follows, in my judgment, that the [contractors'] alternative claim to damages for breach of contract must fail,...

Chapter 13

LIQUIDATED DAMAGES AND EXTENSION OF TIME

A provision relating to liquidated damages is unenforceable if expressed in a manner which is inconsistent with the other contract clauses to which liquidated damages relate.

Bramall & Ogden Ltd v. Sheffield City Council

QUEEN'S BENCH DIVISION   (1985), 1 ConLR 30

In June 1975 the parties contracted on JCT 63 terms for the erection of 123 dwellings for the sum of £1,191,453.07. The date for completion was stated in the Appendix as 6 December 1976, and the Appendix entry relating to liquidated damages read 'Liquidated and Ascertained Damages: (Clause) 22 *at the rate of £20 per week for each uncompleted dwelling*'. Various extensions of time were granted by the architect giving an extended completion date of 4 May 1977.

The parties did not enter into a sectional completion supplement but as the houses were completed they were taken over by the employers, the architect issuing a number of certificates of practical completion in respect of each group taken over. The last houses were not taken over until 29 November 1977 but, between 4 May and 29 November the architect issued various instructions to the contractors.

The employer retained £26,150 (123 × £20) as liquidated damages in respect of 123 dwellings uncompleted, but claimed no liquidated damages beyond 29 November 1977.

The contractors disputed both the extensions of time granted and the employer's right to deduct liquidated damages. An arbitrator issued an interim award in favour of the employer on both points and the matter came before the court by way of an application for leave to appeal under s. 1 of the Arbitration Act 1979.

The contractor's two main contentions were:

(1) 'That by issuing instructions involving variations and extra works

after the date certified in the ... Clause 22 certificate and by failing to issue any extensions of time, the [Council], through their architect, put time at large under the contract and thereby rendered null and void the ... liquidated and ascertained damages' clause.

(2) 'That since the (contract) contained no express provisions for sectional completion ... the ... liquidated and ascertained damages provisions were rendered unenforceable by reason of the operation of clauses 12(1) and 16(e) of the' contract.

Held (inter alia):

(1) Leave to appeal was refused as to the extension point.

(2) The contractor's second contention was correct and the liquidated damages clause was unenforceable.

JUDGE LEWIS HAWSER QC: [Counsel for the applicants] has submitted that there are errors of law on the face of the award in the arbitrator's reasons. His principal argument ... was that the contract does not provide for sectional completion, as indeed it does not and as the arbitrator found it did not. The arbitrator's finding is: 'There was no provision for sectional completion in the Articles of Agreement'. Therefore, according to its terms the respondents would have been entitled to deduct liquidated damages on all the dwellings and indeed on all the works as defined in condition 1(1) and as found by the arbitrator, up to the date of practical completion of the works, irrespective of whether they had taken possession of dwellings during the course of the work. The result, he said, would be that liquidated damages would turn into a penalty, since they could exceed substantially the actual loss sustained. The fact that the respondents chose not to exercise their rights and only to deduct the liquidated damages after the extension date of 4 May 1977 and until the time when they took possession of the last house on 29 November 1977 in respect of houses not taken into possession during that period does not and cannot affect the position in law. He submitted that they may well have operated the provisions in a reasonable manner but this cannot affect their validity. He said that the liquidated damages' provisions here have to be construed strictly and *contra proferentem* – a proposition which seems to be borne out by the decision of the Court of Appeal in *Peak Construction (Liverpool) Ltd* v. *McKinney Foundations Ltd* (1970). There is a passage in the judgment of Salmon LJ which says: 'The liquidated damages and extension of time clauses in printed forms of contract must be construed strictly *contra proferentem*'.

[Counsel's] argument may, I hope, be summarised as follows. The 'Works' cover not only the houses but the other items above referred to. Clause 22 refers to a failure 'to complete the Works' by the extended date. As from that date the employer becomes entitled to liquidated damages until the Works are completed. Clause 16 deals with the consensual taking of possession of part of the Works. Clause 16(e) provides for the sum payable after taking possession in respect of the period during which the Works remain incomplete. The way in which the liquidated damages are dealt with is set out in the Appendix. This does not allow of the calculation to be made which is required by condition 16(e), and one cannot operate the Appendix and condition 16(e) in the circumstances of this case. The inconsistency can only be reconciled if provision is made in the contract for sectional completion of those parts which are taken over and to which specific liquidated damages' provisions are applied. In *M.J. Gleeson (Contractors), Ltd* v. *London Borough of Hillingdon* (1970) Mocatta J came to the conclusion that, as there were no sectional completion provisions in the contract before him, clause 12(1) of the RIBA conditions prevented the provisions of the contract bills from overriding the contract itself and that therefore sectional completion could not be relied upon to justify the deduction of liquidated damages.

The matter is also dealt with on page 330 of Keating's *Building Contracts*, 4th edn. Under the heading '(3) Sectional completion', he says this: 'Until an amendment introduced in July 1973 the side note to this clause was "Sectional completion". This may have misled some people into thinking that it provided, in itself for the work to be completed and handed over in sections if supplemented by provisions in the bills of quantities. In *Gleeson* it was held that this was not so. The contract was for the provision of a large number of houses to be erected in blocks. The form was, for all material purposes, the current form with one date for completion, namely 24 months after the date for possession. There were no amendments to the form. The preliminaries bill provided that the contract was to be completed and handed over in sections, some blocks after twelve months and the rest at three-month intervals up to twenty-four months. The blocks due for completion after twelve months were not completed and the employer applied the liquidated damages clause which provided for liquidated damages at the rate of £5 per dwelling per week. It was held, applying clause 12 (discussed *ante*), that the employer was not entitled to make the deductions because there was no provision in the contract for the completion of the work in sections. Although this case has been the subject of some criticism (see notes to clause 12) it appears to state the law. So if a contract providing for completion of the works in sections is required the form itself must be amended. There are two approaches to the necessary amendments. One is to state the parts of the works and the dates when they are to be completed and to use general words indicating that the contract must be read as if all necessary consequential amendments had been effected. The other approach is to carry out in detail the many amendments necessary. The Joint Contracts Tribunal, in December 1975, issued a sectional completion supplement together with practice note 21.' That case does seem to proceed on the basis that a sectional completion provision was necessary in order to justify the deduction and it seems to me that this is correct. It is of importance to note that the respondents' submissions in the award dealt with this matter in this way: 'If the contract had not included for sectional completion, the respondents could not, in accordance with Clause 16, have taken over the dwellings without the consent of the Claimants and had the respondents not taken over the dwellings as they were completed the Claimants would have made howls of anguish. In their submission Sectional Completion did form part of the contract and Liquidated and Ascertained Damages had been deducted in law correctly'. The arbitrator, of course, held, rightly, that there was no provision for sectional completion in the Articles of Agreement.

[The respondent's Counsel] contended that this argument was based upon a confusion between contractual sectional completion and consensual partial possession under clause 16. He said that paragraph of the award shows that the arbitrator found that there was consensual partial possession and that the liquidated damages were payable at the 'rate of £20 per week for each uncompleted dwelling'. [He] did concede that there might be one difficulty, namely, that arising out of the construction and application of clause 16(e), and, as it has assumed particular importance, I shall read it again: 'In lieu of any sum to be paid or allowed by the Contractor under clause 22 of these Conditions in respect of any period during which the Works may remain incomplete occurring after the date on which the Employer shall have taken possession of the relevant part there shall be paid or allowed such sum as bears the same ratio to the sum which would be paid or allowed apart from the provisions of this Condition as does the Contract Sum less the total value of the said relevant part to the Contract Sum'. He suggested that this might be susceptible to a little tinkering by the court, and that on four grounds it could be held not to operate against his general argument. These four grounds were: (1) the rate in the Appendix is capable of being expressed as £2,460 per week, that is, in respect of all the 123 houses at the rate of £20 per week; (2) that it produced a verbal quirk which could be ignored having regard to the obvious intention of the parties; (3) that one should rely upon condition 22, and not condition 16, and that the sums were recoverable under clause 22 and not subject to or recoverable under clause 16(e); (4) that the court should adopt the principle set out in *Glynn* v. *Margetson & Co.* (1893). There were two passages on which he relied in particular, first, from the opinion of Lord Herschell LC: 'Then is there any rule of law which compels the construction contended for? I think there is not. Where general words are used in a printed form which are obviously intended to apply, so far as they are applicable, to the circumstances of a particular contract, which particular contract is to be embodied in or introduced into that printed form, I think you are justified in looking at the main object and intent of the contract and in limiting the general words used, having in view that object and intent'.

Lord Halsbury said this: 'My Lords, I am entirely of the same opinion. It seems to me that in construing this document, which is a contract of carriage between the parties, one must in the first instance look at the whole of the instrument and not at one part of it only. Looking at the whole of the instrument, and seeing what one must regard, for a reason which I will give in a moment, as its main purpose, one must reject words, indeed whole provisions, if they are inconsistent with what one assumes to be the main purpose of the contract'.

With the greatest possible respect, I do not think that this passage can be applied to the present circumstances, and I do not think that the observations of Lord Halsbury really assist in the point of construction. In my view, the arbitrator was required to construe and give effect to all the provisions in these Articles.

[Counsel for the respondents] described the point raised by condition 16(e) as the only real point put forward by the applicants. There is no doubt that the applicants' argument is a very technical one, but I think that it is correct. I do not find any of the four suggestions put forward by [him] in respect of condition 16(e) convincing. There is no doubt that partial possession was taken by the respondents from time to time of completed houses, and this indeed is how the liquidated damages were computed. Accordingly, it seems to me that the respondents operated under condition 16 and it cannot be said now that it is condition 22 and not condition 16 which applies. I think one must read and give effect to both conditions. I do not think one can avoid the conclusion that condition 16(e) would apply to the present situation, and it does not seem to be consistent with the liquidated damages as set out in the Appendix.

It would of course be open to the parties to have made appropriate provision in the contract itself so as to deal with the situation. My finding does not in any event prevent the respondents from claiming damages for breaches of the contract. It is correct that they have not done so at the present time, preferring to rely upon the liquidated damages provision. I do not myself see any reason why the arbitrator should not allow appropriate amendments if they are put forward. That is, however, a matter for him and not for me.

It seems to me, therefore, that, in the absence of any provision for sectional completion in this contract, the respondents were not entitled to claim or deduct liquidated damages as provided in the Appendix.

Chapter 21

NOMINATED SUB-CONTRACTORS

The problems on re-nomination illustrated on pp. 334–343 were again considered in the following case, which discusses the grounds on which the main contractor is entitled to reject a renomination and the question of what is a reasonable time. It also establishes that the main contractor is not liable for defects in prime cost work which arise before completion where the original nominated sub-contractor drops out.

Fairclough Building Ltd v. Rhuddlan Borough Council

COURT OF APPEAL   (1985), 3 ConLR 38

Fairclough (F) contracted with the Council in JCT 63 terms to construct a complex of swimming pools, a theatre and other amenities at Rhyl. F. took possession of the site on 5 January 1976, and the original completion date was 2 May 1977. The contract sum was just short of £3 million. Gunite (Swimpool) Ltd (G.) were nominated for specialist sprayed concrete work for the sum of £96,700. F. determined the sub-contract for delay on 29 September 1977 and on the same day wrote requesting the architect to make a renomination of a sub-contractor 'to complete the sprayed concrete pool construction ... including making good ... any defects in ... G's work'.

The architect did not renominate for some months. On 20 January 1978 he issued an instruction requiring F. to 'enter into a sub-contract with M. for the sprayed concrete completion work' but the parties agreed that this did not constitute a valid renomination. By an instruction of 13 February 1978 the architect required F. 'to investigate and make good (G.'s) defective workmanship'. The validity of this instruction was disputed by F.

On 24 February 1978 the architect required F. to sub-contract with M. for the completion but not the remedial work. F. rejected the nomination on the basis that (a) they were entitled to an extension of time to accommodate M.'s programme and (b) that the employer's duty was 'to renominate a sub-contractor to carry out all work necessary (whether remedial or not) to complete the work covered by the original sub-contract'. On 23 May 1978, without prejudice, the architect nominated another company to complete G.'s work. G. were in liquidation. The trial judge found in favour of F., and the Court of Appeal dismissed the Council's appeal.

Held:

(1) The architect was entitled to have regard to the employer's interests by seeking lump sum tenders from three possible replacements and was justified in holding to that course until reasonably satisfied that a lump sum was not forthcoming. The architect was not at fault in not renominating before 24 February 1978.

(2) F. were entitled to reject the nomination on the ground that the proposed nominee's programme overran its own, in the absence of an undertaking from the architect that the main contract time-limits would be made compatible with the sub-contract. F were also entitled to so reject the renomination because of the architect's failure to require M. to do the remedial as well as the completion work.

(3) F. were not under any liability in respect of defects in G.'s work which arose before completion, whether those defects were the result of faulty design, bad workmanship or otherwise.

(4) F. were entitled to an extension of time in respect of all such delay as resulted from the employer's failure validly to renominate on 24 February, 1978.

PARKER LJ:
(a) *Did the time provided for in the proposed sub-contract entitle the contractors to refuse the nomination?*

By way of preface it is necessary to state that when the contractors objected to the nomination in February 1978, *Percy Bilton Ltd* v. *Greater London Council* had not been decided, much less reported. This fact needs to be taken into account when considering contentions raised by the contractors in contemporary correspondence on which Mr Keating for the employers sought to base an argument.

The employers' case on this issue is founded on clause 27(a) (ii) of the contract and on the admitted facts (i) that at the time of AI 137 the date for completion of the main contract was fixed at 10 May 1978; (ii) that William Mulcaster Ltd would not enter into a contract which provided for completion of the swimming pool work by that date but required until 1 September 1978 for the completion of such work.

Clause 27(a) (ii), so far as material, provides:

'27 The following provisions of this Condition, shall apply where prime cost sums are included in the Contract Bills ... in respect of persons to be nominated by the Architect ... to supply and fix materials or goods or to execute work.

(*a*) ... all specialists ... who are nominated ... are hereby declared to be sub-contractors employed by the Contractor and are referred to in the conditions as 'nominated sub-contractors'. Provided that the Architect ... shall not nominate any person as a sub-contractor against whom the Contractor shall make reasonable objection or (save where the Architect ... and Contractor shall otherwise agree) who will not enter into a sub-contractor which provides (*inter alia*):

(i) ...

(ii) That the nominated sub-contractor shall observe, perform and comply with all the provisions of this Contract on the part of the Contractor to be observed, performed and complied with ... so far as they relate and apply to the sub-contract works or to any portion of the same.'

The contractors' case is simple. They say that they were under a contractual obligation to complete all the contract work, which included the sub-contract work, by 10 May 1978. Accordingly, unless William Mulcaster Ltd would enter into a contract to perform the sub-contract work by that date, which it is common ground they would not, the architect was not entitled to nominate them and his purported nomination was invalid.

On the plain wording of clause 27(a)(ii) and the admitted facts this contention, even unsupported by authority, appears, at any rate at first sight to be unanswerable. It is, moreover, supported by *Bilton* v. *GLC*.

The facts of that case are somewhat complicated but so far as material for present purposes may be stated in simple form. The main contract was entered into on 25 October 1976 and was in the same form as the contract presently under consideration. The original completion date was 24 January 1979, but it had been extended to 16 March 1979, when on 31 July 1978 a nominated sub-contractor's contract was determined, as in this case, owing to his withdrawal from the site. There then followed a series of abortive attempts to find a new sub-contractor but no possibly effective renomination was made until, at the earliest, 31 October 1978. The new sub-contractor at that time required until November 1979 to complete the sub-contract work but the contractor nevertheless accepted the nomination and entered into a sub-contract which eventually provided for completion in 53 weeks from commencement of work, which was specified to be 22 January 1979. The main contract completion date remained at 16 March 1979 until on 9 May it was extended to 21 June 1979. Thereafter further extensions were granted, finally to 1 February 1980, and on 4 February 1980 the architect issued a certificate that the contract ought reasonably to have been completed by 1 February 1980. In March 1980 the GLC deducted the sum of £24,000 odd from monies otherwise due to the contractor in respect of liquidated damages calculated as from 1 February 1980. This deduction was challenged successfully before His Honour Judge Stabb QC on grounds which we need not rehearse, and the GLC appealed to this court with partial success. Again, apart from one matter we need not go into the reasons.

That matter and its importance cannot be better shown than by quotation of a passage from the judgment of Sir David Cairns (with whom Stephenson and Dunn LJJ agreed) in the Court of Appeal. He said:

'A quite separate argument by Mr Garland is what he described as his 'overshoot' submission; that is to say that, at the time of the application for renomination, the new sub-contractor's date for completion was later than the plaintiffs' date for completion and that, since this would make it impossible for the plaintiffs both to accept the new sub-contractor and to comply with the provision in their own contract as to the time for completion, therefore the time provision must go completely, time will be at large, and the right to liquidated damages will disappear.

I do not accept this argument. The contractor, faced with a sub-contract with such a provision as to completion, would be entitled to refuse to accept that sub-contractor under clause 27; or what the contractor could do would be to say that he would not agree to accept that sub-contract unless at the same time the employer would agree to an extension of time for the completion of the main contract. Or he could simply accept the nomination and rely on the probability that extensions which would later be given would cover the whole of the period which was required by the nominated sub-contractor for completion of his own work. That is what seems to have happened in this case, and with good reason, because already at that time it was probably in every-body's mind that there were going to be such extensions as would make the ultimate completion date for the main contract later than the completion date in the sub-contractor's contract.'

Whether, as we think, this is part of the *ratio* or not, the contractors contention that the nomination in AI 137 was invalid thus has the unanimous support of all three members of the court.

On appeal to the House of Lords the decision of this Court was affirmed. The speech of Lord Fraser of Tullybelton (with whose speech the other members of the House of Lords agreed) contains the following passage.

'There was also another course open to the appellant. It could have exercised its right of 'reasonable objection' under clause 27(a) to prevent the nomination of any new sub-contractor who did not offer to complete his part of the work within the overall completion period for the contract.'

The mention in this passage of the right of 'reasonable objection' appears to be a reference to the alternative ground for refusing nomination under clause 27 rather than that presently under consideration unless, as may well be the case, Lord Fraser regarded an objection on such ground as being necessarily reasonable. Be that as it may, it is clear that the contractors' contention not only has the unanimous support of three members of this Court but also the equally unanimous support of five members of the House of Lords. This is hardly surprising for, whatever may be said of the shortcomings in respect of clarity of other provisions of the contract under consideration, the wording in this particular instance is so clear that no other conclusion appears possible.

How then does [Counsel] seek to escape from what appears to be an inescapable position? He founds his argument, first, on a passage in the contractors' letter to the architects of 2 March 1978 in which they objected to the nomination. This reads:

'With regard to the commencement and completion of the sprayed concrete work, the Sub-Contractor requires a minimum of one week from the date of written acceptance of his tender to commence work and a further 26 weeks to complete the whole of the works.

We are unable to enter into a sub-contract agreement contain-ing the above periods, as these are at variance with clause 27(a)(ii) of the main contract.

To enable us to be in a position to enter into a sub-contract with Messrs William Mulcaster we would require an extension of time. In accordance with clause 23(e) of the main contract, we would request an extension of time of 19 weeks due to the difference between the 27 weeks required for the completion of the works by William Mulcaster from the date of the acceptance of their tender, and the 8 weeks remaining of the 16-week period referred to in the Bills of Quantities for carrying out the whole of the sprayed concrete works contained in the pc sum. We would also confirm that we would seek reimbursement for the above under clause 24(c).'

He founds secondly on the architect's reply to that letter, dated 10 March:

'I would acknowledge receipt of your letter dated 2 March, 1978, and would comment as follows. Regarding Item 1, I would confirm our intention to grant an extension of time in connection with the renominated Sub-Contractor's programme time at such time as the effect upon your overall programme can be ascertained.'

Based on these passages Mr Keating made the following submission, which he set out in writing at the request of the court and handed in at the conclusion or near-conclusion of his argument on the point:

'That the contractor cannot refuse to accept the nomination of any new sub-contractor who does not offer to complete his part of the work within the overall completion period for the contract as a whole if the contractor purports to seek an extension of time under clause 23 in relation to such nomination and the architect indicates that he is willing to grant an extension as if the nomination were a cause of delay under clause 23.'

We cannot accept this submission. The contractors made it abundantly clear that they were not prepared to accept the nomination because the time required by the proposed sub-contractor was at variance with clause 27(a)(ii) unless they first got an extension of time to cure the position. They were thus doing precisely what Sir David Cairns indicated they could properly do. It is true that they referred to clause 23(e), which could not be applicable unless and until the nomination was accepted, but it is not and could not be suggested that the architect or the employers were thereby led to believe that the nomination was being accepted, or that clause 23(e) in fact applied. Indeed the submission itself recognizes that it did not, for it uses the words 'as if the nomination were a cause of delay'.

The importance which it is sought to attach to the mention of clause 23(e) is in our view quite misplaced. It was merely mistaken and the matter would no doubt have been expressed differently had *Bilton* then been decided and reported. What is important is that the contractors made it plain that they would not accept the nomination until they had got such an extension as would bring the main contract and the proposed sub-contract into line.

Furthermore, although (Counsel) described his submission as being a submission on the construction of the contract it does not appear to us to be that at all, except in so far as it is a submission that the contract in some unexplained way should be so construed as to produce the result contended for. That it cannot possibly be so construed appears to us to be clear. The result is contrary to the plain terms of clause 27(a)(ii) and to achieve it would involve reading into clause 27 some such proviso as:

'Provided that if a proposed nominated sub-contractor will not enter into a contract which provides for the sub-contractor to complete the sub-contract work by the time fixed for completion of the main contract but is otherwise willing to enter into a contract containing the provisions of this clause the architect may none the less nominate such sub-contractor if on receipt of the original nomination the contractor seeks an extension of time allowed under the main contract and the architect indicates that he is willing to grant such extension as would be allowable under the main contract if such nomination were included in the causes of delay set out in clause 23'.

This would be to rewrite the contract. In addition it would impute an intention to the parties for which it appears to us there is no warrant whatever. We reject the submission and, like the learned judge, conclude that the contractors were entitled to reject the nomination on the ground that the proposed sub-contractors would not enter into a contract complying with clause 27(a)(ii).

[Counsel] urged upon us that this meant that, if no sub-contractor could be found to complete by the currently fixed date for completion, the employer would be driven, if he wanted the work to be done at all, to vary the main contract, for unless and until there was a valid nomination, there would be no power of extension under clause 23. If this be right we cannot see that it matters, but we do not accept that it is necessarily right. If there is a lacuna in the contractual machinery, as there appears to be, it might, as it seems to us, possibly be filled by an implied term, that if the nomination were accepted, an appropriate extension would be granted.

[Counsel] also urged that where, as here, the anticipated date for actual completion was many months later than the then fixed date for completion, the contractor would be able, if our conclusion is right, to force upon the employer the grant of an extension which deprived him of a potential right to damages as from the currently fixed date, even in a case where the sub-contract work could be carried out without delaying the actual date of completion at all. This may be so but again we cannot see that it affects matters. The contractor through no fault of his own may be without a sub-contractor for months, but is not entitled to an extension for any delay thereby caused until after a reasonable time for renomination has elapsed. In the present instance delay until 24 February therefore falls upon the contractor. If, when his contractual completion date is some $2\frac{1}{2}$ months off, he is asked to do work which will take 6 months to complete, we see no reason for saying that the contract must be so construed that he cannot insist on an extension of time under the main contract to bring it in line with the proposed sub-contract, particularly if the position is that, were he to accept the nomination and then seek an extension in respect of the sub-contract work, any extension granted might be for less than the period required to complete that work.

It was accepted by counsel for both sides that such was the position and in the present case it is unnecessary to decide the matter. It appears to us however to be doubtful. In the present case the date fixed for completion was 10 May 1978. The contractors were therefore, as at 24 February, under a contractual obligation to complete all the contract work by that date, and on the architect certifying that all the contract work ought reasonably to have been completed by that date, to pay liquidated damages thereafter. On 24 February the contractor was then required to enter into a sub-contract for the doing of work which, it is accepted, could not possibly be done until 1 September. It may well be that the doing of such work would not delay *actual* completion of all outstanding work but if the contractor is required on 24 February to do work which cannot be done until September, it appears to us at least arguable that he could not be in breach of contract by reason of failure to do that part of the work until September and thus that he is entitled, if he does not exercise his right to prevent nomination, to an extension to that date ...

(b) *Was the nomination invalid by reason of the fact that the proposed sub-contract did not cover remedial work?*

The contractor's objection on this ground is based on the decision of the House of Lords in *North West Metropolitan Regional Hospital Board* v. *T. A. Bickerton & Son Ltd* (1970). In that case a nominated sub-contractor had gone into liquidation and the liquidator elected not to perform the remaining work. The employer although requested to issue a variation order or nominate a new sub-contractor refused to do so but called upon the main contractor to complete the sub-contract work, maintaining that he was bound to do so. The point for determination was whether the employer was bound to renominate. It was held that he was obliged to do so on the true construction of clause 27 of the contract, which was in the same terms as clause 27 of the contract here under consideration. It was so held on the specific ground that under clause the main contractor was neither bound nor entitled to do any of the sub-contract work himself.

In reaching his conclusion that the contractors' objection to the nomination on the ground now under consideration was valid, the learned judge relied on the following passages in the speeches in their Lordships' House:

Lord Reid:

'I can find nothing anywhere to indicate that the principal contractor can ever have in any event either the right or the duty to do any of the prime cost work himself.'

Lord Guest:

'Upon a proper construction of the conditions of the contract, the contractor is not entitled to do work by himself for which a prime cost sum is allocated in the contract.'

Viscount Dilhorne:

'The effect of a sub-contractor repudiating his contract and ceasing to act is that, unless there is a fresh nomination, there is no one left who can carry out the prime cost work. The contract gives to the contractor no right to do it. In the case of ordinary sub-contractors the contractor may be able to remedy the breach before the date for completion but, when the breach is by a nominated sub-contractor he cannot do that as he is debarred from doing that sub-contract work.'

To those we would add one further quotation wherein Lord Reid sets out the contention which was advanced by the contractors and which was upheld:

'The RIBA form of contract makes no express provision for an event which cannot be very uncommon – that for one reason or another the contract between the principal contractor and the nominated sub-contractor is terminated before the sub-contract work is completed. The problem is what is to happen in that event. There appear to me to be three possibilities and no more were suggested in argument.

The first is that the employer must then make a new nomination, or, if he does not wish the work to be continued, give an instruction to that effect by way of variation. If he does make a new nomination, then as in the case of the first nomination he must make a bargain with the new nominee and then instruct the principal contractor to make a sub-contract in terms of that bargain. Then if the price in the second sub-contract exceeds the price in the first sub-contract which has come to an end, the employer must bear the loss. That is the contention of the respondents.'

It is true that only Lord Dilhorne specifically refers to remedial work, and it was submitted that there is a distinction between remedial work and work which has not been done at all. This distinction was said to be supported by the facts: (i) that under clause 6(3) and (4) the architects can require the contractor to open up for inspection any work covered up and arrange for tests to be carried out and can also issue instructions in regard to the removal from the site of any work, materials or goods which are not in accordance with the contract; and, (ii) that under clause 15 the contractor is obliged to make good defects in all work – including sub-contract work which may appear in the defects liability period under the contract.

For present purposes we are content to accept, without deciding, that clauses 6(3) and 6(4) do apply to nominated sub-contract work, although in our view this is not necessarily so. It was common ground that clause 15 does so apply. Even so they do not avail the employers. Clause 6 does not deal with remedial work. If defects are found in main contract work the contractor will have to put it right pursuant to his overall obligation to perform the contract. So also in the case of work which he has elected to have performed by a sub-contractor. He has, however, no obligation to do nominated sub-contract or prime cost work. (Counsel) submitted that remedial work was not nominated sub-contract work but we do not follow this. The nominated sub-contractor is obliged to produce work which satisfies the sub-contract. If he does some work which does not satisfy the contract he is obliged, if he stays, to cure the position. This may require either relatively minor remedial work or, if the original work is very bad, complete removal and a fresh start. If he leaves the site with neither course taken, there is an outstanding nominated sub-contract obligation still to be performed. This obligation the contractor is neither entitled nor obliged to perform. The employer must therefore either order its omission, or issue a variation order and pay the contractor for it or negotiate a new sub-contract covering such work. In our opinion this follows inevitably from *Bickerton*. If he chooses the renomination route but does not provide for the new sub-contractor to carry out the remedial work he had not carried out his renomination obligation and the contractor can object. It was submitted that this conflicted with clause 15. In our view it does not. It accords with it. The overall intention is that the contractor shall not be liable to perform any part of the nominated sub-contractor work. That is why, under clause 27, he is entitled to demand a renomination if the nominated sub-contractor drops out, and why the nominated sub-contractor can be rejected if he will not enter into a contract containing the provision specified in clause 27(a)(ii). If the renomination includes remedial work and defects in such work are later discovered he can call upon the sub-contractor to remedy them. This is the plain intent of the parties. If the new sub-contract does not cover remedial work that intention is defeated. In the present case all the original sub-contract work had to be removed and the work started afresh. The submission that if the contractor did remove the old work and start afresh he would not be doing work which *Bickerton* decides he is neither obliged nor entitled to do is in our view wholly unsustainable. It is true that in the result the employer will suffer a loss, but that this might happen was specifically recognized by Lord Reid. There is no reason why he should do so if the contract contains no protective provision. It is he who has chosen the sub-contractor who first did the defective work and then quit the site, leaving the contractor no remedy for any delay resulting from the need to find a replacement so long as the delay is not unreasonable. This delay could well be much more serious financially than the cost to the employer of including remedial work in the new sub-contract. If it is, the contractor must bear the burden. That the employer should bear the burden of paying for remedial work does not justify construing remedial work as not including work which the original sub-contractor would have been obliged to do had he not left the site.

(c) *Are the contractors entitled to an extension for the 8 weeks' delay incurred by Gunite before they withdrew?*

Clause 23(g) of the contract included amongst causes of delay,

> 'delay on the part of nominated sub-contractors which the contractor has taken all practicable steps to avoid or reduce *but such delay will only be considered for those reasons for which the contractor could obtain an extension of time under the contract.*'

The words stressed were added in type at the end of the printed provision. Their meaning is in our view plain. It is to ensure that the contractor does not get an extension for *mere* delay on the part of the sub-contractor but can do so only if the sub-contractor's delay is itself due to one or other of the causes of delay specified in the other sub-clauses of clause 23.

Gunite's delay was not so due. Accordingly we, like the learned judge, conclude that the contractors are not entitled to an extension in respect of it.

(d) *Are the employers entitled to charge the contractor with the full cost of remedial work or only obtain the amount which they had already paid in respect of Gunite's work before their withdrawal?*

It is accepted that they are entitled to be credited by the contractors with the lesser amount. Once it is concluded that the obligation to renominate includes the obligation to include remedial work in the work to be done by the renominated sub-contractor and that the contractors are neither entitled nor obliged to do such work, we can see no basis upon which the employers can charge them with the cost of such work. In the course of argument [Counsel] came very near to accepting that this must be so but sought to say that there was a route by which they could be so charged. We can only say that we could not follow any route, however ingenious, which leads to the result, in effect, that the contractors must bear the cost of work the burden of which has been held to fall upon the employers, and which the contractors are not obliged or entitled to do. We therefore reject the employers' contention.

In the result we would dismiss both appeal and cross appeal and affirm the judgment appealed from on all issues.

Liability of a nominated sub-contractor in tort to the employer for defective work includes not only the direct physical consequences of the defective work but the foreseeable economic loss such as compensating the main contractor for delay and disruption and the employer for losses due to delayed completion.

Greater Nottingham Co-operative Society Ltd v. Cementation Piling and Foundations Ltd and others

QUEEN'S BENCH DIVISION   (1986), 4 ConLR 57
(OFFICIAL REFEREES' BUSINESS)

The plaintiffs owned a store in Skegness and in 1979 arranged to build an extension on an adjoining site. The second defendants were the main contractors and the first defendants nominated sub-contractors for piling work. The main contract was on JCT 1963 terms. The employers and nominated sub-contractors entered into a direct employer-nominated sub-contractor agreement.

The site was liable to subsidence because of the silty nature of the subsoil. The original building had been built on a raft but this had not proved entirely satisfactory. In 1971 an earlier extension had been built upon piles which had been driven into the ground. This pile driving resulted in excessive vibration and consequential damage. The plaintiffs therefore decided after consulting their professional advisers to use the augering system of piling which is comparatively vibration-free.

The invitation to tender specified 'bored' piling and drew attention to the need for the piling to be vibration-free. The specialists invited to tender were provided with a geological report on the site of 1971 but not an earlier report of 1958 on subsidence in the main store. The first defendants submitted a tender based on piles of 450 mm diameter; a competing tender provided for piles of 300 mm (the smaller the diameter of the piles, the more are needed).

Piling work started on 17 July 1979. About midday on 19 July cracks appeared in a nearby building and piling work was suspended until October 1979 when it was restarted in a substantially modified form and completed on 20 November 1979. The plaintiffs claimed for the cost of meeting the claim by the adjoining landowner for damage to his building; for the extra cost of the piling operation; (against the first defendants) for the extra sums paid to the second defendants for delay and disruption and for its own losses arising out of late completion.

Held (inter alia): Cementation were negligent in the conduct of the piling operations both by repeated attempts at penetration and by straight-lifting the auger without reverse screwing. They were liable to the plaintiffs for the money paid to the owners of the adjoining property to compensate them and also for the additional cost of the revised piling scheme; for the additional sums paid to the main contractors for disruption and for the plaintiffs' own losses due to late completion.

JUDGE DAVID SMOUT QC: (Counsel) on behalf of Cementation has contended that a finding of negligence would render Cementation liable only in respect of the Ling action, that is in the agreed sum of £82,000. The other claims, namely:

(a) For money spent by the Society on the repairs to the Windsor;
(b) For the additional cost to the Society of the revised piling scheme;
(c) For the additional sums due to Shepherd in respect of delay in execution of the work, and
(d) For the Society's own losses due to late completion

all relate to economic loss claimed by the Society in circumstances where their own property has not been physically damaged. It is

[his] contention that such loss does not come within the ambit of Cementation's duty in tort.

[Both counsel] rely upon different passages and different interpretations of the speeches in the House of Lords in *Junior Books Ltd* v. *Veitchi (1983)*. As Robert Goff LJ commented in *Leigh and Sillivan Ltd* v. *Aliakmon Shipping Ltd* (1985) 'foreseeability of damage is a prerequisite, not the criterion of liability in negligence'.I respectfully adopt his analysis of the authorities culminating in *Junior Books* that:

> 'We have seen the development in the cases, not of a generalised right of recovery in negligence for economic loss, but of recovery in specific instances'.

The ambit of the duty of care must depend upon an examination of the specific relationship between the parties in the factual context, in the light of not only *Junior Books* and Lord Wilberforce's two-stage approach in *Anns* v. *Merton LBC* (1978), but also the observations of Lord Keith in *Peabody Fund* v. *Sir Lindsay Parkinson Ltd* (1984):

> 'The true question in each case is whether the particular defendant owed to a particular plaintiff a duty of care having the scope which is contended for, and whether he was in breach of the duty with consequent loss to the plaintiff. A relationship of proximity in Lord Atkin's sense must exist before any duty of care can arise, but the scope of the duty must depend on all the circumstances of the case'.

The essential facts are shortly recited. A, whilst engaged for reward by B on the instructions of C, performs work designed by A on C's land and thereby accidentally damages the adjoining property of D in circumstances which render C as well as A liable to D. C. spends money towards repair of D's property. A's programme of work is necessarily suspended pending re-appraisal and eventual redesign which involves C in considerable loss and expense, much of which was of a nature which was reasonably foreseeable to A. Taking such facts as a whole, it follows that there is a 'sufficient relationship of proximity' between A and C that in the reasonable contemplation of A carelessness on his part was likely to cause much of the damage that was suffered by C. If it is useful to compare specific instances the proximity in those circumstances seems as close as that which existed between the factory owner and the subcontractor in *Junior Books*. There is thus a *prima facie* duty of care on Cementation not to cause economic loss to the Society of such a type as was reasonably foreseeable. Contractual obligations between the parties are no doubt amongst the leading considerations which may negative, reduce or limit the scope of the *prima facie* duty, as for example in *William Hill Organisation* v. *Bernard Sunley & Sons* (1983). Yet I find nothing in the contractual nexus in the instant case that would diminish Cementation's duty of care to the Society in tort. The existence of any express or implied terms as to good workmanship in the contract between Cementation and Shepherd, and that between Shepherd and the Society, and the nature of the express warranties between Cementation and the Society do not suggest to me that the scope of duty of care in tort is thereby in any way curtailed. Nor so far as I am aware is there any other consideration that would entitle this court to refuse to the Society the remedy of recovery for economic loss.

Chapter 15

FRUSTRATION

Codelfa Construction Proprietary Ltd v. State Rail Authority of New South Wales

HIGH COURT OF AUSTRALIA   (1982), 149 CLR 337

The appellants, a construction company, contracted with the respondents, a railway authority, for the construction of an underground railway. The contract required the work to be completed within a fixed period and the appellants started work on a three shift a day, seven days a week basis. However, the work was noisy and third parties affected by the work obtained injunctions restraining evening and Sunday work.

The appellants commenced an arbitration claiming an addition to the contract price either on the basis that there was an implied term in the contract that if the contractor was restrained by injunction from working the planned shifts, the authority would indemnify it against the additional costs or on the basis that the contract had been frustrated by the grant of the injunction.

The arbitrator found as a fact that the parties had entered into the contract on the basis that work was to be carried out on a seven-day, three-shift basis and in the shared belief that no injunction would be granted to restrain noisy work in the evenings or at weekends.

The High Court of Australia held that no term was to be implied but (by a majority of 4–1) that the contract had been frustrated by the grant of the injunction.

MASON, J: It is not surprising that the cases commonly throw up situations of supervening impossibility caused by a change in the law – they are the more common instances of the unforeseen or unexpected occurrence. But in principle there is no reason why a mutual assumption arising from a mistaken view that an activity is immune from injunctive relief should not attract the principle of frustration. No doubt it is more difficult in such a case to show that the grant of injunctive relief was not foreseen or could not reasonably have been foreseen, but if that can be shown then the doctrine of frustration should apply. The injunction is a supervening event though it does not stem from any alteration in the law.

An unusual element in the present case is that the parties appear to have received, accepted and acted on erroneous legal advice that the contract work could not be impeded by the grant of an injunction to restrain noise or other nuisance, advice which was based on an erroneous interpretation of s. 11 of the City and Suburban Electric Railways Act 1915 (NSW). One might have expected the parties and their advisers to have had reservations about the correctness of the advice and to have given consideration to the possibility that, despite the advice, an injunction might be granted. However, the findings do not reflect the existence of any reservations; indeed, they record Codelfa's acceptance of the representations made by the Authority. Codelfa is a wholly owned subsidiary of an Italian company and this may explain Codelfa's willingness to accept and act on the representation made by the Authority.

The doctrine of frustration is closely related to the concept of mutual mistake. However, in general, relief on the ground of mutual mistake is confined to mistakes of fact, not of law. If the common contractual assumption is of present fact it is a case of mutual mistake; if the assumption is of future fact it is a case of frustration (*Bell*(14), per Lord Atkin), the distinction being that in one case the contract is void ab initio and in the other it is binding until the assumption is falsified. Here the mistake is not one of present fact; it is either a mistake as to future fact or a mistake of law. Even if it be a mistake of law, this is not, I think, fatal to the application of the doctrine of frustration. The unsatisfactory distinction between a mistake of fact and one of law has not so far been carried over into frustration and I see no reason to further complicate the doctrine by invoking this distinction.

The critical issue then is whether the situation resulting from the grant of the injunction is fundamentally different from the situation contemplated by the contract on its true construction in the light of the surrounding circumstances. The contract itself did not require that the work be carried out on a three shift continuous basis six days a week without restriction as to Sundays. But it required completion of the works within 130 weeks. And Codelfa with its tender had submitted a construction programme which involved a three shift continuous basis six days a week. By cl. S.6 of the specifications Codelfa was required to submit a revised programme of work to the Engineer for his determination within thirty calendar days of the issue of a notice to proceed under the contract. This Codelfa did. Again it made provision for the method of operation already mentioned. It was accepted by the Engineer.

It is in this contractual setting that the findings of the Arbitrator have special significance. The relevant findings are set out in pars. 14, 15, 16, 18 and 19 of his award.

'14. The Parties to Contract ESR 1005 each entered into such Contract on the common and mutual understanding and on the basis that:

    (a) the works the subject of the Contract should and would be carried out by the Contractor on a 3-shift continuous basis six days per week and without restriction as to Sundays, and

    (b) the work to be performed was inherently of a noisy and disturbing nature and the work or substantial parts thereof was to be carried out in close proximity to areas of residential neighbourhood, and

    (c) no Injunction or other Restraining Order could or would be granted against the Contractor in relation to noise or other nuisance arising out of the carrying on of the said works on such basis.

15. The matter mentioned in paragraph 14(c) was represented by the Principal to the Contractor and was accepted as the situation by the Contractor prior to and at the time of entering into the Contract.

...

16. The said works could not in fact be carried out by the Contractor in accordance with methods and programmes agreed to by the parties and in acccordance with the contractual stipulations as to time of performance unless the works were carried out on the basis mentioned in paragraph 14(a) hereof.

...

18. Restraining Orders and Injunctions were in fact issued by the Court on grounds of noise and other nuisance arising out of the carrying out of the works by the Contractor on the basis mentioned in paragraph 14(a) hereof the effect of which was (inter alia) to prohibit and preclude the works from being carried out by the Contractor on the said basis and to cause the Contractor to incur additional cost in the carrying out of the works.

19. The said works could not in fact be carried out by the Contractor in accordance with methods and programmes agreed to by the parties without substantial noise and other disturbances arising therefrom or in connection therewith by reason of the inherently noisy and disturbing nature of the work.'

The submission of the proposed programme of work with the tender, its supersession by the revised programme pursuant to cl. S.6 of the specifications, together with the very provisions of cl. S.6 itself dealing with the construction programme, provide a link between the contract and the antecedent discussions so as to enable us, subject to a consideration of specific provisions in the specifications, to say that the contract contemplated that completion would be achieved within the time stipulated by the method of work already mentioned, it being assumed that it could not be disturbed by the grant of an injunction.

I reject the Authority's argument that cl. S.6 is inconsistent with the notion that the contract looked to this method of work as the mode by which the work was to be completed. Certainly cl. S.6(5) envisaged that a major change to the work diagram as determined by the Engineer might be required by 'revisions to the programme'. It seems that the responsibility for initiating such a change lay with Codelfa. But I do not think that this is necessarily inconsistent with Codelfa's case on frustration nor do I think that cl. G.44(7) of the contract, which provides for the grant of an extension of time in case of delays 'owing to causes beyond the control or without the fault or negligence of the Contractor', covers the position. Delay due to the grant of an injunction on the ground of nuisance committed by Codelfa scarcely answers this description, even though it results from performance of the work in the only manner which will enable completion to take place within the time stipulated.

Clause S.8(2)(c) however, poses a greater obstacle. It provides:

'The operation of all plant and construction equipment shall be such that it does not cause undue noise, pollution or nuisance. This may require the use of sound insulated compressor and air tools, silencers on ventilating fans and restrictions on the working hours of plant or such other measures as approved by the Engineer. The Contractor shall not be entitled to additional payment if the Engineer requires that measures be taken to reduce noise and pollution.'

Once the injunctions were granted the Engineer gave notices reflecting the provisions of the injunctions, restricting the hours of work so as to prohibit work at night and thereby inhibited Codelfa from continuing with its three shift operation under the contract.

The first paragraph of cl. S.8(2)(c) contains a promise by Codelfa that it will not operate plant and equipment so as to cause a nuisance. The second specifically looks to the possibility of a restriction on working hours of plant. And the third denies additional remuneration if the Engineer requires measures to be taken to reduce noise and pollution.

Do these provisions support the view that Codelfa was undertaking in any event to perform the contract work, even though the method contemplated by the parties might prove to be unlawful or impossible by reason of its amounting to a nuisance and its being restrained by injunction? I do not think that cl. S.8(2)(c) has such a wide-ranging effect. It involves no subtraction from the language of the provisions to say that it is quite consistent with the contemplated method of work being an essential element of the contract. Indeed, there would be no inconsistency between these provisions and an explicit provision for termination of the contract in the event that the method of work was restrained by injunction. There was plenty of scope for an exercise of the Engineer's power under the second paragraph so long as it did not displace the continuation of that method of work.

I come back then to the question whether the performance of the contract in the new situation was fundamentally different from performance in the situation contemplated by the contract. The answer must, I think, be in the affirmative. Paragraphs 14,, 15, 16, 18 and 19 of the Arbitrator's award go a long way towards establishing this answer. The finding contained in par. 16 proceeds on the footing that the contract work could not be carried out as contemplated by the contract once injunctions were granted, the effect of which was to prohibit the continuous three shift a day operation six days a week. Performance by means of a two shift operation, necessitated by the grant of the injunctions, was fundamentally different from that contemplated by the contract.

There is, of course, no inconsistency between the conclusion that a term cannot be implied and the conclusion that events have occurred which have brought about a frustration of the contract. I find it impossible to imply a term because I am not satisfied that in the circumstances of this case the term sought to be implied was one which parties in that situation would necessarily have agreed upon as an appropriate provision to cover the eventuality which has arisen. On the other hand I find it much easier to come to the conclusion that the performance of the contract in the events which have occurred is radically different from performance of the contract in the circumstances which it, construed in the light of surrounding circumstances, contemplated.

Chapter 18

LIMITATION

LATENT DAMAGE ACT 1986

Time limits for negligence actions in respect of latent damage not involving personal injuries

1.   The following sections shall be inserted in the Limitation Act 1980 (referred to below in this Act as the 1980 Act) immediately after section 14 (date of knowledge for purposes of special time limits for actions in respect of personal injuries or death):

'*Actions in respect of latent damage not involving personal injuries*

Special time limit for negligence actions where facts relevant to cause of action are not known at date of accrual.

14A.   (1) This section applies to any action for damages for negligence, other than one to which section 11 of this Act applies, where the starting date for reckoning the period of limitation under subsection (4)(*b*) below falls after the date on which the cause of action accrued.

(2) Section 2 of this Act shall not apply to an action to which this section applies.

(3) An action to which this section applies shall not be brought after the expiration of the period applicable in accordance with subsection (4) below.

(4) That period is either:

    (*a*) six years from the date on which the cause of action accrued; or

    (*b*) three years from the starting date as defined by subsection (5) below, if that period expires later than the period mentioned in paragraph (*a*) above.

(5) For the purposes of this section, the starting date for reckoning the period of limitation under subsection (4)(*b*) above is the earliest date on which the plaintiff or any person in whom the cause of action was vested before him first had both the knowledge required for bringing an action for damages in respect of the relevant damage and a right to bring such an action.

(6) In subsection (5) above "the knowledge required for bringing an action for damages in respect of the relevant damage" means knowledge both:

    (*a*) of the material facts about the damage in respect of which damages are claimed; and

    (*b*) of the other facts relevant to the current action mentioned in subsection (8) below.

(7) For the purposes of subsection (6)(*a*) above, the material facts about the damage are such facts about the damage as would lead a reasonable person who had suffered such damage to consider it sufficiently serious to justify his instituting proceedings for damages against a defendant who did not dispute liability and was able to satisfy a judgment.

(8) The other facts referred to in subsection (6)(*b*) above are:

    (*a*) that the damage was attributable in whole or in part to the act or omission which is alleged to constitute negligence; and

    (*b*) the identity of the defendant; and

    (*c*) if it is alleged that the act or omission was that of a person other than the defendant, the identity of that person and the additional facts supporting the bringing of an action against the defendant.

(9) Knowledge that any acts or omissions did or did not, as a matter of law, involve negligence is irrelevant for the purposes of subsection (5) above.

(10) For the purposes of this section a person's knowledge includes knowledge which he might reasonably have been expected to acquire:

- (*a*) from facts observable or ascertainable by him; or
- (*b*) from facts ascertainable by him with the help of appropriate expert advice which it is reasonable for him to seek;

but a person shall not be taken by virtue of this subsection to have knowledge of a fact ascertainable only with the help of expert advice so long as he has taken all reasonable steps to obtain (and, where appropriate, to act on) that advice.

Overriding time limit for negligence actions not involving personal injuries.

14B (1) An action for damages for negligence, other than one to which section 11 of this Act applies, shall not be brought after the expiration of fifteen years from the date (or, if more than one, from the last of the dates) on which there occurred any act or omission:

- (*a*) which is alleged to constitute negligence; and
- (*b*) to which the damage in respect of which damages are claimed is alleged to be attributable (in whole or in part).

(2) This section bars the right of action in a case to which subsection (1) above applies notwithstanding that:

- (*a*) the cause of action has not yet accrued; or
- (*b*) where section 14A of this Act applies to the action, the date which is for the purposes of that section the starting date for reckoning the period mentioned in subsection (4)(*b*) of that section has not yet occurred;

before the end of the period of limitation prescribed by this section.'

Provisions consequential on section 1

2. (1) The following section shall be inserted in the 1980 Act immediately after section 28 (extension of limitation period in case of disability on date of accrual of cause of action):

'Extension for cases where the limitation period is the period under section 14A(4)(b)

28A.   (1) Subject to subsection (2) below, if in the case of any action for which a period of limitation is prescribed by section 14A of this Act:

(*a*)   the period applicable in accordance with subsection (4) of that section is the period mentioned in paragraph (*b*) of that subsection;

(*b*)   on the date which is for the purposes of that section the starting date for reckoning that period the person by reference to whose knowledge that date fell to be determined under subsection (5) of that section was under a disability; and

(*c*)   section 28 of this Act does not apply to the action;

the action may be brought at any time before the expiration of three years from the date when he ceased to be under a disability or died (whichever first occurred) notwithstanding that the period mentioned above has expired.

(2) An action may not be brought by virtue of subsection (1) above after the end of the period of limitation prescribed by section 14B of this Act.'

(2) In section 32 of the 1980 Act (postponement of limitation period in case of fraud, concealment or mistake), at the end there shall be added the following subsection:

'(5) Sections 14A and 14B of this Act shall not apply to any action to which subsection (1)(*b*) above applies (and accordingly the period of limitation referred to in that subsection, in any case to which either of those sections would otherwise apply, is the period applicable under section 2 of this Act).'

Accrual of cause of action to successive owners in respect of latent damage to property

3.   (1) Subject to the following provisions of this section, where:

(*a*)   a cause of action ('the original cause of action') has accrued to any person in respect of any negligence to which damage to any property in which he has an interest is attributable (in whole or in part); and

(*b*)   another person acquires an interest in that property after the date on which the original cause of action accrued but before the material facts about the damage have become known to any person who, at the time when he first has knowledge of those facts, has any interest in the property;

a fresh cause of action in respect of that negligence shall accrue to that other person on the date on which he acquires his interest in the property.

(2) A cause of action accruing to any person by virtue of subsection (1) above:

(*a*) shall be treated as if based on breach of a duty of care at common law owed to the person to whom it accrues; and

(*b*) shall be treated for the purposes of section 14A of the 1980 Act (special time limit for negligence actions where facts relevant to cause of action are not known at date of accrual) as having accrued on the date on which the original cause of action accrued.

(3) Section 28 of the 1980 Act (extension of limitation period in case of disability) shall not apply in relation to any such cause of action.

(4) Subsection (1) above shall not apply in any case where the person acquiring an interest in the damaged property is either:

(*a*) a person in whom the original cause of action vests by operation of law; or

(*b*) a person in whom the interest in that property vests by virtue of any order made by a court under section 538 of the Companies Act 1985 (vesting of company property in liquidator).

(5) For the purposes of subsection (1)(*b*) above, the material facts about the damage are such facts about the damage as would lead a reasonable person who has an interest in the damaged property at the time when those facts become known to him to consider it sufficiently serious to justify his instituting proceedings for damages against a defendant whodid not dispute liability and was able to satisfy a judgment.

(6) For the purposes of this section a person's knowledge includes knowledge which he might reasonably have been expected to acquire:

(*a*) from facts observable or ascertainable by him; or

(*b*) from facts ascertainable by him with the help of appropriate expert advice which it is reasonable for him to seek;

but a person shall not be taken by virtue of this subsection to have knowledge of a fact ascertainable by him only with the help of expert advice so long as he has taken all reasonable steps to obtain (and, where appropriate, to act on) that advice.

(7) This section shall bind the Crown, but as regards the Crown's liability in tort shall not bind the Crown further than the Crown is made liable in tort by the Crown Proceedings Act 1947.

Supplementary

Transitional provisions

4. (1) Nothing in section 1 or 2 of this Act shall:

(*a*) enable any action to be brought which was barred by the 1980 Act or (as the case may be) by the Limitation Act 1939 before this Act comes into force; or

(*b*) affect any action commenced before this Act comes into force.

(2) Subject to subsection (1) above, sections 1 and 2 of this Act shall have effect in relation to causes of action accruing before, as well as in relation to causes of action accruing after, this Act comes into force.

(3) Section 3 of this Act shall only apply in cases where an interest in damaged property is acquired after this Act comes into force but shall so apply, subject to subsection (4) below, irrespective of whether the original cause of action accrued before or after this Act comes into force.

(4) Where:

(*a*) a person acquires an interest in damaged property in circumstances to which section 3 would apart from this subsection apply; but

(*b*) the original cause of action accrued more than six years before this Act comes into force;

a cause of action shall not accrue to that person by virtue of subsection (1) of that section unless section 32(1)(*b*) of the 1980 Act (postponement of limitation period in case of deliberate concealment of relevant facts) would apply to any action founded on the original cause of action.

Citation, interpretation, commencement and extent

5. (1) This Act may be cited as the Latent Damage Act 1986.
(2) In this Act:

'the 1980 Act' has the meaning given by section 1; and
'action' includes any proceeding in a court of law, an arbitration and
any new claim within the meaning of section 35 of the 1980 Act (new
claims in pending actions).

(3) This Act shall come into force at the end of the period of two
months beginning with the date on which it is passed.

(4) This Act extends to England and Wales only.

Table of Cases

Note

The following abbreviations of the main Reports are used:

AC – Law Reports Appeal Cases Series
All ER – All England Law Reports
BLR – Building Law Reports
Ch – Law Reports Chancery Series
ConLR – Construction Law Reports
KB – Law Reports King's Bench Series
LGR – Local Government Reports
P – Law Reports Probate Series
QB – Law Reports Queen's Bench Series
Sol Jo – Solicitors' Journal
TLR – Times Law Reports
WLR – Weekly Law Reports

References in heavy type indicate a detailed discussion of the case.

Subject Index